TRANSGENICS AND THE POOR

The genomics revolution in biology has enabled technologies with unprecedented potential; genetic engineering is changing the terrain of development studies.

Societies have reacted with appreciation or indifference to genetically engineered pharmaceuticals, beginning with insulin. Yet applications in food and agriculture have produced globally contentious politics and unprecedented policy dilemmas. Transgenic organisms raise questions of property, ethics, and distributive justice unimaginable a generation ago: what can be owned? with what responsibility? with what implications? Much turns on the science: how publics conceptualize authoritative knowledge, evidence, uncertainty and risk.

The intersection of technological change, human progress, and threats to vulnerable populations is the locus of classic controversies in development studies. This volume brings together social scientists – with expertise in agriculture, economics, area studies, sociology, and political science – and natural scientists to analyze genuinely new questions raised by recombinant DNA technology. Both opponents and proponents of genetic engineering have a poverty story to tell, but with divergent implications.

Analytical attention focuses on:

- yields, incomes and risks
- bioproperty, patents and farmer strategies
- biofortification, nutrition and food safety
- biodiversity, environmental uncertainty and risk
- costs and efficacy of biosafety regimes
- agroecological alternatives.

Transgenics and the Poor begins by disaggregating the needs of the poor – for income, nutrition, environmental integrity – and asks: what does emergent theory and evidence reveal about the conditions under which frontier applications in biotechnology might contribute to meaningful development?

This book was previously published as a special issue of the *Journal of Development Studies*.

Ronald J. Herring is Professor of Government at Cornell University.

TRANSGENICS AND THE POOR

Biotechnology in Development Studies

Edited by
Ronald J. Herring

Routledge
Taylor & Francis Group
LONDON AND NEW YORK

First published 2007 by Routledge
2 Park Square, Milton Park, Abingdon, Oxon, OX14 4RN

Simultaneously published in the USA and Canada
by Routledge
270 Madison Avenue, New York, NY 10016

Routledge is an imprint of the Taylor & Francis Group, an informa business

Typeset in Times New Roman by KnowledgeWorks Global Limited,
Southampton, UK
Printed and bound in Great Britain by MPG Books Ltd, Bodmin, Cornwall

British Library Cataloguing in Publication Data
A catalogue record for this book is available
from the British Library

Library of Congress Cataloging in Publication Data
A catalog record for this book has been requested

ISBN 10: 0-415-38010-3 (Hbk)
ISBN 13: 978-0-415-38010-2 (Hbk)

CONTENTS

CHAPTERS

The Genomics Revolution and Development Studies: Science, Poverty and Politics

RONALD J. HERRING

I. Promethean Science, Pandora's Jug

The intersection of technological change, human progress, and threats to vulnerable populations is the locus of classic controversies in development studies. Ned Ludd contributed his name to one hostile characterisation of opponents of technical change, but raised the critical developmental question: whatever some notional aggregate cost-benefit ratio indicates, there is a socially charged question of disaggregation: whose benefits and whose costs?

The genomics revolution in biology resonates with previous contestations of technical change, but has raised genuinely new problematics: property claims and novel organisms unimaginable a generation ago, with implications for human health, food production, trade regimes and environmental integrity. In medical applications, genetic engineering based on recombinant DNA technologies has brought mostly quiescent or appreciative responses from mass publics. Neither diabetic patients nor their physicians typically resist synthetic human insulin, for example, though it has been produced by genetically engineered organisms since 1978. Transgenics in the pharmaceutical sector have been largely immune to mass mobilisation; as with insulin, life-saving is widely accepted as a legitimate trade-off for uncertainty – or, at a minimum, a political reason to focus mobilisation efforts elsewhere. Yet in agriculture and food systems conflicts are dramatic: field-trial crops are burned, experimental stations are attacked, a remarkable transcontinental caravan of farmers travels from India to Europe to protest the global power of multinationals controlling 'GMOs' (Madsen, 2001). The scope of this global dispute is reflected in titles of recent books: *Gene Wars; Pandora's Picnic Basket; Lords of the Harvest; Politics of Precaution; Seeds of Suicide.*[1]

Science is the fulcrum on which this contentious politics rests. Science as agnostic method for adjudicating truth claims in applied genomics is overwhelmed by a politicised science constructed either as target or legitimation in strategies of corporations, government agencies, evangelical politicians, social movements and NGOs. Science becomes less method than arena. 'Junk science' is a common epithet hurled at Vandana Shiva – the leading Pandoran – who writes of 'imperialist science'. 'Western science' as derogatory identifies a logic of inquiry that claims for itself no geography (Nanda, 2003: 125–181). It is not, as pro-biotechnology forces often argue, a contest between Science and Luddism. Science does not presume to answer questions in normative theory, nor risk preferences. Moreover, there are scientists deeply troubled by genetically engineered organisms.[2] There are specifiable 'known unknowns' – horizontal gene flow, allergenicity from novel proteins –and almost certainly 'unknown unknowns' as well.

To say that biotechnology[3] constitutes 'Promethean' science constitutes a normative stance: there are risks – as with the mythical gift of fire from Prometheus to humans – but also potential for great good achievable in no other way.[4] In gifting fire, Prometheus understood that it would confer novel powers and comforts, but also new hazards for humans and other species. Opponents of genetic engineering argue that we are dealing not with Promethean promise and calculable risk, but with Pandora's jug, with correspondingly darker connotations. Pandora was in Greek mythology the first woman on earth, given by Zeus to Epimetheus, brother of Prometheus; Pandora and Prometheus were related from the beginning. Pandora came with a sealed jug – or box in some tellings – that was not to be opened. Opening the jug released all evils that afflict the human species. Pandora's lesson is that unanticipated consequences of a characteristic human trait – curiosity – may be catastrophic. Inquisitive behaviour beyond some limits is proscribed; in criticisms of biotechnology, the creation of organisms nature cannot make is equated to 'playing God', with severe precautionary implications.[5]

A global debate reproduces cleavages introduced by Prometheus and Pandora: between camps that believe the genomics revolution to be more like fire – a source of human progress, entailing risks but amenable to control and wise use – and those who believe it to be more like Pandora's jug, set to unleash unimagined evils on our species and others: from ecological disaster to bioterrorism.[6] Divergent claims to knowledge reflect and justify widely varying, socially conditioned distributions of risk aversion and risk acceptance (Douglas and Wildavsky, 1982). Most North Americans consume transgenic foods with little thought of allergenicity; Europeans are more likely to reject the same foods. Zambia and Zimbabwe in 2002 rejected United Nations food aid containing some transgenic maize kernels in the midst of famine, terming the shipment 'poison'.[7] The poison in question is corn that American parents feed their children routinely. It is hard to exaggerate this cognitive chasm.

Genomic fault lines have not followed familiar North–South tectonics. Among the leaders in genetically engineered crops are China and Argentina – and more recently Brazil – along with the US and Canada. Significant opposition has appeared in Europe and some African nations, and in metropolitan Asia. Global divisions reflect genuinely new developmental dilemmas. Low-income nations could consider their interests threatened by a global technological revolution in agriculture; this concern was explicit in Brazil's early attempt to compete with the US and Argentina by adopting transgenic soy. In New Delhi, the cliché one hears is: 'we missed the industrial revolution, we cannot afford to miss the information revolution'. From this reading of national interest, the very existence of biotechnology puts pressure on the developmental state to make new choices.[8] Those at the bottom of the global hierarchy could, however, and sometimes do, see transgenics as a new dependency trap – or a risk to exports to Europe and Japan – rather than a source of progress.

Poverty has figured prominently in conflicts over the place of genetic engineering in development strategy. India's then-Prime Minister Atal Bihari Vajpayee sketched the pro-poor developmentalist state 'vision' of 'shaping biotechnology into a premier precision tool of the future for creation of wealth and ensuring social justice – especially for the welfare of the poor'. Biotechnology is to fight obdurate diseases, increase agricultural production, combat nutritional deficiencies and protect the environment (Department of Biotechnology, 2001). Any and all of these outcomes could be pro-poor if realised. Public intellectuals in India have written of 'seeds of death' (Shiva et al., 2000). Both supporters and opponents of transgenics have a poverty story to tell. Proponents have resurrected the Reverend Malthus in a view of aggregate food security that is often global: 'feeding a hungry world' is the corporate expression. Per Pinstrup-Andersen and Ebbe Schiøler, in a book that won the World Food Prize for 2001, conclude: 'once again Malthus's clash between population growth and food production looms threateningly on the horizon'.[9]

In the optimistic scenario, expanding the production possibilities frontier with transgenic crops offers special advantages to the poor. Food scarcity prices poor consumers out of the market. For poor producers, seeds represent a divisible technology, in theory scale-neutral (Lele, 2003). Certain transgenic crops may even alleviate some financial vulnerabilities of the small farm: transgenic seeds such as those containing a gene from the common soil bacterium *Bacillus thuringiensis* – hence the '*Bt*' designation – substitute a plant's own biological processes for cash-intensive inputs such as pesticides. Cash-intensive inputs differentiate farmers along lines of wealth and connections; bad harvests often produce crippling debt through such intermediaries as usurious pesticide merchants, leading to widespread farmer suicides in India in 1998.[10] Typically less able to tap inexpensive credit through social standing or political connections, the poor farmer is especially eager to reduce upfront cash expenditures. If the technology works, the question then turns to the price of seeds, and thus on the intellectual property contained therein.

Opponents of transgenics argue that it is precisely the most vulnerable people who will be most at risk; critiques often begin with intellectual property (Shiva, 2001). Biotechnology is held to enable new potential for monopoly control of seeds; it simultaneously enables global bio-piracy that plunders genetic resources of indigenous peoples and poor nations to make corporate property (Shiva et al., 2000; Shiva, 2001). Poor farmers, in this view, will be crushed by bondage to multinational monopolists, re-subordinating poor nations to neo-colonial control. The most excoriated mechanism has been 'terminator technology', widely (but falsely) charged to Monsanto, the lightening rod of global protest. Critics also see the poor as threatened by environmental degradation, unsafe foods introduced through foreign aid and public distribution systems, or allergenicity from novel proteins (Altieri, 2001; McHughen, 2000: 160–169; Sharma, 2004; Shiva, 2000; Winston, 2002: 107–129).

What unites authors in this collection – scientists and social scientists alike – is conviction that these are questions of great consequence and are amenable to empirical treatment, common practice notwithstanding. Being wrong about transgenics could have adverse consequences for the poor – both poor nations and poor individuals. If opponents of transgenics are correct, but proponents win politically, lives of the poor could deteriorate further. If proponents are correct, but critics win politically, the poor would be denied significant opportunities for improving their lives. Precautionary approaches are therefore not costless: the status quo is hardly risk-free for the world's poor. Proponents of genetic engineering argue that alternatives to some technical advances are more costly, slower, less certain, or impossible.

Honest science is Promethean: there can be no proof that all risks are known, or that current knowledge will not be undermined by new findings. For this reason, science often loses politically to the ominous uncertainty of Pandora's jug: fear of unknown unknowns (Herring, 2001).

II. 'GMOs': The Political Biology of Labels

Almost universally, opponents of genetic engineering label its products 'GMOs' for 'genetically modified organisms'. Thus develops market segmentation and a niche for 'GMO-free' labelling on grocery shelves and export baskets. 'GM-free zones' crop up in southern Brazil, but also in California. The designation 'GMO' posits and reifies a category, and thus a niche for mobilisation and product differentiation, where many biologists would find none – an artificial distinction. Genetic modification is the history of agriculture. All existing crops are genetically modified – that is the purpose of plant breeding, which has been with us in a more or less scientific form for over a hundred years, and with us as a species for at least 6000 years. The current distribution of plant species cultivated for food and fibre has involved radical and purposive reduction of biological diversity for instrumental human ends. We would otherwise be, as a species, unable to feed ourselves.

Plant breeding modifies genomes – there would be no point to it otherwise. Conventional techniques of genetic modification are often inconsistent with an idyll of Gregor Mendel puttering with his peas. In addition to familiar and now naturalised techniques of selection and crossing, the conventional repertoire includes more invasive and radical techniques: emasculation; intergeneric crossing (of unrelated plants of different species – for example, *triticale*, a cross of wheat and rye by conventional breeding); embryo rescue; haploid breeding and induced mutation that produces potentially useful genetic variations, and mostly fatal ones, by means of toxic chemicals or radiation.[11] Recombinant DNA technology – moving a specific sequence of DNA from one place, or species, to another – expands the scope of plant breeding, producing alterations more quickly with less unwanted (and unknown) movement of genetic material. This technology is appropriately called genetic engineering, the product of which is a transgenic organism. 'Transgenic' is a more precise designation of organisms that result from rDNA technology: a biological category, rather than a political one.

Which plant breeding techniques should be considered natural, or dangerously unnatural, is an emotionally charged and non-negotiable vector of global disputes. Whether genetic engineering represents one end of a continuum of plant breeding or a radical departure separates the discourse of Frankenfoods from the stance of the US Department of Agriculture on 'substantial equivalence'. Though the term 'GMO' is now so embedded in policy and political discourse that virtually everyone uses it, it is important to underscore its biological ambiguity and political loading. It is recombinant DNA work that has energised the debate, because of its unique potential and consequent susceptibility: organisms believed to be impossible in nature.

Segmentation of global markets, driven by diverse interpretations of evidence and risk, creates strong interests in maintaining distinctions. A new global market niche for 'GMO-free' food joins organic products. This distinction has consequences for farmers. Papaya, for example, was genetically engineered to stop the ring-spot virus that was devastating crops in all papaya-growing countries. The transgenic was developed by two universities, Cornell and Hawaii, not by corporate giants. Adoption was rapid, and did not follow the three stage S-curve familiar in technology adoption, wherein larger farmers appropriate innovators' rents because they can stand more risk and command more resources. Rather, both in numbers and acreage small farmers dominated large (Gonsalves et al., 2007). Rapid adoption was driven by what might be termed the *insulin effect*: saving the plants took precedence over ideological distinctions. What eventually hurt small farmers who adopted transgenic papayas was not failure of the technology, nor intellectual property, but Pandoran logic: regulatory exclusion of transgenics from their major market, Japan.[12] In a deeply ironic outcome, the beneficiary of Japan's restrictive policy was the transnational Dole, the world's largest producer and marketer of fresh fruits and vegetables.

There is nothing comparable in response to genetically engineered pharmaceuticals: zones for 'GMO-free' medicines. There are both biological and political reasons for this divergence. Genetic engineering creates uncertainties, perhaps risks, in agriculture that are less likely in laboratories. Genes will travel through agro-ecological systems, with unpredictable results. The practice of '*pharming*' – the use of transgenic plants as factories for production of commercially useful chemicals – blurs these boundaries and represents a higher threshold of risk even for techno-optimists (Thies and Devare, 2007).

Some transgenic questions are thus genuinely new. Yet the intense controversies resonate with fundamental developmental questions – the *telos* of societal change, Promethean promises and threats of novel technology, distribution of benefits and risks differentially across segments of society. For many in the development community, the very question of benefits for the poor from transgenics is unthinkable: an instrumental ideological cover for corporate globalisation. Empirics thus far do not bear out their pessimistic scenarios. Yet their concerns warrant our collective attention if we are serious about anticipating consequences and clarifying conditions under which pro-poor outcomes are possible. The public and political discourse around biotechnology has largely taken a dichotomous and generalising form. We collectively seek to disaggregate where needed, while drawing larger pictures when possible.

III. The Emergent Consensus: From Terminator to Toolkit

An optimistically evoked 'international community' repeatedly declares global commitment to poverty reduction. The very existence of absolute poverty constitutes an ethical imperative to apply new knowledge – including plant science – to alleviate limits on human potential (Nuffield Council, 1999). Ethical choice presupposes knowledge, which is difficult to accumulate at the frontier (Leissinger, 2000). Howarth Bouis in this issue of *Development Studies* notes one of the many moral imperatives that make precautionary or obstructionist stances problematic in ethical terms: 'Globally, about 3 million children of preschool age have visible eye damage owing to a vitamin A deficiency. Annually, an estimated 250,000 to 500,000 preschool children go blind from this deficiency and about two-thirds of these children die within months of going blind'. Supposing 'Golden Rice' could reduce that catastrophe, even marginally, what evidence of risk from transgenic rice that alleviates vitamin A deficiency would justify denial of this technology?

Despite continuing opposition to 'GMOs', the terms of debate have clearly changed through acceptance of transgenic crops by more international organisations, farmers, and the epistemic community of agricultural scientists. Endorsement of biotechnology – including positing of benefits specifically for the poor – by the UNDP in its 2001 *Human Development Report* – 'Making technologies work for human development' – stunned some NGOs that had counted on that organisation for support of alternative development thinking (Kothari and Chawla, 2002). The UNFAO (2004) report on the *State of Food and Agriculture 2003–2004* endorsed genetic engineering specifically for the poor, joining the position of the Consultative Group on International Agricultural Research of the World Bank (CGIAR). A recent report by WHO (June, 2005) reviews a great deal of science and concludes there are significant benefits to be obtained, especially in regard to health and nutrition in poor countries. Farmers' positive agronomic experiences would seem to be the reason for continuous expansion of acreage, crops and farmers; these data are tracked in publications of the International Service for the Acquisition of Agri-Biotech Applications (ISAAA) typically authored by Clive James.[13] ISAAA's annual accounting has become the reference of record. Yet even this sharply upward sloping curve significantly understates transgenic adoption: illegal, unregistered, gray-market stealth seeds are cropping up in many settings but cannot be fully documented.[14] The discovery of transgenic crops in fields where they are not supposed to be – either in the bio-safety sense of being approved for planting or in the bio-property sense of violating proprietary claims – indicates substantial undercounting in official sources. We are reminded again that data are social products, the conditions of their production determining their relation to reality (Herring, 2003b). Farmers seem to be voting with their ploughs, whether or not bio-safety committees and Monsanto confer their blessings.

Developmental professionals have increasingly agreed to something like a standard narrative of biotechnology.[15] It is an optimistic but cautious consensus. Per Pinstrup-Andersen and Ebbe Schiøler's *Seeds of Contention* won the World Food Prize in 2001; it is indicative of the dominant narrative. This consensus departs both from the apocalyptic vision of the many NGOs opposed even to testing transgenic crops and from the transparently instrumental propaganda of multi-national firms selling seeds. The metaphor of the 'toolkit' is telling: transgenics will not solve the problem of 'world hunger', but represent a new tool, just as many traditional tools are proving either inadequate or come with too many cumulative externalities – particularly environmental. Whether or not this new tool should be used, and under what conditions, goes to the heart of representation, reflecting the global rift of biopolitics. Participation is increasingly valued in developmentalist logic: who speaks for the poor? Pinstrup-Andersen and Schiøler claim to represent no one, but state that 'too many well-to-do individuals and groups from Europe and North America have taken an unacceptably paternalistic position, claiming to represent the interests of the developing countries and to know what is best for the poor within these countries'. The 'almost silent majority' of people in low-income countries are not being heard (Pinstrup-Andersen and Schiøler, 2000: xi).

Using the right tools is no simple task in this consensus; existing incentive structures, direction of research and property configurations augur ill for the poor. The sharp decline in the ratio of global public sector to private-sector agricultural research means that multinational firms dominate. Yet there is in principle no reason that public sector research could not yield results comparable to those of the private sector. Research and development costs are daunting, but large nations such as China, India and Brazil have public sector institutions that can operate at the technological frontier. The first field trials of transgenics were approved in 1987; globally, more than 11,000 field trials of 81 transgenic crops have been approved. The single most tested crop is maize (corn), followed by canola, potato and soybeans (Pray and Naseem, 2007, Figure 3). Soybean trials rank fourth, tied with cotton at 7 per cent each. Transgenic rice is both officially and illegally undergoing field tests. Yam has not been certified for field trials, but other 'orphan commodities' – for example, bananas, sweet potato, lentils – have all received approval for field trials in at least one country each. The direction of change is not toward consolidation and monopoly, but toward multiplication of actors and crops.

Dominance by multinational firms has been important politically, but has no necessary connection with modern biotechnology. Public investment in biotechnology is held to be justifiable in the emergent consensus because intensification along lines of the *status quo* is unlikely to be sustainable. After weighing logical pros and cons and looking to some evidence, Pinstrup-Andersen and Schiøler conclude that 'the organic approach, while certainly a worthwhile option in regions with the space, the labor and the consumer purchasing power … is not a cure-all' (2000: 79). Biotechnology could bypass some of the worst externalities of the 'green revolution' while avoiding the yield limitations of low-input agriculture (Conway, 1997).

The optimistic scenario is hedged by Promethean concerns: new organisms necessitate surveillance and control. The final chapter of Pinstrup-Andersen and Schiøler's book is tellingly entitled 'Moving forward: handle with care'. The authors underscore problems of concentrated control of technology by unaccountable firms; they endorse ethical scrutiny of each step in evaluating transgenics; they stand for 'free and informed choice' for consumers and farmers, and, emphatically: 'extermination of a terminator' (2000: 135).

'Terminator technology' has been perhaps the most effective weapon for mobilising global opposition to transgenics;[16] it is not surprising that the standard narrative disavows it. The terminator creates a bio-cultural abomination: a plant that cannot produce viable seeds. Opponents charged that gene use restriction technology [GURT] would increase costs to farmers, force them to buy more products (gene activators) and prevent saving of seed for future planting. Though the terminator has been emblematic of the political dramaturgy deployed against biotechnology, the deep irony is that, as Pinstrup-Andersen and Schiøler note, gene use restriction technology would be the only certain means of preventing horizontal gene flow – the major environmental risk. To date there have been no applications for field trials, nor field testing of terminators. Of greatest importance to small farmers, there are compromise technologies that require activation only if the farmer wants to retain the engineered trait; otherwise, the seed reverts to pre-transgenic genome and can be saved, absent the inserted trait.[17] This concept averts the most serious environmental uncertainty and is compatible with concerns for seed choice and seed saving among poor farmers.

The standard narrative then has two major components that temper optimism about pro-poor outcomes: *Biosafety* and *Bioproperty*. With assumptions about adjustments to bioproperty, and establishment of biosafety institutions, the pro-biotech narrative acknowledges major points of critics and answers them. But there is a third assumption less acknowledged or discussed: *Biopolitics*. Recombinant DNA technology introduces a deep ideational divide on the nature of the natural. Benefits for the poor frequently depend on coalitions that use political means to countervail their structural disabilities in markets: minimum wages, social security, nutrition programs, health insurance, transfer payments. These coalitions support public goods especially important to the poor – clean water, education, law and order. Such coalitions are rare and fragile (Moore, 2003); they often divide explicitly on transgenic organisms. Howarth Bouis (2007) in this collection writes confidently that biofortification has the great advantage of not needing to rely 'on behavioural change as a condition for success … [like] … adding fluoride to drinking water in developed countries'. But in Ithaca, New York, home of Cornell University, a leading centre of science research and education, and inhabited by a notably well-educated and progressive community, there is no fluoridation of water, despite widely accepted benefits, particularly for the poor. Even professional associations of dentists, for whom this public good reduces private income, support fluoridation of water. It is politically impossible to put fluoride in the water in Ithaca for fear of poorly specified risks; Pandora has left her mark everywhere.

Citizens encounter biotechnology through a hyperbolic and distorted discourse; scientific literacy is low. The Frankenfish tomato-with-flounder-genes had a robust life in popular junk-science but was a hoax (McHughen, 2000: 14–16). 'Terminator technology' cotton did not exist in India but frightened both public and some farmers, though the discourse ultimately departed too far from farmers' experience and interests to be sustainable (Herring, 2005).

Public caution is thus understandable. And there are good reasons for mass publics to lack confidence in science. The Merck Vioxx (rofecoxib) episode is illustrative. Merck is charged with suppression of data, in collusion with regulators, causing tens of thousands of unnecessary deaths; the firm now faces about 7000 court cases in the US. One scientific study Merck cited in its defence of Vioxx has been discovered to contain falsified data: the prestigious *New England Journal of Medicine*, in an unusual step, posted an editorial disclaiming the article, which it had published five years earlier.[18] Dr David Graham, associate director for science in US Food and Drug Administration's office of drug safety, testified in Congressional hearings that the number of additional heart attacks and strokes could be 139,000, with 55,000 deaths. Graham explicitly argued that data were available to predict this catastrophe but were ignored by the FDA. Senate Finance Committee Chairman Charles Grassley, Republican of Iowa, rather quaintly termed the relationship between regulators and regulated too 'cozy'.

Critiques of biotechnology posit hasty, self-referential science dominated by symbiotic relationships with regulators and producers (Winston, 2002: 58–82). Moreover, there is no assurance that research driven by the profit needs of corporations or wealthy farmers will coincide with the needs of the poor. There is little private incentive to produce for small markets of poor people, especially when the political climate for acceptability of transgenic crops in low-income countries is uncertain or hostile (Potrykus, 2004). Orphan crops could thus well join orphan drugs as instances of market failure. Addressing market failures that produce human misery is a central legitimation of development policy. The standard narrative thus converges on redirection of biotech research and development as necessary conditions for reaching the poor. Getting the institutions right is then a necessary condition for purposive pursuit of poverty-reducing outcomes (Cohen et al., 2003). Where both Promethean and Pandoran discourses mislead is in failure to disaggregate both biotechnology and poverty: across economies, social stratification systems, crops and agrarian structures.

IV. Disaggregating Pro-Poor Claims

Poverty is defined variously, but a biological construct seems least subject to measurement noise and subjective entanglements. Early treatment of 'absolute poverty' defined a human condition so limited that the genetic potential of the species is denied to individuals. This is an outcome notion of poverty: do people have the means of realizing the potential of the genes with which they are universally born?

Much of the debate on biotechnology has been about farmers, but most of the world's poor do not control food-producing land; their relation to the genomics revolution is through food prices, labour markets, biofortification of staples, bioremediation of environmental hazards, and possibilities still on the drawing board. Though the poor constitute a heterogeneous category, some primary desiderata are universal:

(a) *Incomes*: For farmers, the question is net returns from new seeds, which depend on an interactive effect of yields and costs. Net employment and wage effects (shadowing productivity gains) relative to food prices are most important for the most vulnerable poor: field workers who have only labour to sell.
(b) *Nutrition*: The poor need more affordable and more nutritious food to improve their health and to live longer and more productive lives. Low-cost food is obviously important; yet the poor consumer's gain can be the poor farmer's loss unless total factor productivity on farm rises.
(c) *Environmental integrity*. More often than for the rich, livelihoods of the poor depend on ecological integrity; environmental degradation affects most quickly and seriously those with the least flexibility in life choices.

This simple accounting does not exhaust the needs of the poor – one thinks of land, shelter, medical services, political access, cultural acceptance, and personal security among others. Nor should consideration of transgenics obscure more fundamental levers: the international regime of agricultural subsidies and protectionism in rich countries, for example, has a much larger impact on incomes of the rural poor than any transgenic crop. Yet technology may matter fundamentally.

Income

Analytically simplest, at first blush, is the question for farmers: under what conditions does genetic engineering allow scale-neutral deployment to raise incomes of small and marginal farmers? Scale-neutral technical change can in theory lower the size threshold of a viable farm; the declining size of holdings worldwide underscores the urgency. The most marginal farmers are most at risk from potentially remediable biotic and abiotic stresses on crops, as they are driven to the most difficult land with the fewest resources for amelioration; water is illustrative (Nuffield Council, 2004: 3.42).

Most evidence from the field is limited to a few crops developed for use in high-income countries: glyphosate[herbicide]-resistant soybeans, corn (maize), canola; insect-resistant corn and cotton. Even so, there is evidence for scale-neutral transgenics giving superior net-returns. The clearest evidence is probably from *Bt* cotton, where small farmers have increased net income through two mechanisms: less cash expenditure on insecticides and better protection from pests, hence better yields.[19] David Zilberman and his co-authors (2007) explain theoretically why this outcome is predictable, and widespread. Unlike lumpy capital investments such as tractors or tube-wells, transgenic seeds – both in theory and on the ground – exhibit what Rogers (1995) calls 'trialability': experimentation with an innovation may be done on a limited basis. Florence Wambugu (2000) stresses that the technology package is in the seed itself, allowing existing skills to be applied without stratifying extension infrastructure and knowledge hierarchies. The empirical evidence that small farmers can take advantage of biotechnology to avoid debts for inputs, provide some insurance against crop failure and raise production has been vetted by development organisations such as UNDP (2001) and the UNFAO (2004).

Evidence to date does not indicate widespread resistance of small farmers to biotechnology where it is available and affordable; quite the contrary, smuggling and *creolization* crop up where transgenics are unavailable or unaffordable. Gray-market seeds spreading underground, farmer to farmer, in Brazil and India, for example, eluded both state managers of biosafety and corporate enforcers of property rights (Herring, 2007). Unless one assumes small farmers to be irrational, this evidence suggests income-positive effects. The worst-case scenario would be for improved yields from transgenic crops to depress farm-gate prices for poor farmers who lack access to the technology but suffer from backwash price effects; this outcome was one of the failures of the 'green revolution' in poverty reduction (Lipton forthcoming). The outcome for poor producers could then be negative even as the outcome for poor consumers would be positive. Should commercially driven export economies continue to adopt cost-saving biotechnology, farmers in resisting nations will face new competitive disadvantages piled on top of the subsidy and protectionist market-rigging of wealthier states.

The exception to scale neutrality must be when the cost-of-entry barrier is too high for poor farmers in imperfect credit markets: for example, expensive seeds from profit-maximising monopolists. Concerns with dominance of private capital in bioproperty are thus warranted, but critics have exaggerated the burden of patents, private-property claims and technology fees. There is a here a curious rejection of simple prudence: farmers will not buy seeds that do not increase margins. Even those farmers who pay royalties are expanding acreage, suggesting positive income effects. Bioproperty has proved much more fluid, contingent and variable than opponents of the technology had feared, as discussed in the section on institutions below.

Global segmentation of markets reflects differential reach of Pandoran logic, and has poverty implications. If the United States Department of Agriculture is correct in its stance on 'substantial equivalence', market discrimination is artificial and wasteful. Though there is softening of European hostility to transgenics, it is not yet clear how product identity preservation and labelling in the global market will affect opportunities for poor farmers. Labels such as 'organic' provide market niches for labour-intensive farmers to sell at premium prices in rich countries. But 'organic' in the US legally precludes transgenic seeds, whatever the agro-ecological practices on farm.

The most obdurate problem of rural poverty is that of landless workers who must find wage employment on whatever crops need labour. They are put at risk by crop choice, but have no voice. What is a livelihood for the labourer is a cost for the farmer. In high-wage agriculture, labour-saving technologies are profitable and will draw investment and development. Herbicide-tolerant transgenics account for a majority of the global transgenic acreage; farmers save money and labour under certain agronomic conditions (e.g., Nuffield Council, 2004: 40–42). Reduction in aggregate demand for labour under many agrarian conditions either destroys livelihoods or puts downward pressure on wage rates or both. Moreover, the rural poor who depend on weeding for a livelihood are frequently those cumulatively disadvantaged across dimensions of social stratification: women, depressed castes, ethnic minorities, migrants. Under those conditions, if herbicide-resistant crops are desirable on other grounds – soil conservation, for example, or use of less toxic herbicides – a pro-poor strategy would necessarily begin with simultaneous discussion of land reforms, rural public works, food subsidies, and other mechanisms to avoid making the poor pay for technology-induced profits (Herring, 2003a).

Labour-displacement is caused by a trait of a variety, not of the method of plant breeding. Herbicide-resistant crops can be bred by traditional means as well as transgenic. For poverty analysis, what matters is the effect of the variety on demand for labour, not whether the varietal characteristic was achieved through transgenic or conventional breeding.

If transgenic crops do reduce demand for labour, poverty effects would be differentiated by agrarian structure. Where holdings are small and relatively equal, as in China, smallholders would be saving their own labour, not depriving others of employment; this seems to be the case with *Bt* cotton. Poor farmers improve their position by freeing up time for other crops or other jobs, as few small farms provide subsistence on their own. Where holdings are larger and less equal, as in India, labourers would lose work applying pesticides.[20] Then the critical question becomes: are wage losses in chemical applications compensated by more harvest labour if yields increase, and by safer ground water and less exposure to toxins? If net wages are lost, but health improves, a difficult trade-off arises. This trade-off depends on variations in wage systems: for example, when wages are based on weight harvested – rather than a daily sum – income increases with yield and ease of harvest, as sometimes occurs in *Bt* cotton (see discussion in Roy, Herring and Geisler, 2007). Moreover, whatever the effect on demand for spraying labour, protection from crop loss has implications for labour: there are no harvesting wages if crops are destroyed by bollworms. It was only fields of *Bt* cotton that survived the 'bollworm rampage' of 2001 in Gujarat (Visvanathan and Parmar, 2002). To the extent that transgenics reduce risk of crop failure, they reduce risks for the landless poor, as well as farmers.

Where farm labour is either scarce, or mostly supplied by farmers on their own land, ability to control weeds enhances yields, returns to labour, and opportunities to take on more land for cultivation when available. One should not assume that poor households have endless supplies of labour, nor that they need little cash. Increasingly, neither is true. Florence Wambugu (2000) has stressed labour constraints as limiting factors in African agriculture. The objective interests of the rural poor in India and Africa may then diverge in herbicide-resistant technology. But even in India, demand for rural labour is highly seasonal: an aggregate surplus of labour, indicated by insecurity and poverty among workers, does not rule out labour shortages in times of peak demand. When farmers complain of short labour supply, they often mean that they cannot hire in discrete bundles of time separated by enforced idleness at a wage that gives them a profit but leaves labourers below the poverty line in annual income (Herring and Edwards, 1983). Disaggregating agrarian structure and crop traits indicates the kind of patient empirical research that will be necessary for meaningful poverty analysis for particular agrarian systems and particular crops.

Health and Nutrition

Most of the global poor are not farmers. The overwhelming facts of poverty are insecurity and restricted options: food comes first, and consumes a larger share of expenditures the poorer one is. Moreover, food expenditures of the poor tend to be weighted towards staples rather than fruits, vegetables and animal protein. As deadly as protein-calorie malnutrition is, nutritionists increasingly recognise that micronutrient deficiencies may be equally or more debilitating. The often imperceptible 'hidden hunger' of micronutrient deficiency reduces energy, stunts growth, impairs cognitive development, and raises morbidity and mortality rates (WHO, 1999, 2001). The potential of bio-fortification of staple food crops – of which pro-vitamin A Golden Rice is the poster plant – figures heavily in claims for the life-saving potential of biotechnology. Engineering staple plants to make bioavailable nutrients for those who cannot afford diversified diets may be superior – in cost and sustainability – to alternatives such as supplementation or fortification of processed foods. The situation in the South reverses that in the North. Consumers in the global North benefit little if at all from transgenic foods, but have been the guinea pigs for testing allergenicity and other putative health risks. In the global South, malnutrition and poor health are more common, corrective health care less available, ability to alter diets less robust (Bouis, 2007).

Biofortification may prove to be the most significant contribution for the poor; crops can be destroyed by natural catastrophe, recessions reduce incomes; international market rigging can reverse gains from higher productivity. But so long as adequate entitlements to food staples can be maintained, nutritional enhancement of those staples contributes directly and significantly to welfare. Spatial disaggregation is important as well. Conventional supplementation and fortification programs may not reach the isolated poor in remote areas, or do so only at great expense and without regularity. Biofortification of staples specifically targets those most at risk. Yet much is unknown. How practical are nutritional enhancements in different agronomic regions and crops? Will consumers accept transgenic foods? Will farmers grow bio-fortified crops? Are there dangers in over-dosage of specific micronutrients for specific individuals?

Howarth Bouis analyses the potentials of biofortification in three breeding substrategies that may be deployed separately on in combination: (i) reducing the level of antinutrients (for example, phytic acid) in food staples, which inhibit the bioavailability of minerals and vitamins; (ii) increasing the levels of nutrients and compounds that promote the bioavailability of minerals and vitamins; and (iii) increasing the mineral and vitamin content. The effort to improve the micronutrient content and/or bioavailability of commonly eaten foods does not substitute for urgent efforts to provide the means for a better and more varied diet desired by but too expensive for the poor.

Claims that transgenic crops are themselves a threat to health are persistent but without validation. For its *State of Food and Agriculture 2003–2004*, the UNFAO (2004) evaluated a large body of scientific literature and found no evidence of any danger to human health from transgenic food plants currently being grown. Similar findings characterise the WHO's recent (June, 2005) report. *Modern Food Biotechnology, Human Health and Development: An Evidence-based Study*. The Nuffield Council report (2004: 4.47) concluded from a survey of existing literature and that 'there is no empirical or theoretical evidence that GM crops pose greater hazards to health than plants resulting from conventional plant breeding'. That report goes further: biotechnology may reduce known dangers in conventional foods from substances such as mycotoxins, glucosinolates, and alkaloids. Because of higher standards of testing, it is fairly certain that no one will die of *Bt* maize, but an analysis published by the US Centers for Disease Control finds that 'foodborne diseases cause approximately 76 million illnesses, 325,000 hospitalisations, and 5,000 deaths in the United States each year' (Mead et al., 2004). Regulatory attention to vague and hypothetical risks of new crops has opportunity costs for attention to clearly documented threats from conventional foods.

Environmental Integrity

The poor depend more on the environment for livelihoods and have fewer exit options than the rich. Critics of biotechnology have conjured ecological disasters; Vandana Shiva and colleagues (1999) wrote that the 'terminator' would 'terminate biodiversity' if India allowed even test trials of *Bt* cotton. Predictions of gene flow and super-weeds conflate to biological Armegeddon (Winston, 2002: 235 ff; McHughen, 2000: 261–262).

The normative question for any technological change is always: compared to what? Modern agriculture poses well-known, and almost certainly unknown, risks; the struggle against pests and weeds has reached desperation levels in many settings (Pimentel et al., 1998). Widespread cultivation of transgenic plants in North and South America has produced no evidence of even incipient environmental disasters (Thies and Devare, 2007). Against the Armageddon narrative of critics, some scientists posit environmental benefits from transgenics (Horsch and Fraley, 1998). Pinstrup-Andersen and Schiøler note that increases in aggregate agricultural production have historically come from two sources other than improved seeds – conversions of landscapes and application of chemicals to fields. Both have ecological consequences. Conversions fragment and destroy habitats and disrupt eco-system services. They argue that 'without the scientific breakthroughs associated with the Green Revolution, the increase in India's wheat production alone between 1966 and 1993 would have necessitated ploughing another 40 million hectares of land' (2000: 20). Raising production requires either more per acre or more acres; the source of more acres is typically ecological destruction.

All agriculture carries risks for ecological systems (Conway 1997: 86–107, *passim*). Mark Winston notes that 'there are hundreds of examples of genes from domesticated crops jumping to feral weeds ... twelve of the world's thirteen most important food crops have sent genes into wild plants through pollen transfer'. One of the banes of my teenage years was Johnson grass (*Sorghum halepense*), an invasive and destructive species that exchanges genes with cultivated sorghum (*Sorghum bicolour*).[21] Franken-weeds resulting from plant gene flow were theoretically possible long before genetic engineering, though most gene flow reduces rather than enhances fitness of subsequent generations of feral plants. The relevant question is of *incremental* risk: how much more risk, if any, does introduction of a transgenic variety entail in comparison with a variety modified by other means?

The poor are especially vulnerable to hazards of modern agriculture; they cannot afford to be selective about what jobs they take. If someone is going to don a backpack sprayer and walk unprotected and often half naked while spewing toxins, it is likely that he will be poor. If anyone is going to drink contaminated surface water, or water from shallow wells, she is most likely from the bottom of the social hierarchy, not the top. Here the claim of *Bt* technology, especially in cotton – a crop particularly prone to pests and toxins – is compelling. Surveys of 400 farmers in northern China found that adoption of *Bt* cotton led to dramatic reductions in use of dangerous organophosphate pesticides and an FAO class II pyrethroid, resulting in significant reduction in incidence of pesticide poisonings (Pray et al., 2002). There is evidence indicating improved farmer health in *Bt*-protected fields as opposed to sprayed fields (Huang et al., 2002; James, 2002: 138–147, *et passim*). Reduction of pesticide spraying can be expected to help curb the epidemic of pesticide poisoning (Jeyaratnam, 1990), conserve some water, and reduce the wholesale destruction of both beneficial insects in the fields and wildlife that depend on those agro-ecological niches. Insect-resistant crops also measurably lower contamination of food crops by mycotoxins produced by opportunistic fungi that infect wounds made by feeding insects (Munkvold et al., 1999).

Though some critics expect biotechnology to 'terminate biodiversity' (Shiva et al., 1999), no mechanism is discernible. The purpose of conventional agriculture is to reduce diversity in the fields, removing competitive plants from those the farmer wants to thrive. Zilberman and his colleagues (2007) find that there are reasons to expect crop genetic diversity not to decline with adoption of transgenics; they cite the evidence of Klaus Ammann, who 'finds that genetic uniformity of crop varieties has decreased by 28 per cent during the period of introduction of transgenic varieties in the US. The evidence suggests that introduction of GMVs around the millennium did not have significant effect on genetic uniformity'. It is possible that the net effect of rDNA breeding would be greater crop diversity because of faster results, more precise breeding for niche applications, and illicit on-farm breeding associated with *creolisation* and stealth seeds.[22]

Generic toxins kill indiscriminately; non-target fauna – some of which prey on pests – suffer high casualty rates – as do their predators, spreading costs up the food chain. *Bt* crops target only a narrow range of *Lepidoptera*: the insect must attack the plant to receive a dose of toxin. There is a kind of natural poetic justice in targeted endotoxins. The selectivity possible in transgenic endotoxins should improve biodiversity of fauna in comparison with spraying of generic toxins. There is evidence for this outcome in China: increased number of insect species in *Bt* cotton fields (Pray et al., 2001; Pray and Naseem, 2007).

Disaggregation produces complex analytics that require significantly more empirical research than exists; in this sense, the serious literature on transgenic crops and poverty is just beginning.[23] Thies and Devare, in analysing one of the poster animals in the gene wars – the monarch butterfly [*Danaus plexippus* L.] as affected by *Bt* toxins – emphasise that there will be variance across transgenic crops modified by different genetic events [particular gene insertions]. Effects can be understood only by disaggregating time: plants flowering at different times have less risk of gene flow in comparison with simultaneously flowering plants, for example. Ecological effects then depend on varietal differences. It was for this reason that one cultivar of *Bt* maize (event 176) was not re-registered with the US Environmental Protection Agency in 2001. The variety contains a pollen-specific promoter and consequently exhibits significantly higher pollen toxin levels than *Bt*11 or MON810 *Bt* corn. Only what Thies and Devare call 'the dialectics of normal science' will sort out these effects, crop-by-crop, event-by-event.

Pitting certain benefits against uncertain dangers presents a difficult public-choice dilemma. When probability distributions are known, risks can be assessed rationally – for example, whether to fly or drive. With transgenics and environmental change we have uncertainties, not probability distributions. It is not helpful to say, as techno-optimists often do, that science should decide; there is no scientific means of placing values on uncertain outcomes. Risk aversions are not subject to refutation; some people fear airplanes, others fear rare diseases – no data will determine correct preferences. The only solution to this public-choice problem is interaction between democratic processes and biosafety institutions. The poor are the least likely to be effective in these deliberations under existing institutional arrangements.[24]

V. BioInstitutions: Property, Regulation, Safety

Property should figure prominently in any poverty discussion. Property enables the capture of income streams from innovation. If private firms are to invest in biotechnology, they expect returns, presupposing enforceable property rights. This functionalist argument for strong property rights is widely accepted in the standard narrative. Not surprisingly, the concept patent on the so-called 'terminator' was for a 'technology protection system'.[25] But there is little money for technology fees on subsistence farms. Moreover, without the terminator, inability to control the technology and to enforce property rights on the ground limits commercial interest in biotechnology for the poor. Biosafety confounds problems with bioproperty. Regulatory delays tie up capital and introduce uncertainty. Limited market potential, fear of piracy and high regulatory costs are typically cited as disincentives for capital. The answer in the standard narrative is more investment by public sector institutions, where public interest – not private return – guides allocations. But as support for agriculture declines in aid budgets, and international institutions' budgets are stretched, would a transgenic initiative divert funds from higher priorities?

Pro-poor opponents of biotechnology argue that monopoly power will soak the poor for super-profits (Shiva et al., 2000), but the picture is more complicated. First, 'humanitarian-use' transfers of intellectual property, and segmentation of markets to permit poor producers to avoid fees charged to commercial producers provide a workable model (Lybbert, 2002). Extremely complex property claims in Golden Rice – 70 patented technologies controlled by 30 institutions – were consolidated to privilege poor producers while allowing for fees on rich farmers (Kryder et al., 2000), under the control of a 'humanitarian board'. This outcome is for many proof of concept in plant breeding driven entirely by needs of the poor. There is also the public sector. China's public sector *Bt* cotton successfully competes with Monsanto's version locally and entered India by way of partnership with Nath Seeds. In India itself, public sector research is increasing significantly, though still far behind that of China. The analogy to pharmaceuticals seems clear: market-driven distribution with strong property rights does not serve the poor, but market segmentation and humanitarian intervention are not impossible, as with retroviral drugs in Africa. In addition to agreements for humanitarian use of proprietary technology, public–private partnerships, and public sector transgenics, farmer-created transgenics have emerged in opposition to global proprietary claims of multinationals. Property rights are not self-enforcing. Monsanto expends great energies trying to collect technology fees in Latin America, with spotty results; high prices of Monsanto's *Bt* cotton in India spurred development of the stealth-seed market, and led to demands to ban Monsanto's varieties – with success in one state. Some transgenes have spread so widely underground that they resemble open-access or open-source technology more than monopoly, more Linux than Microsoft. The transgenic genie is out of the bottle.[26] Even in strong property regimes such as the US, Monsanto is forced into undesirable publicity – suing modest farmers even to bankruptcy – to increase compliance with otherwise unenforceable claims. Since it is impossible to catch everyone who violates contracts prohibiting replanting of transgenic seeds, Monsanto seeks to make examples of a few farmers for deterrence (Liptak, 2003: 18). Such strong manifestations of intellectual property have not proved practicable on a global scale for reasons of transactions costs, politics and law. To date, biotechnology has invigorated a vigorous anarchic and artisanal agrarian capitalism through the spread of stealth seeds, whereas global monopoly power of multinational property in biota is difficult to discern on the ground (Herring, 2007).

Bioproperty and biosafety regimes interact. The weak nature of bioproperty regimes on the ground means that if there is to be monopoly it will be through a strong biosafety regime. Strict control and testing regimes raise costs of seed development beyond what is affordable by small firms, enhancing the power of deep-pocket corporations. If Monsanto cannot enforce its property claims in India, could Delhi's biosafety regime create de facto monopoly by prosecuting producers of underground *Bt* seeds? In theory, yes; in practice, no. India farmer and seed organizations have charged that biosafety officials had precisely this objective, and demand regularization of illegal transgenics, especially Navbharat 151 – the original stealth seed – and especially in Gujarat. Strong bio-property rights and demanding bio-safety regimes drive high prices of official seeds, and thus invigorate the underground market.

There are unresolved property questions spawned by genetic engineering: what can be owned? Where is the line between common and acceptable private property? (Kloppenburg, Jr. and Kleinman, 1987; Hilgartner, 2002). Firms recognize that these issues are not settled. Evidence does not support a super-profit gold-mine interpretation of biotech dominance: Pray and Naseem (2007) show that private firms are decreasing their investments in agricultural biotechnology, whereas public sector institutions in low-income countries are increasing investment. But the public-private distinction in research is often overdone in any event. Even conceptually, it is not clear that private investment precludes collective benefits from biotechnology. Michael Lipton notes that critics have a point in their claim that the poor are unlikely to benefit from the 'focus, so far ... on "cottonseed and chicken-feed" (soy, yellow maize)'. But does the basic science not have implications beyond those seemingly humdrum products?

The standard narrative recommends redirection of and public participation in both research and property. This is happening, but is it not possible that the science itself will enable pro-poor outcomes, whatever the incentives directing research? *Bt* cotton, for example, helps free farmers from pesticide merchants and debt, independently of Monsanto's objectives. Who benefits from the global spread of transgenic crops? Pray and Naseem (2007) conclude from their analysis that the primary beneficiaries of increased revenues to date are not multinationals but farmers and consumers, even in countries that enforce strong intellectual property rights.

Is it possible that 'farmers' in general could benefit but poor farmers lose in the shift from public to private sector dominance of seed property? If transgenics require more upfront cash for technology fees than alternative seeds, wealthier farmers will be advantaged. The poor are less able to afford cash payments from savings and are excluded from or disadvantaged by credit institutions. They often pay more for credit. Precisely the same probabilities of lower ranking in the social hierarchy that make poor farm labourers especially vulnerable to income insecurity and nutritional crisis afflict small farmers in stratified agrarian systems (Herring, 1977). For these same reasons, to the extent that transgenes substitute for upfront cash costs, as in *Bt* substitution for pesticides, they are of special benefit to the poorest farmers; yield increases may well be highest on the poorest farms, reflecting prior inability to purchase sufficient inputs for want of cash or credit (Lipton, 2007). Creative credit institutions are in general of special importance to the poor, especially under conditions of technical change.[27] In the more common scenario, lacking political power, connections, credit and cash, the poorest farmers would be better served by a cropping regime that substituted biological controls for purchased chemical controls.

Technology fees and their enforceability then matter greatly. Poor farmers would certainly lose if technology fees prove to be prohibitive – and enforced – and yields improve on farms of those who can afford the fees. 'Farmers' as a class could still benefit, but poor farmers would be caught in a backwash of lower output prices because of increased yields on adopter-farms, but with no reduction in input costs or increased yields on their own farms. Technical change in this scenario would accelerate agglomeration of ownership and the ruin of small farmers.

The final institutional caveat in the standard narrative is effective regulatory capacity for biosafety, linking far-flung villages to a global regime of the Cartagena Protocol on Biosafety of the Convention on Biological Diversity. Discussion of these institutions often conveys a certain magic-wand quality. The most difficult questions are without answers: How much regulation? Will regulation work? Are the results worth the costs? Thies and Devare note that in the US 100 times more is spent in technology development than in risk assessment and monitoring. Is this too high a ratio or too low, or, in the Goldilocks formulation, 'just right'?

What are the real threats to biosafety? Gene flow among plants is common; more critically, whole genomes are continually inserted into agricultural and other ecologies through invasive plants (Pimentel et al., 1998) and conventionally bred varieties. Invasive species are almost certainly a greater danger than flow of a single gene, which would have to increase the fitness of some wild relative significantly to be of any concern. The Center for Biological Informatics of the US Geological Survey estimates the costs of invasive species to be $138 billion per year in the US (http://invasivespecies.nbii.gov/). Where should the biosafety dollar be spent? There are many competing priorities; the poorer the country, the more poignant the trade-off. Should India tie up microbiologists in *Bt* cotton surveillance while biofortification research goes begging or invasive species wreak ecological havoc?

Biosafety regimes will certainly consume resources; will they work? Seeds are not only divisible as working capital – contributing to scale-neutral characteristics (Lele, 2003), and hence to poor-farmer income – but sufficiently divisible to evade the Panopticon assumed in official biosafety discourse. The original import of *Bt* cotton seeds into India was one-hundred grams; there are now millions of acres under dozens of unauthorised transgenic cotton varieties in the field – no one knows precise numbers.[28] *Refugia* set-asides are recommended to prevent development of resistance to endotoxin-producing transgenes such as *Bt*; it is clear that small farmers would be especially disadvantaged by these requirements – *and* that many, perhaps most, ignore them. However, the science is imcomplete: if there are alternative hosts for the insects, as seems to be the case for bollworms in India, any *refugia* requirement is an unnecessary cost imposed on farmers and society alike.[29] Identity-preservation measures for segregating and labelling transgenic crops are neither costless nor foolproof. In 1999, Aventis Corporation's StarLink corn hybrid, containing the *Cry9C Bt* gene for resistance to European corn borer, was discovered in corn products within and exports from the US. No ill effects from Starlink have been demonstrated, but the variety was approved only for non-human food uses. Nevertheless, food processors and distributors throughout the marketing chain were required to remove all products containing StarLink DNA, at an estimated cost of $1 billion (Moose, 2003). This entire operation may well have been a complete waste of food and money.

Discursive reifications of both seed and state – or 'patents' and 'biosafety regime' – sit uneasily with ground experience. Seeds are highly portable. Saved transgenic seeds and farm-to-farm exchanges undermine both bioproperty and biosafety regimes – a useful corrective to the presumption of monopoly power of multinational corporations, but troubling for the optimistic discourse of biosafety. In both Brazil and India, farmers have crossed transgenes from Monsanto technology to fine-tune local varieties, and save seeds to cut costs. Unauthorised and creolised *Bt* cotton seeds spread underground in India, undetected by the Genetic Engineering Approval Committee in Delhi (Jayaraman, 2001, 2004). Once it discovered these seeds through commercial rivalry – not its own devices – the GEAC proved essentially impotent. As a result, a kind of genetic anarchy evolved in India's cotton regions: packaged farmer-generated *Bt* crosses, F2 seeds from earlier crosses, and officially approved Bollgard varieties (21 as of June 2005), all competing for space in the fields at different price points (Roy et al., 2007). Oppositional discourse in both Brazil and India posited monopoly and dependence – premised in both cases on 'seeds of death' – and targeted Monsanto specifically. On the ground, Monsanto looked more like Gulliver than Goliath; biosafety law appeared as more aspiration than constraint. 'GMO-free zones' declared by popular organisations and governmental institutions seem quaintly fantastical.

Brazil and India present special difficulties for environmental regulation: authority is divided structurally between centre and states in expansive federal political systems with long porous borders; significant power resides in multi-tier courts; farmers are courted by democratic regimes; bureaucracies are not immune to special pleading. Yet one can find comparable characteristics militating against enforcement in many political systems: penetrated or captured regulatory agencies, corruption, bureaucratic incapacity, local power of the nominally regulated. Surveillance of nature is no mean task, either macro or micro, whatever the claims of the 'high-modernist' state of James Scott (1998). It will be difficult given the skill repertoire of modal bureaucrats to distinguish transgenic plants from isogenic varieties. Moreover, gene police will be no more popular in villages than forest guards.

In the optimistic narrative, state technicians will be able to make for society the cost-benefit analyses necessary to decide which transgenics, when, and where. But there are also no probability distributions from which a true risk assessment can be derived and few means of stopping the flow of seeds that farmers want – short of the much-maligned but un-tested terminator. The regulatory regime of the Cartagena Protocol on Biosafety of the Convention on Biological Diversity will be costly and difficult to implement, particularly in the poorest countries, and perhaps ineffective. If benefits are captured by a subset of farmers and seed companies, but costs are spread to society generally, the case for transgenics is proportionately weaker on developmental grounds. If transgenics represent progress, caution or 'regulatory excess' has high costs for the poor. In the foreword to the Nuffield Council on Bioethics report (2004), Chairman Bob Hepple argued: 'It cannot be responsible to render a technology unavailable to those whose needs are urgent. Nor can it be responsible to be partisan in a debate where empirical evidence should be decisive in settling the question'. It is hard to argue with this sentiment, but unlikely that mere empirics will be decisive drivers in biopolitics.

VI. Biopolitics

For transgenics to be incorporated into pro-poor development strategies depends on widely shared assumptions: an appropriate intellectual-property regime, pro-poor research priorities and bio-safety controls for environmental protection. Another assumption, seldom explicitly stated, is that pro-poor strategies for biotechnology are politically feasible. Because the poor are unlikely to be politically powerful, coalitions become critical (Moore, 2003; Herring, 2003a); these coalitions are mediated by politicised science in transgenic policy.

NGOs speaking on behalf of the poor have intermittently blocked or destroyed field trials designed to test transgenic crops. This phenomenon illustrates absence of even the most basic epistemological and methodological grounds for resolution of the politicised science of 'GMOs'. The political vulnerability of genetic engineering is acknowledged uncertainty. Uncertainty is the most powerful weapon of the anti-transgenic movement. The elision of and escalation from uncertainty to anxiety meets little cognitive resistance. Fearing the unknown is not only a first response, but a rational response. There cannot be, for example, conclusive evidence that transgene flow will not cause ecological disruption somewhere, sometime. Real science is inevitably, often radically, incomplete. High anxiety with low information is the condition most likely to generate the powerful effects of symbolic politics (Edelman, 1964: 91–92, *passim*). Where science is uncertain, opponents of transgenics have had ready answers.

There is thus introduced a deep indeterminacy into biopolitics: science is a powerful cognitive filter that stands between structure and interest. Science continually presents new challenges to the way interests are understood by citizens and political classes that control states; the sea change in redefinitions of interests introduced by the atmospheric science of ozone holes and climate change is archetypal. Counter-intuitive links between refrigerator gases and skin cancer were not on the agenda of citizens in pre-Montreal Protocol days; there was a time when climate change was on no one's political horizon. Transgenic organisms represent a particularly compelling instance of these dynamics. Ecosystem health, biodiversity, allergenicity, gene flow – all are sufficiently removed from ordinary conceptual and practical knowledge of most people that reliance on experts, congealed knowledge, or persuasive framing, becomes inevitable. Effects take a long time to become apparent; causality in over-determined chains of dynamics is difficult to parse. And the science keeps changing. Interests are knowledge-dependent, but the knowledge is constantly in flux at the margins, and contested.

It is not surprising that opponents of biotechnology are suspicious of the science. Because science is expensive, what will be found out depends in part on who funds what research, and how much of whatever is found out is suppressed or ignored. Allocative decisions within big science will be driven by markets and politics, not free-floating curiosity or epistemic necessity. In court, each side has its scientists. There are potentially high stakes in small findings – hence restrictions on international field-work for fear of biopiracy. The value of these new commodifications of nature is dependent on social acceptability of the enabling science that certifies safety of products and procedures. Pharmaceuticals have created a spectacle of politicised science for mass publics uninterested in science, but deeply interested in health. NGOs in India demanded field-trial data on *Bt* cotton vociferously while simultaneously telling anyone who asked that no one would believe the results if released – data were produced by a state that was promoting biotechnology, by scientists whose careers depended on the results and by commercial interests with profits at stake.[30] It is equally true that many critics would lose livelihoods if 'GMO' controversies died out.

The genomics revolution created possibilities for conversion of nature to property unimaginable a generation ago (Tanksley and McCouch, 1997). There is a profound question of social justice in allocating returns from property in nature: whether or how those who have forgone benefits from destruction of nature by leading impoverished but low-impact lives might benefit from new knowledge. This potential seemed to alter the biopolitics of conservation, giving nature itself tangible value. So long as biodiversity is valued only in normative terms, as a desirable thing, its political base is everywhere fragile. But valuable genetic information may depend on actually existing biodiversity. Biotechnology potentially valorises, perhaps remunerates, both biodiversity and local knowledge (Gupta, 1998; Weiss and Eisner, 1998). Yesterday's pest could harbour tomorrow's miracle gene; who is opposed to curing cancer? In the normative spectrum from 'biopartnerships', to 'bioprospecting' to 'biopiracy', there is a common assumption that variable relationships between value and new forms of property are worth contesting (Svarstad and Dhillion, 2000). Through this knowledge-based revaluation of nature, potential monetary incentives are injected into the political struggle over micro-nature. For some years, the Environment Minister had something to say to the Commerce Minister when opportunity costs of conservation were questioned.[31] Bioprospecting largely failed to materialise, but the very potential left behind opposition to biopiracy. When the BBC characterised the Indian firm Navbharat's appropriation of Monsanto's *Bt* cotton gene as 'biopiracy', the rhetorical tables were turned; the assumption that genetic flow can move only from South to North is clearly problematic.

Anti-transgenic ideologies have a tactical political advantage in biopolitics, derived from core characteristics of science-as-method – sceptical agnosticism, tentative conclusions, replicability, validation in epistemic communities – commitments at odds with the simple messages and certainty privileged by politics. Junk science forgoes these complications in favour of reductionist explanations that evoke anxiety in spheres of low information or cognitive complexity: 'the terminator will terminate biodiversity' was ringing through rallies even as farmers saved, propagated and replanted the 'terminator' seeds in India. A claim in real time that UN food shipments to Africa are 'poison' will elicit a precautionary response; it is quite a different matter to assess the probabilities that some novel proteins may prove allergenic to some people at some time. As science is inevitably a work in progress, and therefore incomplete, uncertain if honest – mis-recognition of interests through obfuscating ideologies renders real science politically vulnerable in biopolitics.

VII. Development Frontiers

Despite Michael Lipton's caution that biotechnology's 'hype-line' often over-runs the 'pipeline', there is a great deal at the frontier. Thies and Devare discuss how rDNA technology is being used to create crops resistant to abiotic stresses such as drought, soil acidity, and salinity. Advances in resistance to abiotic pressures could stabilise and increase yields of crops grown in marginal, low-productivity areas in which the poor are often trapped. Increased storage stability and delayed ripening would disproportionately benefit those with little market holding power or resources to invest in storage and transport facilities.

Because of the seemingly limitless scope of genetic engineering, development discourse has placed too much emphasis on the seeds themselves: whether 'miracle seeds' or 'suicide seeds'. Alleviation of the life-limiting conditions of poverty cannot rely on technical change. The primary causes of poverty, and therefore solutions, lie in much larger structural forces and policy choices. Seeds cannot carry too much of the load when so much is biased against the poor.

Development policy hinges on alternatives: compared to what? The emerging consensus among development professionals is that more public sector funds should go into biotechnology, in part to mitigate poverty. But what returns at the margin would justify more international funding of transgenics? At the cost of what other projects? It is easy to argue that biofortification research is more likely to aid the poor than any number of amorphous and trendy sinkholes for monies that satisfy cyclical imperatives of funders and fundees. But what of research on invasive species and ecological decline, water conservation – or agroecology? The frontier issue is opportunity costs at the margin: the best use of the additional dollar.

The first ethical dictum of development policy is to do no harm. Yet there is no consensual metric to determine how much caution is warranted, how much is too costly. Consider the problem of scale. Agro-ecologies vary from district to district in India, even from village to village, but are not fully mapped. Does a biosafety regime require testing across all ecological variations, or is rough comparability enough? What is an adequate sample? The Nuffield Council on Bioethics recommends sharing methods and data among countries that have similar agroecologies (2004: 73–75). This recommendation begs the question: how similar? Rajasthan is something like Texas, sharing acacias and low rainfall, but is quite different in other ways. Few would argue that Texas data are proxies for Rajasthan, but does one test district by district, village by village? For how long? If we knew what exactly in the ecology needed to be matched up, we would have powerful enough theory to do away with much testing. There is institutional uncertainty as well. Mechanisms for biosafety are almost certain to be at least partially ineffective: is the certainty of gene-use restriction technology – aka 'the terminator' – a more cost-effective answer to gene flow? A terminator that could ensure some biosafety desiderata could simultaneously render property claims of commercial firms much more enforceable than is currently the case. Under what conditions would societies accept terminator technology?

Equally unsettled are the comparative returns on different avenues of research. Compared to spraying toxins, *Bt* crops fare well in terms of biodiversity in the fields, human health and crop protection. But is the relevant comparison conventional agriculture – which many believe unsustainable – or alternatives from agro-ecological approaches? Norman Uphoff's contribution to this collection argues that research monies for improving genetic potential compete with superior pay-offs from agro-ecological research. Uphoff's data from multiple test sites across several continents on the system of rice intensification – itself subject to scientific controversy (Surridge, 2004) – suggest that there are novel ways to increase the percentage of genetic potential realised by plants in the field by improving management and technique.[32] Is research in this field more productive at the margin than research designed to increase the nominal genetic potential of the plant itself? Uphoff's eco-agricultural techniques for rice are certainly compatible with rice that makes beta carotene in its grain, not just its leaves; the result would presumably be a double benefit for the poor: higher yields of more nutritious rice. Yet Golden Rice required an extraordinary feat of institutional cooperation to become viable and is yet to reach actual farmers (Kryder et al., 2000; Potrykus, 2004).

One could have Golden Rice with high or low yields; one could have high SRI yields with no pro-vitamin A in the endosperm, or only in the non-edible plant parts as with conventional rice. From an agro-ecosystem perspective, use of herbicide-tolerant crops allows reduced and zero-till practices to work more effectively, reducing soil disturbance so as to retain more organic matter, improve soil structure, reduce soil compaction and improve soil water relations. In both cases, combination of agro-ecological advances with transgenics seems optimal, as some agro-ecologists recognise (McNeely and Scherr, 2003: 153). Bambawale et al. (2004) found that the best performance of participatory integrated pest management in their field study in

India was with transgenic cotton – Bollgard 162. Devparna Roy's unpublished field investigation of organic farmers in India finds that some use *Bt* cotton – a transgenic crop very much contested nationally – with chemical-free cultivation techniques in accord with Gandhian values.[33] This strategy of walking on two legs is the essence of Gordon Conway's (1997) 'doubly green revolution'.

There is but one problem with this sensible conclusion: the unresolved issue between Prometheus and Pandora. In the US, 'organic' is a standard that defines a rapidly expanding and advantageous market niche. 'Organic' officially excludes 'GMOs'. Yet the objectives of transgenic engineering and organic farmers in reducing soil erosion, synthetic chemical inputs and destruction of helpful insects are congruent. To the extent 'GMO-free' remains a meaningful market category, the threat of gene flow escalates in financial terms. It is precisely poorer farmers who have the lowest opportunity costs of labour and can engage in labour-intensive practices associated with organic agriculture. This market niche is thus of some importance to poor farmers, so long as rich consumers prefer organic produce and are willing to pay the premium. In this case, the externalities of transgenic crops through gene flow ('genetic pollution') are anti-poor, but only so long as the 'organic' label categorically rejects transgenic plants. This bright-line distinction between recombinant DNA and other methods of moving DNA around plants undermines an otherwise intuitive complementarity.

Whether or not genetic engineering constitutes a fundamental break in manipulation of plant genomes represented by thousands of years of breeding depends on a more basic cognitive divide impervious to data. This is a divide between what we might call an *organismic* view of nature and a *molecularist* view. From an organismic perspective, putting a fish gene into a tomato – a hoax used by opponents of genetic engineering, but not an inconceivable outcome – violates some threshold of the unnatural. For the pure molecularist, there are no fish genes or tomato genes, just variable organisations of bases in DNA, arranged in different networks: all life is composed of the same stuff. Different arrangements of this genetic variation are largely accidents of evolution. In the organismic view, species constitute the natural world; to disturb this order is 'playing god' (Herring, 2001). Political contestations of transgenic crops, like those in embryonic stem-cell research and cloning, depend more on theological and ontological dispositions than on science. Whether or not transgenic technology for the poor is in the public interest depends on how one conceptualises the public, how one couches the alternatives, the normative position one takes on uncertainty and risk, and the projections one makes from an inevitably incomplete science.

Acknowledgements

The editing services, patience and persistence of Kristen Ebert-Wagner made this collection possible; we all owe her a deep debt of gratitude.

Notes

1. Respectively, Dawkins (1997), McHughen (2000), Charles (2001), Paarlberg, (2001), Shiva et al. (2000). On the global divide, Winston (2002: 235–8), Pinstrup-Andersen and Schiøler (2000: 1–31).

2. Ho and Ching (2003); e.g., The Union of Concerned Scientists in the US and the Independent Science Panel in the UK.
3. Defined as use of living organisms or their products for commercial purposes, biotechnology begins at least as early as written history in brewing beverages and making bread. Contemporary biotechnology is much broader than genetic engineering, including such techniques as tissue culture and marker-aided selection.
4. Serageldin and Persley (2000). Claims of the unnatural nature of biotechnology reproduce – through references to *Frankenfoods* and even *Frankenpants* (made from *Bt* cotton) – the original Promethean connection established by the monster's creator: Mary Wollstonecraft Shelley subtitled her 1818 classic *Frankenstein, 'The Modern Prometheus'*.
5. The most notorious statement is perhaps that of Prince Charles, to the effect that scientists working on transgenic foods had strayed into 'realms that belong to God and God alone'. Though the Bush administration presses transgenics upon Europe and Africa, evangelicals in their coalition oppose genetic engineering on theological grounds.
6. Sharma (2004: 17–23) conjures the 'demon seed'; on the common attributions 'seeds of death' or 'suicide seeds', Herring (this issue).
7. *Scientific American* August 2004, 291(2): 8. The UNFAO warned that the 'war of rhetoric' might be more dangerous for small farmers than the transgenics themselves. Denying food to starving people to protect them from risk was almost certainly motivated by concern about European markets for agricultural products more than welfare of poor people.
8. Paarlberg (2001: 6–8 and Chapter 4); on developmental state theory, Woo-Cumings (1999).
9. Pinstrup-Andersen and Schiøler (2000: 31). The authors understand obvious problems in the Malthusian narrative, but there remains a wide-spread conception – echoing corporate public relations – that biotechnology means more food and more food means less poverty. To take but one (large) example, India is beset by staggering costs of managing 'surplus' food production, but the poor remain malnourished.
10. It is not known whether or not farmer suicides were abnormally frequent, nor what percentage were connected to economic distress (Karnataka, 2002), but the tragedy was attributed by some public intellectuals to transgenic seeds (Stone, 2001; Herring, 2005).
11. McHughen (2000: 63–66), Winston (2002: 11–34, *et passim*), Nuffield Council (2004: 22–5), Shelton (2004).
12. World Trade Organization regulations permit such technical barriers when 'information about health, hedonistic, or ethical attributes of agricultural products is either unknown or asymmetrically distributed between producers and consumers, and the transaction costs of obtaining this information are prohibitively high for consumers'. Roberts et al. (1999).
13. See his annual *Global Status of Commercialized Transgenic Crops*, ISAAA Briefs, Ithaca, NY. For 2004, James reports approximately 8.25 million farmers in 17 countries growing transgenic crops on about 81 million hectares; one-third of this area fell in low-income countries. The preponderance of area is in the US, Argentina, Brazil, Canada, and China.
14. Brazil was counted as zero in transgenic production until illegal movement of glyphosate-resistant soy seeds across international and internal borders assumed such proportions and raised so many political and juridical conflicts that the myth of 'GMO-free' could not be sustained; it is now a leading producer. Likewise, before 2002, India officially had no transgenic crops, but farmers were growing *Bt* cotton in several states. Paarlberg (2001: Chapter 4), Visvanathan and Parmar (2002), Herring (2005).
15. A 'development narrative' is a story both normative and causal, that simplifies complexity and through its simplifications suggests some course of action. Roe (1991), Dryzek (1997) for examples of useful application.
16. Genetic-use restriction technologies (GURT) were dubbed 'terminator technology' by RAFI, a Canadian NGO. It is a telling characteristic of contemporary global politics that a few people and a server could have launched so powerful a representation.
17. Curiously, the Nuffield Council Report (2004: 51–2) does not investigate this potential, despite its concern with the consequences of GURT.
18. On 18 November 2004, in hearings before the Senate Finance Committee in the United States, an FDA report was cited as estimating that 27,785 heart attacks and sudden cardiac deaths would have been avoided if patients had not used Vioxx (rofecoxib), an arthritis drug, from 1999 through 2003. Text relies on CSPAN telecast Senate committee hearings and reportage by the *Washington Post* and

New York Times. The Editorial of the *New England Journal of Medicine* was posted to their web site, 8 December 2005.

19. James (2002). Evidence is discussed in this issue in contributions by Herring, Roy et al., Lipton, Pray and Naseem, Zilberman et al.
20. Ruifa Hu and Carl Pray, personal communication, find in their survey of about 400 *Bt* farmers in China, self-labour constituted 96.5 per cent of the total labour used, hired labour only 1.7 per cent.
21. Winston (2002: 95) states that Johnson grass 'is a hybrid between cultivated and wild sorghum species'; the more common view holds the origin to be importation from the Mediterranean region. He notes that 'between one quarter and one third' of all domesticated crops are known to have transferred genes to wild relatives. See also Ammann et al. (2004).
22. On Indian farmers' cottage-industry hybridisation of transgenic cotton, Jayaraman (2004), Gupta and Chandak (2005), Herring (2007), Roy et al. (2007).
23. For early compendia of pro-poor potentialities, many of which seem born out in later empirical work, Qaim et al. (2000), Persley and Lantin (2000).
24. For results from a major project testing conceptual and empirical dimensions, see 'Democratizing biotechnology: genetically modified crops in developing countries,' Institute for Development Studies, University of Sussex, Brighton, UK, www.ids.ac.uk/biotech.
25. US Patent 5,723,765 granted 3 March 1998, held by Delta and Pine Land Company, in collaboration with the United States Department of Agriculture's Agricultural Research Service. Monsanto's attempt to purchase Delta and Pine Land failed, though this fact did not change the global protest focus on 'Monsanto's terminator'. Terminator technology was not commercialised, due in part to vigorous international protests and intervention of then-President of the Rockefeller Foundation, Gordon Conway (personal communication).
26. Pray and Naseem (2007) note that descriptions of many proprietary laboratory technologies have been published. Moreover, 'Some genes are in commercial use and can be obtained through reverse engineering, and some techniques have made their way to developing countries by way of unauthorised routes'. Patents either cannot be or have not been obtained in many – perhaps most – low-income countries, and are unenforceable in others.
27. For example, debts at usurious rates to pesticide firms were a significant source of cotton-farmer financial crisis and widely publicised farmer suicides of 1998 in India. It is now clear that the Cry1Ac endotoxin in cotton substitutes for pesticides in a very cost effective way, more so when technology fees are avoided than when they are paid.
28. Conversations with seed producers in Gujarat, June 2005; they too do not know the precise acreage, as the farmers are producing *Bt* hybrids on their own farms and some still use transgenic F2 seeds. Data from Navbharat Seeds, progenitor of the first and most successful of the underground *Bt* lines, and parent to most, puts sales at 52.45 *lakh* packets of illegal *Bt* cotton for *kharif* 2005, enough seed cotton to plant 5.245 million acres, or roughly 25 per cent of India's cotton acreage (personal communication). Legal *Bt* sales are increasing rapidly as well. See Herring (this issue).
29. There was 'lively debate' on this question at a recent meeting of a scientific board of the US Environmental Protection Agency on this topic, with no resolution; the 'louder voices' supported more research to improve databases – not surprisingly. Personal communication from Dr Janice Thies, who attended.
30. My experience with *Bt* critics in various parts of India; see also Bharathan (2000). Mark Winston, a biologist, found that secrecy of the firms dealing in transgenics was a major source of distrust, but was experienced as an imperative by holders of inside information (2002: 55–7).
31. After the famous Merck-INBio deal's novelty wore off, this prospect decidedly dimmed. On the optimistic scenario, see Reid (1996), Gupta (1998), Weiss and Eisner (1998). For an economic explanation of failure, Simpson et al. (1996).
32. Also Uphoff (2002, 2003), Uphoff et al. (2002); for a broader view, McNeely and Scherr (2003).
33. They do this, Roy finds, of economic necessity: returns are higher and 'they have to feed their children'. Roy, personal communication; see also Roy et al. (2007).

References

Altieri, M. A. (2001) *Genetic Engineering in Agriculture: The Myths, Environmental Risks and Alternatives*, (Oakland, CA: Food First).

Ammann, K., Jacot, Y. and Al Mazyad, P. R. (2004) Weediness in the light of new transgenic crops and their potential hybrids, Botanical Garden, University of Bern, Switzerland, accessed at: http://www.botanischergarten.ch/Biotech-Biodiv/Report-Biodiv-Biotech12.pdf.

Bambawale, O. M., Singh, A., Sharma, O. P., Bhosle, B. B., Lavekar, R. C., Dhandhapani, A., Kanwar, V., Tanwar, R. K., Rathod, K. S., Patange, N. R. and Pawar, V. M. (2004) Performance of *Bt* cotton (MECH-162) under integrated pest management in farmers' participatory field trial in Nanded district, Central India, *Current Science*, 86(12), pp. 1628–33.

Bharathan, G. (2000) Bt-Cotton in India: anatomy of a controversy, *Current Science*, 79(8), pp. 1067–75.

Bouis, H. E. (2007) The potential of genetically modified food crops to improve human nutrition in developing countries, *Journal of Development Studies*, 43(1), pp. 79–96.

Center for Biological Informatics, US Geological Survey (n.d.) Invasive species information node, accessed at: http://invasivespecies.nbii.gov/

Charles, D. (2001) *Lords of the Harvest: Biotech, Big Money, and the Future of Food*, (Cambridge, MA: Perseus Publishing).

Cohen, J., Komen, J. and Zepeda, J. F. (2003) Innovation, regulation and capacity for bioengineered crops – sustaining a pro poor agenda?, paper presented at the Conference on Transgenics and the Poor, November 7–8, Cornell University, Ithaca, New York.

Conway, G. (1997) *The Doubly Green Revolution: Food for All in the 21st Century* (New York: Penguin Books).

Dawkins, K. (1997) *Gene Wars: The Politics of Biotechnology* (New York: Seven Stories Press).

Democratizing biotechnology: genetically modified crops in developing countries, Institute for Development Studies, University of Sussex, Brighton, UK, accessed at: www.ids.ac.uk/biotech.

Department of Biotechnology Ministry of Science and Technology, Government of India, (2001) *Biotech News*, 2(2).

Douglas, M. and Wildavsky, A. (1982) *Risk and Culture* (Berkeley: University of California Press).

Dryzek, J. S. (1997) *The Politics of the Earth: Environmental Discourses* (Cambridge: Oxford University Press).

Edelman, M. J. (1964) *The Symbolic Uses of Politics* (Urbana: University of Illinois Press).

Gonsalves, C., Lee, D. R. and Gonsalves, D. (2007) The adoption of genetically modified papaya in Hawaii and its implications for developing countries, *Journal of Development Studies*, 43(1), pp. 177–91.

Gupta, A. (1998) Rewarding local communities for conserving biodiversity: the case of the honey bee, in L. D. Guruswamy and J. A. McNeely (eds) *Protection of Global Biodiversity: Converging Strategies*, pp. 180–9 (Durham: Duke University Press).

Gupta, A. K. and Chandak, V. (2005) Agricultural biotechnology in India: ethics, business and politics, *International Journal of Biotechnology*, 7(1/3), pp. 212–7.

Herring, R. J. (1977) Land tenure and credit-capital tenure in contemporary India, in R. E. Frykenberg (ed.), *Land Tenure and Peasant in South Asia* (New Delhi: Orient Longman).

Herring, R. J. (2001, 3 Dec.) *Promethean science, Pandora's Jug: conflicts around genetically engineered organisms in India*, Mary Keating Das Lecture, Columbia University, New York. Transcribed and web-published.

Herring, R. J. (2003a) The political impossibility theorem of agrarian reform: path dependence and terms of inclusion, in P. P. Houtzager and M. Moore (eds) *Changing Paths: The New Politics of Inclusion* (Ann Arbor, MI: University of Michigan Press).

Herring, R. J. (2003b) Data as social product, in R. Kanbur (ed) *Q-Squared: Combining Qualitative and Quantitative Methods in Poverty Appraisal* (New Delhi: Permanent Black).

Herring, R. J. (2005) Miracle seeds, suicide seeds and the poor: GMOs, NGOs, farmers and the state, in R. Ray and M.F. Katzenstein (eds) *Social Movements in India: Poverty, Power, and Politics*, pp. 203–32 (Lanham, MD: Rowan and Littlefield).

Herring, R. J. (2007) Stealth seeds: bioproperty, biosafety, biopolitics, *Journal of Development Studies*, 43(1), pp. 130–57.

Herring, R. J. and Edwards, R. M. (1983) Guaranteeing employment to the rural poor: social functions and class interests in the employment guarantee scheme in Western India, *World Development*, 11(7), pp. 575–92.

Hilgartner, S. (2002) Acceptable intellectual property, *Journal of Molecular Biology*, 319, pp. 943–6.

Ho, M.-W. and Ching, L. L. (2003) *The Case For A GM-Free Sustainable World*, (London: Independent Science Panel).

Horsch, R. B. and Fraley, R. T. (1998) Biotechnology can help reduce the loss of biodiversity, in L. D. Guruswamy and J. A. McNeely (eds) *Protection of Global Biodiversity: Converging Strategies*, (Durham: Duke University Press).

Huang, J., Pray, C. and Rozelle, S. (2002) Enhancing the crops to feed the poor, *Nature*, 418(6898), p. 678.

James, C. (2002) Global review of transgenic crops: 2001 feature: *Bt* cotton, ISAAA Briefs No. 26, ISAAA, Ithaca, NY.

James, C. (2003) Preview: global status of commercialised transgenic crops: 2003, ISAAA Briefs No. 30, ISAAA, Ithaca, NY.

Jayaraman, K. S. (2001) Illegal *Bt* cotton in India haunts regulators, *Nature Biotechnology*, 19(12), p. 1090.

Jayaraman, K. S. (2004) India produces homegrown GM cotton, *Nature Biotechnology*, 22(3), pp. 255–6.

Jeyaratnam, J. (1990) Acute pesticide poisoning: a major global health problem, *World Health Statistics Quarterly*, 43, pp. 139–44.

Government of Karnataka (2002) Farmers' suicides in Karnataka: a scientific analysis, report of the Expert Committee for Study on Farmers' Suicides, Bangalore.

Kloppenburg, J., Jr. and Kleinman, D. L. (1987) Seed wars: common heritage, private property, and political strategy, *Socialist Review*, 95(September/October), pp. 7–41.

Kothari, S. and Chawla, M. (eds) (2002) U N *DeePer in the Techno-Corporate Mire*. Kalpavriksh, Lokayan et al. New Delhi.

Kryder, R. D., Kowalaski, S. P., and Krattiger, A. F. (2000) The intellectual and technical property components of Pro Vitamin-A Rice (Golden Rice™): a preliminary freedom-to-operate review, ISAAA Brief No. 20, ISAAA, Ithaca, NY, USA.

Liptak, A. (2003) Saving seeds subjects farmers to suits over patent, *New York Times*, 2 November, p. 1,18.

Lipton, M. (2007) Plant breeding and poverty: can transgenic seeds replicate the 'green revolution' as a source of grains for the poor?, *Journal of Development Studies*, 43(1), pp. 31–62.

Lybbert, T. J. (2002) Humanitarian use technology transfer: issues and approaches, *Intellectual Property Strategy Today*, 2, pp. 17–24. Revised version in *International Journal of Food, Agriculture & Environment*, 1(2003), pp. 95–9.

Leissinger, K. M. (2000) Ethical challenges of agricultural biotechnology for developing countries, in G. J. Persley and M. M. Lantin (eds) *Agricultural Biotechnology and the Poor* (Washington, DC: Consultative Group on International Agricultural Research).

Lele, U. (2003) Biotechnology: opportunities and challenges for developing countries, *American Journal of Agricultural Economics*, 85, pp. 1119–25.

Madsen, S. T. (2001) The view from Vevey, *Economic and Political Weekly*, 29, pp. 3733–42.

McHughen, A. (2000) *Pandora's Picnic Basket: The Potential and Hazards of Genetically Modified Foods* (Oxford: Oxford University Press).

McNeely, J. A. and Scherr, S. J. (2003) *Ecoagriculture: Strategies to Feed the World and Save Wild Biodiversity* (Washington: Island Press).

Mead, P. S., Slutsker, L., Dietz, V., McCaig, L. F., Bresee, J. S., Shapiro, C., Griffin, P. M. and Tauxe, R. V. (2004) Food-related illness and death in the United States, *Emerging Infectious Diseases* [US Centers for Disease Control and Prevention], 5(5).

Moore, M. (2003) Arguing the politics of inclusion, in P. Houtzager and M. Moore (eds) *Changing Paths: The New Politics of Inclusion*, (Ann Arbor, MI: University of Michigan Press).

Moose, S. (2003) StarLink™: fallout from a shooting biotech star. University of Illinois, accessed at: http://agronomyday.cropsci.uiuc.edu/2001/tours/Starlink/.

Munkvold, G. P., Hellmich, R. L. and Rice, L. G. (1999) Comparison of fumonisin concentrations in kernels of transgenic *Bt* maize hybrids and non-transgenic hybrids, *Plant Disease*, 83, pp. 130–8.

Nanda, M. (2003) *Prophets Facing Backward: Postmodern Critiques of Science and Hindu Nationalism in India* (New Brunswick, NJ: Rutgers University Press).

Nuffield Council on Bioethics (1999) *Genetically Modified Crops: The Ethical and Social Issues* (London: Nuffield Council on Bioethics).

Nuffield Council on Bioethics (2004) *The Use of Genetically Modified Crops in Developing Countries* (London: Nuffield Council on Bioethics).

Paarlberg, R. L. (2001) *The Politics of Precaution: Genetically Modified Crops in Developing Countries* (Baltimore, MD: Johns Hopkins University Press).

Persley, G. J. and Lantin, M. M. (eds) (2000) *Agricultural Biotechnology and the Poor* (Washington, DC: Consultative Group on International Agricultural Research).

Pimentel, D., Tort, M., D'Anna, L. and Krawic, A. (1998) Ecology of increasing disease, *Bioscience*, 48(10), pp. 817–26.

Pinstrup-Andersen, P. and Schiøler, E. (2000) *Seeds of Contention: World Hunger and the Global Controversy over GM Crops* (Baltimore: Johns Hopkins University Press).

Potrykus, I. (2004) Experience from the humanitarian Golden Rice project: extreme precautionary regulation prevents use of green biotechnology in public projects, BioVision, Alexandria, 3–6 April.

Pray, C. E., Huang, J., Hu, R. and Rozelle, S. (2002) Five years of *Bt* cotton in China—the benefits continue, *The Plant Journal*, 31(4), pp. 423–30.

Pray, C. E. and Naseem, A. (2007) Supplying crop biotechnology to the poor: opportunities and constraints, *Journal of Development Studies*, 43(1), pp. 192–217.

Qaim, M. and Zilberman, D. (2003) Yield effects of genetically modified crops in developing countries, *Science*, 299(5608), pp. 900–2.

Qaim, M., Krattiger, A. F. and Von Braun, J. (eds) (2000) *Agricultural Biotechnology in Developing Countries: Towards Optimizing the Benefits for the Poor* (Boston: Kluwer Academic Publishers).

Reid, W. V. (1996) Gene co-ops and the biotrade: translating genetic resource rights into sustainable development, *Journal of Ethno-Pharmacology*, 51, pp. 75–92.

Roberts, D., Josling, T. E. and Orden, D. (1999) A framework for analyzing technical trade barriers in agricultural markets, USDA Economic Research Service, Technical Bulletin Number 1876, March.

Roe, E. (1991) Development narratives, or making the best of blueprint development, *World Development* 19(4), pp. 287–300.

Rogers, E. M. (1995) *Diffusion of Innovations* (New York: Free Press).

Roy, D., Herring, R. J. and Geister, C. C. (2007) Naturalising transgenics: official seeds, loose seeds and risk in the decision matrix of Gujarati cotton farmers, *Journal of Development Studies*, 43(1), pp. 158–76.

Scientific American (2004) Editorial, 291(2, August), p. 8.

Serageldin, I. and Persley, G. J. Consultative Group on International Agricultural Research. (2000) *Promethean Science: Agricultural Biotechnology, the Environment, and the Poor* (Washington, DC: Secretariat Consultative Group on International Agricultural Research).

Scott, J. C. (1998) *Seeing Like a State* (New Haven: Yale University Press).

Sharma, D. (2004) *GM Food and Hunger: A View from the South* (New Delhi: Forum for Biotechnology and Food Security).

Shelton, A. M. (2004) Risks and benefits of agricultural biotechnology, in F. E. Ahmed (ed) *Testing of Genetically Modified Organisms in Foods*, pp. 1–53 (Binghamton, NY: Haworth Press).

Shiva, V. (2000) *Stolen Harvest: The Hijacking of the Global Food Supply* (Cambridge, MA: South End Press).

Shiva, V. (2001) *Patents: Myths and Reality* (Delhi: Penguin).

Shiva, V., Emani, A. and Jafri, A.H. (1999) Globalization and threat to seed security: case of transgenic cotton trials in India, *Economic and Political Weekly*, March, 6–12, 13–19.

Shiva, V., Jafri, A. H., Emani, A. and Pande, M. (2000) *Seeds of Suicide: The Ecological and Human Costs of Globalisation of Agriculture* (New Delhi: Research Foundation for Science Technology and Ecology).

Surridge, C. (2004) Feast or famine?, *Nature*, 428, pp. 360–1.

Svarstad, H. and Dhillion, S. S. (eds) (2000) *Responding to Bioprospecting* (Oslo: Spartacus Forlag AS).

Simpson, R. D., Sedjo, R. A. and Reid, J. W. (1996) Valuing biodiversity for use in pharmaceutical research, *Journal of Political Economy*, 104(1), pp. 163–85.

Tanksley, S. D. and McCouch, S. R. (1997) Seed banks and molecular maps: unlocking genetic potential from the wild, *Science*, 277(5329), p. 1063.

Thies, J. E. and Devare, M. H. (2007) An ecological assessment of transgenic crops, *Journal of Development Studies*, 43(1), pp. 97–129.

Stone, G. D. (2002) Biotechnology and suicide in India, *Anthropology News*, 43(5), p. 5.

United Nations Development Programme (2001) *Human Development Report: Making Technologies Work for Human Development* (New York: Oxford University Press).

UNFAO (2004) *The State of Food and Agriculture 2003–2004: Agricultural Biotechnology: Meeting the Needs of the Poor* (Rome: UNFAO).

Uphoff, N. T. (ed) (2002) *Agroecological Innovations: Increasing Food Production with Participatory Development* (London: Earthscan Publications).

Uphoff, N., Fernandes, E. C. F., Yuan, L. P., Peng, J., Rafaralahy, S. and Rabenandrasana, J. (eds) (2002) *The System of Rice Intensification: Proceedings of an International Conference, Sanya, China, April 1–4, 2002*, Cornell International Institute for Food, Agriculture and Development, Ithaca, NY, accessed at: http://ciifad.cornell.edu/sri/.

Uphoff, N. (2003) Higher yields with fewer external inputs? The system of rice intensification and potential contributions to agricultural sustainability, *International Journal of Agricultural Sustainability*, 1(1), pp. 38–50.

Visvanathan, S. and Parmar, C. (2002) A biotechnology story: notes from India, *Economic and Political Weekly*, 6 July, pp. 2714–24.

Wambugu, F. (2000) The current and future situation of agricultural biotechnology in Africa, in M. Qaim, A. F. Krattiger and J. von Braun (eds) *Agricultural Biotechnology in Developing Countries: Towards Optimizing the Benefits for the Poor* (Boston: Kluwer Academic).

Weiss, C. and Eisner, T. (1998) Partnerships for value-added through bioprospecting, *Technology In Society*, 20, pp. 481–98.

Winston, M. (2002) *Travels in Genetically Modified Zone* (Cambridge, MA: Harvard University Press).

Woo-Cumings, M. (ed.) (1999) *The Developmental State* (Ithaca, NY: Cornell University Press).

World Health Organization (1999) Malnutrition worldwide, accessed at: http://www.who.int/nut/malnutrition_worldwide.htm.

World Health Organization (2001) Nutrition/micronutrient deficiencies: combating vitamin A deficiency, accessed at: http://www.who.int/nut/vad.htm.

World Health Organization, The Food Safety Department (2005) Modern food biotechnology, human health and development: an evidence-based study, Geneva, June.

Zilberman, D., Ameden, H. and Qaim, M. (2007) The impact of agricultural biotechnology on yields, risks and biodiversity of low-income countries, *Journal of Development Studies*, 43(1), pp. 63–78.

Plant Breeding and Poverty: Can Transgenic Seeds Replicate the 'Green Revolution' as a Source of Gains for the Poor?[1]

MICHAEL LIPTON

I. Introduction

Most of the world's poor depend mainly in farming for their income. Most poor farmers cannot readily farm more land, or obtain more water. Yet the farm population is still growing, and the farm workforce is growing faster. In such circumstances, the conditions for an improved farm technology to benefit all major groups of the working poor[2] – those earning income mainly from farming, farm labour, and non-farm rural and urban labour – are tight. The improved farm technology must 'walk two tightropes' (Lipton, 2005: 11). It must:

- Raise labour productivity, so poor farmers get more reward for effort; *but*
- Raise land and water productivity *faster*, so employment, on unchanged land and water, rises.
- Cut staples prices through increased output (so the non-farm poor gain); *but*
- Raise total factor productivity (conversion of inputs to outputs) *faster*, so poor farmers gain.

Section II traces six main routes from GR (or GM) to poverty change: via consumption; nutrition; cereals production; employment income; risk; and sustainability. For each route, we confirm that the GR tended to reduce poverty, and ask what this suggests for GM. Section III enquires why the GR, though aimed at raising food supply, also raised the poor's income and food entitlements; why it missed many areas; and why its impact has slowed – and explores whether, in a changed situation, GM might help. Has the agricultural, demographic or nutritional environment changed between GR and GM, in ways affecting the poverty impact of improved seeds? Does change (e.g., privatisation) in supply, demand and 'industrial structure' of seed research between GR and GM mean changed poverty outcomes? Or is science – its continuity or exogenous change – more important? Section IV sketches some evidence about GM and poverty reduction so far. Section V explores issues for GM, arising from earlier conclusions.

A key issue turns out to be: what mainly determines whether new seeds have the features most conducive to poverty reduction, 'property' or 'science'? If it is mainly whether research *institutions and outputs* are in the public-purpose sector (as with the GR), then, for GM seeds – researched and sold mostly by private companies – to help the poor more than incidentally may depend on new public-purpose incentives to private GM research, e.g., a shift from royalties to contracts. If the poverty impact of modern seed research – GR or GM – is mainly driven by the options opened (or closed) by the basic science, then what matters most is that GM both widens and sharpens GR's tendency to seek *broad-range* changes in plant architecture or chemistry. Genes, with identified effects transferred to one GM crop, may produce similar effects in others. Hence GM advances, developed for crops of main interest to rich farmers and consumers, may prove applicable to other crops, eaten or grown mainly by the poor. Further, transgenics offers hope of yield enhancement for crops whose pre-GM field varieties were selected, by farmers, less for yield than for survival in ill-watered, high-risk environments (where the poor mainly live and work (IFPRI 2001)) – in Africa, sorghum, millet, and root and tuber crops.

Major gains from GM crops for the world's dollar-poor[3] require staple-crop research focus on yield, water-efficiency and robustness, via employment enhancement. Details depend on initial situations (changed by GR) and local agro-ecologies; methods depend partly on whether science or property is the main driver of GM research focus. These issues have not been much addressed. So we question most claims that gains for the poorest from GM are very close. However, exceptions (Section IV) suggest potential medium-term (10–30 year) poverty reduction from GM comparable with that in the GR – with attainable changes, not just in research institutions and incentives, but in scientific and economic criteria within GM research.

II. From Higher Yields to Less Poverty? Lessons From Six Paths in the Green Revolution

The GR is the spread of radically improved non-transgenic[4] staples in developing countries. The new plant types were maize hybrids from the mid-1950s and rice and wheat semi-dwarfs from the early 1960s. These have spread across high proportions of cropland in Asia and Latin America, and small but significant proportions in Africa. The new varieties, from the start, were much more fertiliser-responsive and better yielding than the old, in areas with good water control and pest management. From about 1970, new varieties focused on resisting successive pest biotypes, and on less secure soil-water environments. They were mostly initiated in the international public sector, and crossed for local use by public-sector national research. The GR more than doubled food supply in Asia in 25 years, with an increase of only 4 per cent in net cropped area (Rosegrant and Hazell, 1999). The main routes from the GR to poverty change were through consumption; nutrition; cereals production; labour income through employment; risk; and ecological sustainability (Lipton with Longhurst, 1989; Kerr and Kohlavalli, 1999; Hazell et al., 2000).

Consumption

The GR raised staples output and therefore restrained prices. It raised output most where wheat or rice was grown with adequate, controlled water. Extra output restrained prices most in big, protected countries, where international trade had little effect on prices. Both applied in huge economies – Bangladesh, Brazil, China, India, Indonesia, Mexico, Pakistan – where before the GR the dollar-poor were massively concentrated. Hence poverty fell sharply in 1970–90: most sharply when, and where, the GR was raising staples yields fastest.

Staple food consumption typically uses 45–70 per cent of income for the dollar-poor, much more than for others. So staples price restraint, due to the GR, helped the poor much more than the rich. On the consumption side, the GR was not only pro-poor but also redistributive. Even in Latin America, where land inequality limited the poor's gains as farmers, this consumption effect reduced poverty (Pinstrup-Andersen et al., 1976). For GM similarly to cut poverty via consumption requires much more focus on enhancing staples, yields and robustness.

Nutrition

The poor are the most harmed by protein-energy malnutrition. Almost always, ending calorie deficiency is the best route to protein adequacy. Hence GR or transgenics best attacks malnutrition by raising staples yield, and thus calories, per hectare and per litre,[5] on small farms and employment-intensively (that is, so the poor get more income and can afford more calories).[6]

However, by focusing on calories, GR bypassed widespread, serious shortfalls in iron, zinc, and vitamin A. Dietary diversification is a long-run solution; now, many dollar-poor cannot afford this *and* sufficient staples. Given the huge damage due to deficient bio-absorbable micronutrients (FAO, 2000), can one increase them in key staples? Further, iron, zinc and vitamin A deficiencies probably worsen the nutritional harm from a given calorie inadequacy. The new prospects opened by transgenics for micronutrient staples breeding are treated in the contribution of Howarth Bouis (2007). Micronutrients apart, the nutritional issue facing transgenics (as was the case for GR) is: will it raise food entitlements for the malnourished?

Production

In Asia, one would expect production effects of the GR on rural poverty, by way of small farmers, to loom larger than in Latin America, compared with consumption effects. A much larger proportion of cropland in Asia is in smallholdings. In Asian GR areas – though at first poor adopters faced severe problems of access and management – from the late 1960s smaller and poorer farmers used GR innovations, *where they were adopted*, as widely, and at least as intensively, as other farmers (Hossain, 1988; Lipton with Longhurst, 1989).

GR greatly helped poor staples farmers where it worked, but in non-GR areas created problems for them. If 'stuck' in staples production and immobile, the GR brought them no higher productivity, but a lower output price. Indeed, rural areas of India with least GR-based productivity growth around 1965–85 also had least poverty reduction. Yet four factors helped *some* poor farmers in *some* non-GR areas. Poverty impact of transgenics in these areas – usually the poorest – will improve if scientists and economists review these four 'lessons from GR' before prioritising traits, target regions, and research lines.

First, many poor farmers outside GR zones were net buyers of food staples and thus gained from lower prices. Second, some areas without a GR in cereals were able to shift land profitably into crops such as sugar, mustard and cotton that had been abandoned in areas benefiting from the huge cereals productivity improvements of the GR (Hazell and Ramasamy, 1991). Third, many non-target areas benefited from GR varieties. For example, second-generation and later rice and wheat semi-dwarfs spread widely in South Asia even where they brought little yield gain, due to their increased robustness (Lipton with Longhurst, 1989). In India, it was in these later stages of GR that its poverty reduction impact was greatest (Smith and Urey, 2002). Fourth, in India and China, returns to seed research are now higher in some supposedly non-GR areas than in the classical GR lead areas (Fan et al., 2000; 2000a).

The message from GR to transgenics is that, for poor farmers outside initial target areas to gain from new seeds, research policy response is key. Sri Lanka's came early in the GR, switching rice breeders' emphasis from the Dry Zone to the lagging Wet Zone around 1975. India and China implemented similar shifts later. Faster poverty reduction followed. In African countries with less explicit research policies, improved varieties were little help to poor farmers outside target areas.

What happened to farmers, often the poorest, growing staples less affected by the GR: millet, maize, sorghum, cassava? They lost as sellers; GR, by making rice and wheat cheaper, also cut the price of competing staples. They derived modest offsetting productivity gains for millet and sorghum in Asia, but far less in Africa (Lipton, 1994). What will transgenics do for such farmers? Will it be profitable to research, say, transgenic millet? The contrasts between the GR in Africa and Asia suggest that good, carefully prioritised *national* adaptive breeding is crucial, if transgenics is to help poor farmers stuck in 'backward' crops or areas.

Employment and Wage Income

In rural India in 1999–2000, one-third each of households had farming and farm labour as main income source. Of farmers, 24 per cent were below the national poverty line; of farm labour households, 47 per cent.[7] In India, and almost certainly much of South Asia, Latin America and Southern Africa, ratios of poor farm labourers to poor farmers are steadily increasing. So the poverty effects of GR, and even more transgenics, may operate by way of impact on farmworkers (through wage-rates and employment) more than on farmers (through output and price).

Ensuring transgenic varieties are suitable for small farmers helps poor labourers too. Small farms use more labour per hectare than large farms – even, usually, more *hired* labour. Also, as new varieties – GR or transgenics – make *family* labour more rewarding, small farmers work more on the home farm, leaving the hired labour market to poorer near-landless labour.

However, while the GR raised rice and wheat yields – and hence employment – tractors and weedicides reduced employment. Staples employment rose about 40 per cent as fast as yields in the early 1970s, but only 15 per cent as fast in the mid-1980s, and perhaps less now.[8] The fall is partly due to labour-saving bias in new varieties, for example, their being designed increasingly to accommodate combines or herbicides, or to suit larger, less labour-intensive farms. The fall is not always bad news for poor workers.

(a) In some areas, for example, Malaysia, it happened because labourers were doing well; labour displacement by new GR varieties responded to their increasing scarcity and higher wage.
(b) Where farmland is 'very equal' (for example, much of China), almost all the rural poor *choose* whether they wish to save labour (largely their own) by using such varieties.
(c) Where there is decent unused land (parts of West Africa), the poor can break it in, if they can save labour on their existing holdings.
(d) However, falling employment content of farm output – aggravated by glyphosate-resistant crops, for example – *is* bad for the growing majority of rural poor who rely on farm labour more than farming, where farmland is unequal and extra farmland scarce: in most of Latin America and Southern Africa, and much of South Asia. There, poverty is best reduced by steering transgenics research towards labour-intensive uses and farms.

Overall policy, not just breeding strategy, affects the poverty impact of GR or transgenics. Policy makers have 'agency', and can turn almost any breeding strategy in favour of – or against – the rural poor. If policy *shifts* sharply towards getting them more land, the poor may even come to gain from labour-saving varieties that reduce their effort, especially if they can use it to earn outside farming.[9] However, it is seldom realistic for breeders to tell policymakers: redistribute land, so our labour-saving varieties will benefit the poor. Breeders' and funders' 'agency' – crop and trait choice in research objectives – is therefore a key influence on poverty.

Our ignorance of the effects on labour-demand, and hence labour poverty, of alternative strategies for plant breeding, and for agricultural research more generally, is a gap in our understanding of both GR and transgenics impact. Also, such impact on *total* employment income[10] can differ from impact on *local staples* employment income. Public-purpose research can 'learn' to respond. For example, IRRI shifted from initial focus on maximal-yield but risky varieties such as IR-8, requiring high inputs and management and best suited to big farmers, towards more robust varieties such as IR-20. *For a private researcher, cutting unit costs of labour is as valid an aim as cutting any other type of unit cost. From a social viewpoint, that is false in a poverty-ridden world.*

Risk

The very early GR varieties were more management- and water-sensitive than older staples varieties, and therefore were seen in the 1960s as increasing farmers' income risk. New GR varieties increasingly aimed to reduce *farm-level risk*. After the first thrust of yield enhancement through semi-dwarfing, successive varieties steadily sought improved resistance to a widening range of pests and pest biotypes. This partly explains the spread of GR varieties in Asia to areas where their yield impact was small.

It is less clear that the GR reduced *consumption variability*. Rising staples production variability accompanied the GR in some countries for some 15 years.[11] In India, where national staples consumption is mainly from domestic production, that increased consumption instability. For most Indians, GR nevertheless cut *nutritional risk*; a greater likelihood of a 5 per cent fall below mean calorie intake was outweighed because GR had also raised that mean, so a 5 per cent shortfall did less harm. However, for Indians who did not raise mean consumption (for example, because they worked in non-GR areas), more variable consumption meant more nutritional risk. This danger is now less; since the mid 1980s, new varieties have increasingly *defended* yields against higher variability, rather than maximising them in lead areas. A more serious variability issue is ironically due to the excellence of individual GR varieties, which become dominant over a wide area; if a new pest biotype 'likes' something about a dominant variety,[12] widespread disaster can result. In those cases, skill, speed and luck led to a resistant replacement variety very soon. Until this happens – or if it fails – it is the poorest who suffer most.

Agriculture, however, is a natural enemy of biodiversity: profitable crops and varieties drive out others through farmer choice. Formal research systems tend to make matters 'worse', because they tend to seek varieties offering, over large areas, dramatic rises in yield, and short-term immunity to pest attack: good varieties drive out bad, and that's bad.[13] The GR has been an extreme case: especially for rice, where almost all GR varieties depend on a single dwarfing gene, further selection for multiple pest resistances often means that, over quite large areas, even distinct varieties are genetically close. But we should not assume that breeders ignore this obvious point. They attempt, with some success, to preserve biodiversity, through varietal collections and to some extent in situ. However, there are limits. SE Asian rice in the field is disturbingly un-diverse genetically.

Transgenics, contrary to its reputation, promises substantial increases in field biodiversity. The John Innes Institute has introduced a new dwarfing gene into rice.[14] If rice varieties incorporating it are developed, field-tested and widely adopted, they will add diversity to the rice varieties available. More generally, transgenics increases both the precision and the crop range of gene insertions. To oppose transgenics is not to *defend* biodiversity (with its concomitant long-run reduction of pest-disaster risk), but to *attack* a possible path to biodiversity which one dislikes on other grounds.[15]

The privatisation of much varietal improvement – absent public-purpose incentives to private breeders – may bias, in an anti-poor way, some of the ways in which transgenics approaches risk. For instance, transgenic *Bt* insect resistance slashes, for now, risk from bollworm or corn borer in 'surge' years for those pests. But the resistance is single-gene, or (in some varieties) from two genes expressing similar proteins. If a new pest biotype overcomes the resistant gene, the rich can access and afford pesticide, but the poor may not be able to do so. They therefore prefer horizontal (polygene), moderate resistance or tolerance to biotic stresses. This is well suited (though difficult) for transgenics, but unattractive to breeders whose incentive is to serve large, not very risk-averse farmers.

Ecological Sustainability

The poorest can do least to pre-empt environmental degradation, yet are most harmed – having fewest viable options – once it degrades their land-soil-water systems below some threshold. Many people believe that (a) sustainability of land-soil-water systems is threatened by the shift from traditional, low-input, 'organic', or localised farming to modern, intensive, pesticide/inorganics-heavy, or commercialised (globalised, homogenised) agriculture; (b) the GR, and especially transgenics crops, are an inextricable part of that threat. Some of the appurtenances of modern farming are unaesthetic and arguably, or potentially, unhealthy. However, plant breeding is a tool – of companies, states or societies – not an inherent part of an overall agricultural trend. This paper claims that:

(a) Agricultural intensification *in a world of overall population and income growth* is more environment-preserving than the alternative, extension of farm area.
(b) Plant breeding, if directed responsively to scientific and public[16] concerns, is more environment-preserving than alternative means of intensification, such as more pesticide use.
(c) The GR illustrates these facts, though it meant more use of fertiliser (and, more gravely, water).
(d) Transgenics can enhance the environment-preserving aspects of plant breeding.

(a) Agricultural intensification. Population growth leads to roughly proportionate growth, other things equal, in demand for staples and other farm products. Income-per-person growth leads to less than proportionate growth in demand for staples, but (approximately balancing this) more-than-proportionate growth in demand for animal products, which typically use three to six times more staples per calorie available to humans. Hence 'stylised reality' in 1950–2000 – world population growth and income-per-head growth each at 2 per cent yearly – requires, to match expanding demand without dearer food, a *7.2-fold* rise in staples output. With little good unused cropland, the options are: expand cropping into increasingly unviable marginal lands; or accelerate yields through intensification. In fact, world staples growth *outpaced* demand growth. In 1950–2000, the GR helped push up staples yields fast enough for real prices to fall *and*, GR regions, for staples-based employment income to rise. That allowed unprecedented poverty reduction, without 'help' from cropped-area expansion that degraded marginal lands elsewhere, especially in Africa. Also, the GR

slowed sharply from 1985, leaving much of Africa, and ill-watered parts of Asia, little affected. These times and places also saw faster degradation of marginal lands and/or slower poverty reduction. Though population growth will be slower in coming decades than in 1950–2000, consumption of animal products – and hence demand for feedgrains – is rising faster (because incomes are higher); and workforce growth remains around 2 per cent yearly. How, without further massive degradation of marginal lands,[17] can the required extra food *and* 'entitlements' for the poor (through employment income) be provided? Clearly, further big staples yield increases are needed. Extra workplaces are most cheaply generated in rural areas, and are most cost-effective in improving food intakes if they make local calories cheaper and more. Worsening water shortages and unreliability, bearing most harshly on the poor (International Fund for Agricultural Development [IFAD], 2001: 94–7), increase the obstacles to cropped area expansion, and the need for more 'crop per drop' – as well as more resistance to moisture stress – alongside higher yields.

(b) Plant breeding. To restore and spread growth in staples yields, a major new impetus is required. The staples yield growth trend in developing countries, around 3 per cent in 1975–85, has steadily fallen to about 1 per cent now. More fertilisers can revive the trend, but without substantial improvement in germplasm they have diminishing returns; higher applications are economically unattractive at current prices in the poverty heartlands: much of Africa, West China and East–Central India. Progress requires, in some areas, more irrigation, and almost everywhere better progress in plant breeding. Conventional breeding has rightly and increasingly become 'defensive', seeking yield *maintenance* in face of new pest biotypes and water depletion: hence slower yield *enhancement*.

(c) The GR. The GR illustrates the environmental prospects – but also the traps, and the need for economic analysis – in transgenics policy construction. Both GR and transgenics normally raise output per litre, per extra unit of water, fertiliser and pesticide; how will this affect the use of these resources? Successive new pest-resistant varieties have almost certainly reduced pesticide use per ton of output. However, GR varieties greatly raised the marginal product of nitrogenous fertiliser around its pre-GR use levels, and hence its use. In water-scarce wheat areas, it did the same for water. These incentives led to severe water depletion – sinking water-tables, overused surface-water systems – and some nitrate and nitrite pollution of drinking-water sources, in many parts of India. The environmental damage from achieving comparable output increases through area expansion would have been far more; but varietal and other input choices in research can and should be predicted and planned to improve environmental impact.

(d) Transgenics. Transgenics – even in its deeply flawed manifestations and commercial origins so far, and much more so potentially – illustrates the environment-preserving role of plant breeding, under conditions discussed below. Alongside new water science, functional crop genomics and transgenics must be part of any solution to the water crisis of farming, complementary to the water reforms more widely advocated (World Water Council, 2000).

III. Green Revolution Puzzles and Transgenics-Era Parallels

To explore the scope, limits and prospects of transgenics in reviving and spreading poverty reduction, we ask four questions.

(a) Why was the GR – while and where it happened – so good at reducing poverty?
(b) Why were there big regional gaps, and a sharp slowdown, in the GR – and can transgenics help?
(c) Has the poverty impact of staples seed improvement been affected by major change, from GR to transgenics eras, in 'demand' for seed research?
(d) As for changes in 'supply' side of such research, does poverty impact of new seeds depend more on change in the 'industrial structure' of seed research, or on change (or continuity) in the science?

How did GR, with no 'Food Entitlements' Agenda, Slash Poverty and Undernutrition?

A now-classic 1959 study diagnosed the need for yield-enhancing strategies, as prospects to expand cropped area dwindled (*India's foodgrain crisis*, 1959). An emergency was signalled by India's near-famines in 1965–66 and China's in 1960–63. The GR seemed a credible response to emergency, seeking to raise staples *availability* by enhancing yields of the most promising crops (rice and wheat), initially on reliably irrigated land, for larger farmers, in institutionally better-served areas.

Now, however, we know (Sen, 1981) that low or reduced food *entitlements*, not availability, cause most famine and hunger. They exist in a world, countries, and even small areas that – even in bad years – grow more than enough to provide ample calories for all. If the designers of GR overestimated the role of increased local food *production and availability* in overcoming hunger, and underestimated the role of increased *entitlements* to food, why did the GR do so well in cutting poverty? The answer is partly luck; the lesson for transgenics is threefold.

- The GR, where it happened, raised staples availability in most countries in ways that also raised food entitlements. This involved walking two tightropes. In GR areas, it raised total factor productivity (so poor farmers gained) *faster* than it lowered the price of staples output (so poor consumers gained). And the GR raised the average and marginal products of labour (so employed workers got more income per hour), but in land- and/or water-scarce areas it raised the average and marginal products of land and/or water *faster* (so farm unemployment fell) (Lipton, 2005).
- Where the poor at first gained little or lost, the (public-purpose) institutions and (science-based) incentives of most GR research made it responsive to criticism (for example, the shifts of breeding priorities towards robust varieties and neglected regions). Private research lacks pro-poor incentives.
- The science underlying GR and transgenics alike – as adapted and applied by farmers where there is abundant labour – may, however, bring innovations that 'walk the tightropes' above.

The greater the extent to which the internal organisation of research (rather than the nature of modern seed research however it is done) was the main reason why the GR so successfully raised staples *entitlements* of the poor, the more radical is the change needed to make transgenics structure and organisation produce results as pro-poor as the GR's generally were.

Why the Gaps and the Slowdown in the GR – and is Transgenics Well Placed to Help?

Gaps. High-yielding GR rice and wheat semi-dwarfs and maize hybrids cover more than 80 per cent of areas sown to these crops in South, East and West Asia and North Africa; over 70 per cent in Latin America; but below one-third in Sub-Saharan Africa. Despite major efforts in leading developing-country and international research institutes, offtake was weak in uplands and unreliably watered lowlands. Also, yield gains in farmers' fields from varieties embodying GR approaches were much smaller, and spread more slowly, for sorghum, millet, cassava and yams – alongside maize, the main staples grown and eaten by Africa's poor.

The gaps are due partly to national policy priorities constraining, for example, local research and fertiliser use, but mainly to limited prospects for applying the GR concept to the unaffected areas. National policy is in part a response to such limitations: why spend scarce resources on GR research, if you believe it will achieve little in your country? The three main limitations are (a) scanty or uncontrolled farm water; (b) few, or badly-absorbed, plant nutrients from fertiliser; (c) a crop-mix locked into low-value staples that are selected (by nature and by farmers) for hardiness, not high yields.

The three features are linked. If water is unreliable, returns to fertiliser are lower and less secure, and farmers are driven to prefer hardy, lower-input staples. Yet the features also have exogenous causes. The recency of land scarcity in parts of Sub-Saharan Africa has delayed incentives, to governments as well as farmers, to recognise a need for intensification, and hence for better water control and nutrient supplementation.

Water-control issues will have to be faced, before GR-type varieties can make a large and rapid impact in Sub-Saharan Africa, and probably also in many little-affected areas in Asia. Gross irrigated area comprised 1.5–3.8 per cent of gross land cropped in Sub-Saharan Africa in 2000, compared with 49 per cent in Bangladesh, 37 per cent in China, 32 per cent in India, 14 per cent in Indonesia, 23 per cent in Mexico and 82 per cent in Pakistan.[18] It is on water-controlled land that GR varieties have transformed staples yields.

However, while plant breeding cannot now raise yields at 'Punjab rates' in unirrigated semi-arid or upland farming, transgenics may make a big difference. If the genetic mix of field varieties of main African staples is adapted to good performance when water is scarce, then DNA from other crops may be needed to generate new varieties offering substantially higher yields without unacceptable water risk. For example, conventional plant breeding has had little success either in raising African millet yields, or in introducing latency, especially at the time of anther formation, into maize. Many farmers grow low-yield millets or maize landraces, rejecting high-yielding maize hybrids because they are believed to raise unacceptably the risk of crop failure if the rains start late. Transgenics may be necessary and sufficient to introduce latency into high-yielding maize varieties – or dramatically higher yield into millet – across species barriers. There are, as always, efforts in the pipeline,[19] but if company profits are driven by royalties rather than contracts, their focus is bound to be, as usual, on breeding for soil-water conditions of large, mainly North American, farmers from whom royalties are readily recoverable.[20]

Water scarcity or unreliability constrains fertiliser use.[21] Together, in many areas they constrain returns from conventionally bred varieties, and shift farmers towards crops where the GR has achieved less. Transgenics probably has the *potential* to transform yields in the latter crops, and to make standard GR crops less prone to moisture stress (Nuffield, 2004: 3.42). To slash poverty in some recalcitrant non-GR areas, they ultimately must get water control or quit agriculture – just as nutrient deficiency must ultimately be tackled by diversified diets. But meanwhile the poor must live, and transgenics offers hope. The case for more irrigation, major as well as minor, in Africa is stronger than often claimed (IFAD 2001), but it will be a long haul, offset by the shift of water to domestic uses, and by faster evaporation due to global warming. Until the long haul is well under way, the gaps in the GR – and the overlapping gaps in rural poverty reduction – will not be addressed unless transgenics turns towards enabling higher yields *in the conditions of poor rural areas* despite scarce or unreliable water.

Slowdown. The growth of staples yields in developing countries in the 1990s was about a third as fast as in the 1970s (Lipton, 2000). Combined with the falling responsiveness of farm employment to yield growth, the growth downtrend has made rural poverty reduction harder. The downtrend has several causes, but fundamental is the falling GR contribution. Can transgenics attack its causes?

The GR's falling effect in yield growth partly reflects its success. GR varieties first spread through the best-suited areas; if similar varieties spread to less well-suited areas, yield impact falls. Transgenics could address this, if gene insertions could do things, infeasible within a crop's genome, to make it less sensitive to the abiotic constraints that create 'less-favoured areas'. On matters such as drought resistance, there are proof-of-concept examples (Nuffield, 1999: 2.49; 2004: 3.42), but incentives to develop them are reduced by another cause of yield slowdown. That is the shift, with transgenics and otherwise, of agricultural research resources towards the private system. Companies have other priorities than reaching small, dispersed, poor farmers, especially with self-pollinating crops. Further, even within the public system, resources have shifted sharply away from crop productivity enhancement, towards environmental and policy research (Lipton, 2000). Perhaps above all, even the depleted public-sector research on crop productivity has been forced (by ever-changing pest biotypes, falling water-tables and soil micronutrient depletion) to shift from yield enhancement to yield maintenance, that is, to become increasingly defensive.[22]

Are there funds to develop such transgenic seeds for profitable small-farmer use? There are currently two impediments. First is the cost of clearing regulatory hurdles, arbitrarily imposed on just one sub-class of new seeds (transgenics) that is no more, and probably less, risky for health or environment than other sub-classes. Second is the reluctance of funders to divert resources from seed development or poverty reduction to a tiring, boring struggle against pseudo-environmentalist PR and scientific ignorance.

Great inventions – such as new techniques of plant dwarfing to raise yield through a higher harvest index and reduced lodging – produce a 'watershed' of applied innovation and adoption. The flow inevitably peaks and then declines, unless the source is replenished through new inventions. Despite important advances in conventional breeding, transgenics seems the only substantial new source to re-start the GR process of rapid poverty reduction, but only if changed public resources and private incentives greatly change the current directions of transgenics research.

Changes in Farming, Nutrition, and Demography

Both demand for staples research and the poor's need for its products depend on farming, nutrition and demography. All have changed between 1955–70 and 1995–2010, the formative years for GR and transgenics research respectively. Does that alter the impact of research decisions on the poor?

Changes in farming. In low-income countries, staples demand continues to be increased by rising population, income per person, and staples requirements per calorie as diets switched to animal products.[23] However, incentives to farmers and researchers behaved contrarily. Support to rich-country farmers surged,[24] while repression of farm prices and production in poor countries declined. Both these trends have increasingly glutted farm markets, four main considerations are relevant.

First, to maintain developing-country farmers' competitiveness, research has to cut their unit costs faster than the GR did – but with a more degraded land-water environment: 'to produce a doubly green revolution at half the price' (Maxwell, 2004). Adopting transgenics further raises market supply and cuts price. This will make farming in countries that reject transgenics ever more uncompetitive[25] – presaging futile[26] efforts to create a world 'half-transgenic and half-nontransgenic'. These price effects are mitigated if transgenics research to raise staples output is focused on people who tend to eat it themselves: small farmers, their employees, and poor areas.

Second, perversely, the focus tends to be on big and rich farms, where demand for farm products is being transformed by new institutions of globalisation: supermarkets, public and private product grades and standards, and export horticulture (Reardon et al., 2003). This transformation is fast and far advanced in some developing areas (Latin America, SE Asia, Southern Africa). Where there is 'intermediation failure' between the new institutions and small farmers, their cash-cropping may well become unviable. These 'unmediated' poor will focus more on staples production, and will place distinct demands on transgenics for crop-mix and traits.

Third, the poor's need for sustainable income sources is much more threatened by water shortage than in the GR's formative years. The GR raised crop-per-drop, but thereby also incentives to water use; on balance, demand for water rose. Further, associated higher fertiliser use reduced the recyclability of farm water. Transgenics has potential to raise crop-per-drop further, but only if 'steered' to identify, test and spread the complex multi-gene insertions probably required.

Fourth, public-sector capacity to meet the poor's needs from agriculture has been undermined since GR days by the collapse of aid to the sector. It fell sharply in the 1980s, and again by 65 per cent in real terms in 1988–98 (IFAD, 2001: 41).

Deteriorating farm infrastructure (irrigation, roads) requires revived investment, including aid, to benefit from transgenics and other research.

Small farms are smaller, yet more productive, than in the GR era. They face new, and differently distorted, environments, governments, institutions and incentives. So do the researchers who seek to serve these farmers. Planning for transgenics research needs to be aware of how these things affect specific local farm contexts. Hence agricultural economists and other social sciences need far more direct linkage with natural scientists in research planning.

Nutrition. The GR helped East Asia and Latin America to slash protein-energy malnutrition, which remains widespread in Sub-Saharan Africa, where it has not lessened, and South Asia. In those areas, it remains the first public-purpose goal of breeding to raise the poor's command over calories.[27]

The GR's partial success against undernutrition has exposed a second problem: micronutrient deficiencies. Conventional plant breeding here has limits. Bioabsorbable iron sources in rice can be increased successfully by conventional breeding; but only transgenics can introduce provitamin A into the endosperm of rice. Nutritional 'diseases of affluence' (obesity, coronary heart disease, diabetes) increasingly plague poor countries. The needs of the poor from transgenics are relevant. Child *under*nutrition increases exposure to diseases of affluence,[28] that is, of adult *over*nutrition, just as child obesity does. All can be tackled with growing, parallel understanding of human and plant functional genomics. This may permit a transgenic transition from 'diseases of poverty' to greater wellness.

Transgenics and the new demographics. Developing countries' rapid growth rates of population in the 1960s have slowed substantially, even in sub-Saharan Africa. Working-age population, however, is still growing at over 2 per cent annually in much of the developing world. So the workforce/dependant ratio, as compared with 1960–90, is rising much faster in 2000–30. In 2000, there were 101 people of prime working age for every 100 dependants in Ethiopia; the projection for 2030 is 139. For Nigeria the rise is from 101 to 150; for Bangladesh from 126 to 181; for India from 140 to 172 (UN [ECOSOC], 2005). After 2030–50, ageing populations start to reverse these gains. So the next thirty years form a 'window of opportunity' – and danger – for poverty reduction. Will more workers and savers, with fewer extra dependants, slash poverty – or employment and wage-rates? The choice depends heavily on whether staples breeding strategy creates enough productive workplaces.[29]

Agricultural, nutritional and demographic differences between the early GR and early transgenics eras may transform (a) how seed researchers' strategy affects poverty, (b) what they can do if they prioritise poverty reduction. GR evidence indicates that fast technical progress for small-scale and/or employment-intensive farms seems necessary and sufficient to *initiate* sustainable mass poverty reduction in low-income areas. How much, if at all, will transgenics help?

Supply of Transgenics, Changing Research Economics and Seed Science

In the GR, supply of seed research to developing countries came mostly from public-sector institutions, under pressure to follow, in part at least, a public-purpose agenda.

Perhaps 90–95 per cent of applied transgenics research, however, is managed by a few large private companies, seeking appropriable profit opportunities. Profits are less, and/or less appropriable, when seeds are self-pollinating (for example, rice, wheat); when they go to numerous dispersed small farmers (for example, millets); and when they embody traits prioritised by such farmers (for example, risk reduction, rather than labour-saving). Why should companies pay to develop such seeds?

One answer might be 'because the public sector decides to make it pay'.[30] It is too pessimistic to say that this cannot be done: that anti-poor, labour-saving bias in agricultural research outputs is an inevitable response to growing labour shortage (and rising wage-rates) in rich countries.[31] Public action influences research rewards. Also, the 'supply side of science', from Crick and Watson through to plant breeders, is not *only* a response to such rewards. True, rewards respond to peer pressures and career incentives, and (especially at the sharpest end of applied science) these are partly created by firms that *do* respond to factor and product prices. But that is not the whole story.

Modern seed research has been informed by insights from basic science, generating 'watershed' innovations – sequences of improved varieties – that raise the conversion efficiency of land, water, sunshine, plant nutrients, and pest management into crop. This basic-science impetus is enhanced in transgenics, because gene insertions, even if designed for crops grown and consumed mainly by the rich, are often suitable also for those of the poor.

Another reason why big-company motivation need not drive out poverty-reducing priorities, is the still-large public seed-research sector.[32] True, it is small compared with private companies, and (China apart) probably cannot *alone* achieve a big shift in transgenics outcomes. However, it can greatly *change incentives* for private transgenics. Large private firms must seek profits, but need not do that solely by charging final customers, with consequent bias towards the crops, traits and regions of interest to large, rich farmers. The public sector, after identifying key poverty-reducing priorities and within an agreed IPR regime, can buy in private R & D for particular tasks, for example, developing high-latency transgenic maize, attractive to small farmers in identified areas. Fee-for-service is feasible, either company-specific or on competitive tender. There are other ways: subsidising 'blue skies' research on company time and equipment, offering substantial prizes, and so forth. Success would draw aid and charitable funds in support. Transgenics companies' need to improve their public image makes this a promising time to develop and explore such options.

IV. Transgenics and The Poor: The Evidence

Of 67.7 million hectares planted to transgenic crops in 2003, 73 per cent were selected for herbicide tolerance, 18 per cent for *Bt*-gene insect resistance, and 9 per cent for both. Herbicide tolerance was induced almost entirely by inserting a gene conferring resistance to damage from glyphosate into soy (41.4 million ha), canola (3.6 m), maize (3.2 m) and cotton (3.1 m). The *Bt* gene in maize (9.1 m ha) and cotton (3.1 m) has induced resistance to corn borer and cotton bollworm. A further 3.2 m ha of maize and 2.6 m ha of cotton have both *Bt* and herbicide-tolerance gene insertions (James, 2004).

Most field evidence is about the impact, not of transgenics as such, but of glyphosate-resistant maize, canola or soy; corn-borer-resistant maize; and bollworm-resistant cotton. These or similar traits have also been induced by conventional plant breeding. Farmers choose a transgenic variety if, and only if, they judge it safer or more profitable. Most arguments, allegedly for or against transgenics, are in fact for or against (a) traits that transgenic varieties embody; (b) the cost, and effects (on the poor, health, environment, and GDP), of obtaining such traits by transgenics vis-à-vis other means: conventional breeding; agrochemicals; agronomic, biological or manual controls; integrated pest or weed management; or crop-mix shifts.

We also need to distinguish *traits* for which transgenic crops now in the field were selected, and the *routes* through which breeders sought those traits. Both may discriminate between poor and non-poor. If a pro-poor *route* to bollworm control in cotton were being designed for poor farmers, a public-purpose agency would attend to their special and extreme vulnerability if resistance breaks down, due to their difficulty in affording or finding extra pesticides quickly, and in bearing risk. Even before gene stacking, *Bt* cotton's resistance to bollworm proved much more stable than expected. But such single-gene, high-level resistance challenges bollworms to develop new biotypes. A more pro-poor cotton breeding strategy would be to seek stable, polygene, moderate resistance or tolerance to bollworm. Biotechnology to understand the functional genomics of both cotton and bollworm, followed by appropriate multiple gene insertions into cotton, seems suited to such an approach, but it is a long haul. Both science and the poor suffer if it is made longer by regulatory restrictions that multiply alongside gene insertions.

Poor and rich alike want better, cheaper, or less poisonous control of insect pests. However, it is usually anti-poor to develop transgenic (or other) varieties for herbicide resistance. This allows chemical, rather than manual, weed control, reducing the demand for labour and hence the wage-rate and perhaps employment. There are exceptions. In some West African uplands, the switch to herbicides, by freeing up family labour, allows poor farmers *who already control land* to farm more of it. Even then, better herbicide resistance is seldom a high priority for pro-poor transgenic research, compared to (a) higher yields per hectare and per litre; (b) greater robustness under moisture stress; (c) focus on areas neglected by agrotechnical progress in the past 50 years; (d) attainability of these advances for small farmers, staple crops and employment-intensive methods.

How well, if at all, will transgenics, through its supplying institutions, address these goals? That is the main factor affecting its long-run poverty impact. However, transgenic outcomes for overarching, polygene traits such as moisture-stress resistance is some way off. Meanwhile, what is the poverty impact of what transgenics mainly does now: inserts *Bt* and glyphosate-resistant genes?

Transgenic Insertion of Bt Insect (Bollworm) Resistance in Cotton, and Analogues

'*Bt* cotton' is code for a range of varieties with a *Bt* gene insertion. These were designed for, and first spread to, large growers. From nil in 1996, *Bt* cotton spread to

4.8 m Chinese farmers (almost all small) on over 2 million hectares in 2001, that is, 43 per cent of China's cotton area. In India, South Africa, Mexico and Argentina, similar varieties have reached millions more small farmers and many large ones (Frisvold et al., 2006). *Bt* cotton has affected the poor through:

- Less pesticide use, with lower costs for adopter-farmers and better health for labourers and farmers: such effects are substantial only where pre-*Bt* pesticide use was *high*.
- Higher yield: important only where pre-*Bt* pesticide use was *low*, or in 'bollworm surge' years.
- Higher net farm income for poor adopter-farmers, due to these two factors combined.
- More employment where few pesticides were used pre-*Bt* (*smallholder Bt* adoption in South Africa and Argentina) and/or bollworm 'surged', so *Bt*-induced rises in harvest labour outweighed falls in spraying labour.
- Otherwise, *less* employment, harming poor labourers (Mexico).
- For similar reasons, less employment in China,[33] but without loss to the poor, because (due to land equality) the fall was in voluntary self-employment, hired farm labour being minimal.

Where pre-*Bt* pesticide use was very high, as in China and Mexico, yield gains from small-farm adoption of *Bt* cotton are small; income and health gains, both mainly due to reduced pesticide use, are large; and employment effects are negative. Further, despite health and income gains to poor farmers and labourers, '*Bt* cotton's success has attenuated its benefits. Rising yields and expanding area has begun to push cotton prices down. As a result, some of the gains that accrued previously to producers are now being enjoyed by consumers'.[34] Where small farmers' pre-*Bt* pesticide spraying was lower (as in Argentina and South Africa), *Bt* adoption raises yields, only modestly reduces the already low pesticide use levels, and on balance increases employment, because the effect of higher yield in raising demand for harvest labour outweighs the effect of modest cuts in demand for spraying labour.

In India there are competing claims, NGO-mediated and company-mediated. However, there is little objective field evidence on transgenic cotton spontaneously managed by farmers (as opposed to field trials). Poor and other farmers are 'voting with their feet' in successive years for *selected Bt* cotton varieties. In October 2002, Qaim and Zilberman (2003) reported field trials showing that 'the amounts of pesticides applied... were reduced to one-third of what is used in conventional cotton, while – under severe pest pressure – yield gains were 80 per cent and higher'. Consequently, 'medium-term projections show sizeable welfare gains for the overall economy, with small-scale farmers being the main beneficiaries'. Qaim and Zilberman[35] 'looked more closely at 2001 data from 157 farms in Maharashtra, Madhya Pradesh and Tamil Nadu. *Bt* cotton yielded 80 per cent better than identical cotton without the *Bt* gene, and 87 per cent better than the local hybrids'. Roy et al., in this issue, explore a wide range of outcomes, dependent on variety and farmer, in a low-infestation year; *Bt* varieties, legal and illegal, were favoured by farmers.

Several distinct *Bt* cotton varieties, some with Indian research and/or public-sector involvement, are at various stages of multi-location field trials, clearance for release, and commercial use in many states of India. Effects vary with experience, location, pre-*Bt* pesticide type and level, weather (does *Bt* cotton outperform competing varieties most when pesticides wash off them?), and numbers and biotypes of bollworms.[36] In Gujarat in 2002–03, smallholders and breeders illegally crossed Mahyco/Monsanto *Bt* cotton with local varieties, yielding cheaper seeds that they believe to be better suited to local conditions.[37] Gujarati farmers seek appropriate *Bt* cotton varieties, especially if they can pay non-monopoly prices, but some are willing to pay the Monsanto price as well, depending on their experience and local conditions, as Roy et al. (2007) document. This might seem fine: the market is working, and environmental effects have been tested in a wide range of *Bt* varieties across millions of smallholder hectares from China to Mexico. However, if varieties – *transgenic or other* – wholly escape regulation or validation, might poor, remote farmers fall victim to snake-oil salesmen?

Virulent opposition to *Bt* cotton in India is not simple anti-transgenics or anti-market foolishness. There are genuine uncertainties. Loss of seed quality can lead to significantly lower yields than are available from 'official' varieties. Absence of refugia can cause loss of biodiversity (Western India is the world Centre of Biological Diversity for cotton). If royalties are avoided, private transgenics inputs may be withdrawn. These are not trivial side-effects, for the poor, of desirable democratisation and marketisation of small-farm seed access – what Herring (2007) calls 'agrarian anarchocapitalism'. Keeping professional breeders in the game may require shifts: from 'smallish royalties for ever' to 'one-off fee for service'; and from (bogus) law enforcement to efficient extension and collection, as the means to monitor environmental impacts.

The price regime and trajectory are crucial for small farmers. Globally, *Bt* cotton benefited rich farmers in rich countries first, rich farmers in poor countries next, and poor farmers in poor countries last. When poor farmers lacked *Bt* crops, rising world production (due partly to rich farmers' *Bt* varieties) depressed prices and incomes for poor farmers in the developing world. These, therefore, lost from price falls, both when others adopted *Bt* varieties, but also by benefiting less when they eventually did get *Bt* seeds. This denial of economic rent to latecomers – much the same as poor farmers – is familiar from the GR. What follows?

First, then as now, the cure is to get technical progress to the poor quickly, and to tailor traits to their needs. The alternative is perverse: to restrict technical progress. Then, the poor get not even the crumbs from the rich man's table.

Second, to tailor transgenics or other researched traits to poor people's needs, one has to look at the timed *work environments* of *local* groups of the poor. *Bt* cotton reduces demand for spraying labour and raises demand for harvest labour. Potentially, the former harms, and the latter benefits, the rural poor to the extent that (a) (cotton) land is fairly unequal; and (b) farmwork is a major component of potential extra income for the poor.[38]

Third, a pro-poor transgenics planning process should also look at the timed *risks* facing local groups of the poor. *Bt* cotton's resistance to bollworm, like *Bt* maize's resistance to corn borer, has lasted many years longer than expected, and has been 'refreshed' by gene stacking. However, *Bt*-resistant bollworm biotypes will develop eventually. Small, poor farmers, who are most information-deprived, and whose risk situation makes them least able to mobilise cash, are least likely to find appropriate reserve pesticides swiftly should this happen. The answer is *not* to delay the spread to poor farmers of *Bt* cotton. It is to create public-private partnership actions to alert small farmers to early signs of *Bt* resistance breakdown and to provide the means for rapid remedy.

More important is the long-run lesson. Concentration on highly resistant crop varieties that maximise challenge to the pathogen, and on single-gene resistance that maximises plant loss if the pathogen successfully develops a new biotype, is risky, and thus anti-poor. The alternative strategy of seeking polygene, moderate resistance to (or tolerance of) pathogens has, however, so far been slower, harder, and less matched to the professional incentives of the plant sciences. But might biotechnology (not just transgenic plants) change this? Public-sector and Monsanto-based efforts have provided a near-complete, public-domain rice genome; work at the John Innes Institute has established that this accurately models most other cereal plants; and almost every month more (though still little) is known about the functions of genes, and proteins, in cereal plants. Feasibility and timing are a matter for the scientists; but economists can point to risk-reducing advantages for the poor, whose influence on these decisions is small.

The Case of Transgenic Glyphosate Resistance

Despite real problems and cautions, resistance to a small set of very damaging insects by way of the *Bt* gene helps poor farmers. So, as a rule, does a shift towards pest control through affordable seeds, rather than costly and dangerous pesticides. However, the poor's 'top 50' priorities, if compiled by them prior to transgenics research, would not have included better weed control through resistance to the herbicide glyphosate (Roundup).

There are non-transgenic varieties that resist various herbicides too. But, since glyphosate resistance is inserted into over 80 per cent of transgenics crops in the field, it raises the issue: what are the effects of transgenics research, managed (for non-public purposes and large farmers) so as to select possibly labour-displacing goals? This is a bad use of transgenics resources from the standpoint of poverty reduction relatively, but not necessarily absolutely. Labour-displacing seeds will not be adopted by farmers unless profitable, that is, normally, unless they cut unit costs. That tends to push down prices and help consumers. However, glyphosate-resistant seeds, and linked replacement of labour by herbicides, can harm the poor by cutting demand for labour. That effect was not significant in Argentina or even Paraguay, where mainly big-farm adopters previously used little weeding labour (costly in a middle-income country), preferring chemical or mechanised weed control, with or without glyphosate-resistant varieties (G. Traxler, personal communication, 2004). However, in South Asia (with lower wages and hence much more handweeding), glyphosate-resistant crops – while raising yields and cheapening food – would also significantly displace weeding labourers, usually women and among the poorest of the poor.

In Colombia, an *ex ante* analysis of the impact in 2016 of transgenic herbicide-resistant cassava production reveals substantial gains in producer and consumer surpluses, but a loss of 25 per cent (over two million person-days per year) of employment.[39] Consumers of cassava, even more than those of soybean, tend to be poor. Pachico et al. (2002) comment that, though glyphosate-resistant cassavas:

> ...might appear to risk jeopardising the welfare of workers in an economy (with) high unemployment, (there would also be a rise in) labour productivity, and thus in principle wages... the welfare issue (of employment creation) should perhaps be seen as part of the overall macroeconomic performance of the economy instead of a matter that can be resolved through a single line of production like cassava.

A similar defence can be made for computerisation of banking in India: major gain to consumers is worth more to GDP, and perhaps to the poor, than the cost to workers of shifting employment. There is, indeed, a case against 'loading labour absorption or employment generation on plant breeding'[40] – *if* the losers (as in the banking case) are not the most poor and vulnerable, or if government mounts (or buys in the marketplace) measures to compensate, or retrain and resettle losers. However, Latin American cassava weeders are often landless, resident in remote and marginal uplands, and/or female *indigenos* with a non-Spanish first language (Psacharopoulos and Patrinos, 1993), and as such among the poorest, least educated, and hence least mobile of workers. With such vulnerable losers – unless they are compensated, for example, with reform lands – one cannot so readily absolve labour-displacing innovation. This is normally a *cause of falling* rural wage-rates. In other circumstances – where herbicide-resistant varieties are a *response to rising* rural wage-rates and urbanisation in a booming economy; where weeding is anyway largely mechanised; where weeding employment provides little of the income of the rural poor; or where rates of cropped area expansion exceed rates of reduction in labour-demand per hectare – there may be net poverty reduction effects from herbicide-resistant varieties.

In much of Asia and in some areas elsewhere, farmers often hire as weeders near-landless or landless women, often from ethnic groups or castes with mobility hampered by language, poverty or illiteracy. Reduced demand for their labour, and hence cuts in their wage-rates, harm the most vulnerable of the poor. This makes a case, not for banning such varieties (infeasible anyway), but for introducing public-purpose incentives to future transgenics breeding choices.

What about arguments that, in some developing rural areas, labour rather than land is scarce, so that innovations that lower the per-hectare demand for labour are desirable? Such arguments are suspect. First, labour scarcity is often, maybe usually, a claim made by employers seeking to justify even lower wage-rates. Second, many parts of Africa that had spare croppable land twenty years ago do not have it now. Third, most of the few remaining places that have spare croppable land now – as research priorities for transgenics are set – will no longer have it when that research delivers. Fourth, much remaining spare land is croppable only at high risk to biodiversity (Amazonia), or of overfarming and exhausting land-water systems better left in extensive grazing (parts of East and Southern Africa).

Finally, special cases for relieving a labour limitation are almost always seasonal, or to permit timely, early planting despite a workforce depleted by HIV/AIDS. Seasonal labour shortages, where labour still earns little, require farming methods (including new seeds) to shift the labour peak, not to cut overall labour demand and thus push down employment and wage-rates. As for HIV/AIDS, it can severly cut labour supply, but this is short-term and often local. It does not mean that plant breeders should prioritise cutting crop demand for labour and hence the income of the poor, including working relatives of HIV/AIDS sufferers. Also – with 5–15 year lags from new breeding strategy to widespread field adoption – labour-saving strategies cut against big medium-term rises in workforce.

If rich farmers cut costs by selecting transgenic varieties for herbicide resistance (or other labour-displacing traits), must poor farmers imitate or die? That is too gloomy, for three reasons.

- It takes too homogenising a view of agricultures. Public policy for research and innovation, as well as for employment, can run with the tide of factor scarcities, steering research into more socially desirable channels.
- It is too limiting a view of science. Can national research do more than adapt, screen, or at best locally cross varieties with gene insertions from elsewhere? Yes, but only in the few developing countries with strong enough research capacity. Such countries are, almost by definition, big; but most of the world's dollar-poor live there.
- It downplays the role of the public sector: in sponsoring applied research (transgenics-based and other) corresponding to its own poverty priorities; in changing incentives to affect private-sector research aims; and in regulating varieties and imports both before and after field trials.

The latter is a controversial area. In general, governments should *reduce* the regulatory delays and costs that keep the benefits of transgenic crops away from poor farmers. However, plants such as *Lathyrus sativus* (the lentil *kesari dal*, known to cause lathyrism) are rightly restricted in many countries. Could it be right to discourage the planting of varieties, whether conventional or transgenic, that cut employment and increase poverty?

Key Crops, Pipelines and Hypelines: Reflections on Golden Rice

Vitamin A deficiency, a major cause of death and blindness among under-fives, is prevalent among poor people with rice as their main staple. They can seldom afford well-diversified conventional vitamin A sources. Vitamin A precursors cannot be inserted into rice *grains* except by transgenics. The 'Golden Rice' programme is a credible use of transgenics to help people too poor to afford conventional vitamin A sources, and is more cost-effective than supplementation or fortification (H. Bouis, this issue). Before we explore what 'Golden Rice' tells us about the poverty impact of transgenics regulation, biosafety and priorities, there are two prior questions. First, what role have staples, vis-à-vis cash crops, in poverty-reducing transgenics research? Second, how can non-specialists – including most research planners and economists – make sense of the welter of PR hype supporting, and NGO hype opposing, ongoing transgenics research?

Staples and transgenics. Though the poor can gain from applicable research into main smallholder cash crops, the poor lose from the focus, so far, of transgenics on 'cottonseed and chicken-feed' (soy, yellow maize) instead of main staples. Despite steady real price falls and some income rises, the dollar-poor still typically spend 50–60 per cent of income on staples alone, and derive 70 per cent of calories (and most proteins and micronutrients) from them. So transgenics making staples cheaper offer the poor, as consumers, potentially big gains in income and nutrition.[41] However, most of the world's poor – mainly farmers in Africa and China, and mainly farm-workers in South Asia – obtain their main income from small farms,[42] largely growing staples. Increases in such income are increasingly constrained by shortage of land and water. It is therefore important that transgenics-led innovations walk the second tightrope (pp. 31–32), so gains for the poorest as consumers are not offset by losses as producers or employees.

It is seldom in the interests of transgenic-seed companies to focus on staples grown by small farms in low-income countries. Regulatory costs and barriers are high.[43] Unfavourable public relations have been created, by unwise company actions and by some NGOs. It is in any event hard for companies to get a normal return on such R & D investments. It is hard to collect seed or technology charges on food crops, often self-consumed or sold outside formal markets, from millions of tiny farmers. It is hard to stop them retaining transgenic seed to replant or sell to neighbours, especially for self-pollinated crops such as rice and wheat; from crossing transgenic seeds with their own varieties; or from otherwise cutting 'monopoly' rent (on large and risky research costs) by breaching what firms regard as their intellectual property rights (IPRs). Monsanto, far from foisting transgenics on the poor, has pulled out of transgenic rice development (after first putting its knowledge of the rice genome in the public domain).

Nevertheless, there is scope for transgenics research to improve staples performance on small farms and for poor people. First, once a company has profited from a successful gene insertion into a non-staple crop, it can test the same gene's impact after insertion into staples. Even if not profitable, the PR may do more for the bottom line than IPRs, which are often unenforceable; it is in the interest of private companies to be seen to act in the interests of the world's poor. Second, private-sector scientists, directors and shareholders often prefer moral to immoral action. Third, as for public-sector crop research institutions, their income depends partly on impact on the millennium development goal of halving dollar poverty in 1990–2015. Personal commitment apart, public agencies may find it cost-effective to stimulate private researchers to shift towards that goal, through public-private partnerships, fee-for-service, prizes, or other means. Finally, there is a growing not-for-profit sector supported by philanthropists through foundations (compare the 2003 announcement of $25 million support by the Gates Foundation for the CGIAR 'challenge grant' work on staples bio-fortification: Bouis, this issue).

Hype for and against staples transgenics. Contributions from transgenics to poor people's staples have been announced, sometimes many times. Nuffield (1999, 2004) reported the claims: that:

> . . . inserting genes from two wild rice relatives into the best performing Chinese rice hybrids has raised yields by 20–40 per cent. Research . . . has produced a transgenic rice variety resistant to the tungro virus . . . potato varieties bred in

> Peru with stable multi-gene resistance to late blight, a wild wheat cross yielding 18 tonnes/ha. (Nuffield, 1999: 4.29)

> ...improved resistance to environmental stresses such as cold, moisture-stress and high salt levels in the soil...achieved in transgenic rice; transgenic bananas [showing] the possibility of achieving protection against serious fungal diseases and reductions in applied pesticides [with] a more diverse range of varieties, which would allow for additional protection against the impact of pest infestation. (Nuffield, 2004: 43)

What happened to such work-in-progress?[44] In 2003, below 0.05 per cent of transgenics area was planted to any main food staple,[45] including those recorded above (James, 2004).

Committed researchers innocently oversell the potential, perhaps especially when their motives are more moral than commercial. This is well known in natural resource management research (micro-watersheds) and conventional plant breeding (the gap between claims, spread and performance of high-yielding cassava in West Africa). The transgenics examples have proof of concept, and usually plants and seeds with the characteristics claimed, but little in trials and less in farmers' fields. One reason for pro-transgenics hype is reaction against anti-transgenics hype. Inflated, pseudo-scientific health and environmental bugbears have led to delay, over-regulation and even destruction of transgenics research, trials and commercial releases, especially for food and above all for staples. Researchers overclaim partly because they feel beleaguered by anti-science, innuendo and suspicion. But that does not justify the extent of 'transgenic overclaim': the pervasive, counter-productive, and long lag (or permanent gap?) between claim and gain (or grain). It is normal, even with no hype, for 5–10 years to elapse between the start of research and the release of a commercial variety. Yet faster pre-selection is a strongly (and credibly) claimed advantage of transgenics.

Provitamin-A-enhanced rice. The main target for transgenics or other staples breeding is to raise and stabilise sustainable yield, water productivity and employment income. Micronutrient enhancement, while important to the health of the poor, is secondary, and is often best achieved by non-transgenics breeding: consumers of IRRI's pioneering ferritin-enriched rice showed greatly improved iron status (Haas et al., 2005). Sometimes, however, transgenics is the best way to enhance a plant as a source of a bioabsorbable micronutrient.

With vitamin A in rice, it is the only way (Bouis, 2007). And vitamin A deficiency (VAD) overlaps strongly with poverty in rice regions. In 1995, three million children had xerophtalmia, the main cause of child blindness; a further 11 million had clinical VAD; and a further 250 million had sub-clinical VAD, raising the risk of infections such as measles (ACC/SCN with IFPRI, 2000). At least a third of VAD is found in Asia, mainly among poor consumers for whom rice is the main staple. Rice is unique as a staple (a) eaten by most of the dollar-poor (most are Asian); (b) absent transgenics not a source of vitamin A for humans; (c) especially in flooded flatlands, often a monoculture for agronomic reasons, so non-staple sources of vitamin A, such as amaranth, are often unavailable.[46] But is transgenic β-carotene-enhanced rice feasible; attractive to farmers, sufficient as a source of bio-absorbable vitamin A, likely to reach those with VAD, or a cost-effective way to cut it?

On feasibility, Beyer et al. (2002) experimentally transferred four relevant genes. After negotiating the necessary patent clearances and material transfer agreements from the companies involved, they inserted these genes in 2002 into IR-64 – the most widespread conventional semi-dwarf rice, which at present has adequate resistance to all major pests of rice. Photographs of small piles of 'goldenised' IR-64 have been published in peer-reviewed journals (Paine et al., 2005).

Farmer acceptance will depend on profitability and risk. Testing for these means crossing, selection and field-testing. National regulatory clearance is needed for the four gene insertions separately and jointly. Field tests also need to monitor environmental risk, while laboratory and other tests check allergenicity. Deregulation for countries participating in the Golden Rice Network (India, China, Indonesia, Vietnam, Bangladesh, the Philippines, and South Africa) has proved onerous. Influenced by the European debates, regulatory agencies have been hesitant to grant licences for field trials. Companies have removed constraints on smallholder planting, though they can, in theory, charge rich farmers (tightly defined) a royalty. Rice is self-pollinating, so 'Golden Rice' seed can be replanted or sold by farmers without effective constraint. However, the legal situation is cloudy in countries having most rice staples consumers with VAD: China, India, Indonesia, Bangladesh. In early 2004 key researchers *hoped* to complete field tests in the Philippines by 2007; by early 2005 they were estimating a further delay of five to eight years before significant quantities of Golden Rice reach consumers.

On sufficiency of vitamin A from transgenic rice, there is both rapid change due to work in progress, and much controversy due to extreme estimates by both supporters and opponents. Three issues determine how useful 'Golden Rice' will be to those with VAD:

(a) Per kg of rice, how many micrograms of β-carotene?
(b) Per microgram of ß-carotene in rice, how much humanly bioabsorbable vitamin A?
(c) Per unit of bioabsorbable vitamin A, how much 'gain', at different degrees of VAD?

On (a), 160 micrograms of β-carotene per 100 g of cooked rice were found in the earliest tests, but successive transgenic varieties claim dramatic improvements. On (b), on a gloomy view, β-carotene from Golden Rice converts to vitamin A at 12:1, the same as from leafy greens; Bouis (this issue) suggests a better ratio and the optimists assume 4:1, and the optimism seems to be confirmed by recent and ongoing Tufts-USDA studies (Bouis, personal communication).[47] On (c), the usefulness of extra vitamin A probably *increases* as initial VAD rises; the definition of VAD is such that VAD-related diseases in the average child are prevented by *half* the recommended daily allowance.

By combining the gloomiest assumptions on (a), (b) and (c), Greenpeace claimed that a child must eat an impossible 3 kg a day of Golden Rice to make a useful impact on VAD risk.[48] On more optimistic assumptions, the fathers of 'Golden Rice' estimate 200 g, giving enough vitamin A *even with no other source* to prevent most

VAD-related diseases.[49] Since then, there have been very substantial increases in the provitamin-A content of transgenic rice, though new gene insertions which still require regulatory approval.

Whether or not transgenic rice is the most cost-effective approach to VAD depends on acceptance by farmers and consumers. Howarth Bouis (2007) compares alternative approaches. *Commercial fortification* worked well in the case of iodised salt, and cost-effectively cut iodine deficiency for large populations, but badly for VAD. *Vitamin A megadoses* were strongly advocated for schoolchildren once, but (save for acute or hospitalised cases) no longer, suggesting that the approach was not very cost-effective, and/or that those most needing help are pre-schoolers or school-age non-attenders, both heavily overrepresented among the poor. *Dietary diversification* into non-staple sources of β-carotene is ruled out in the short term by poverty. *Nutrition education* is promising only in the small minority of cases of VAD due to parental ignorance. *Non-transgenic breeding* is infeasible for vitamin A enhancement of rice. Other paths may have their place in a strategy against VAD, but it is implausible that the mass of cases among the rice-eating poor will be tackled rapidly or cost-effectively by such methods alone.

It has been estimated that each year of delay in 'Golden Rice' brings the Philippines 9000 more blind children and at least 60 extra deaths (Zimmermann and Qaim, 2002). The Asia-wide numbers would be at least 30 times higher (scaling up from the Philippines to Asian rice-staple populations and allowing modestly for the latter's generally worse poverty and undernutrition). The large majority of VAD sufferers, and hence potential beneficiaries, are poor.

V. Lessons for Transgenics from the GR: Some Tentative Conclusions

It has become clear than most poverty and hunger are linked to inadequate staples *entitlements*. Yet the GR successfully brewed poverty reduction out of seed improvements designed to raise staples *availability*. Also, GR did less well at cutting poverty by raising the poor's entitlements when (since 1990) and where (most of Western China, East–Central India and Africa) it did less to raise staples availability. Sections II–IV suggest two alternative assumptions underlying this paradox, implying distinct lessons for transgenics.

- Assumption 1: *The GR worked, and helped the poor, only when and where there was adequate support for plant-productivity research in the international and developing-country*[50] *public sectors, because these are more responsive than profit-motivated companies to the poor's needs and criticisms.* Transgenics companies seldom find it profitable, given the costs of research and regulatory management, to prioritise the poor's central trait requirements: higher yields and greater robustness for staples in water-insecure environments. Risk-averse or royalties-motivated scientists and companies will seldom prioritise these traits. Though probably amenable to transgenics, such traits are *recalcitrant*: complex, polygene, and with high risk of research failures and difficulty in appropriating royalties from millions of tiny farmers. Mainly private transgenics research may yield significant incidental gains to the poor (Section IV), but on Assumption 1 the central, recalcitrant traits needed by the poor will not attract much private-sector research.

- Assumption 2: *The GR addressed the entitlements problems of the poor –'walked the tightropes' – due less to its public-sector, publicly accountable origin than to the yield- and robustness-enhancing bias of modern seed research and innovation.* If so, the poor's central trait requirements will be helped, in the long run, by almost any new impetus from basic science to seed research. This is especially likely with the impetus that produced transgenics, because it is not specific to the host crop. On Assumption 2, whether the *motive* of transgenic seed research is profit or the public good, its *long-term outcome* is likely to benefit the poor, in adopting areas, substantially and fundamentally (as did the GR), not modestly or by chance side effects (as with transgenics so far). The poor may still lose from specific transgenic innovations (for example, better herbicide-tolerance that displaces weeding labourers), or from innovations heavily concentrated on competitors against poor farmers. That need not invalidate the pro-poor potential of any powerful new tool of seed science. The tool, as a source of new products, cannot be monopolised for long, or confined to rich beneficiaries, even if particular manifestations of its product may be.

Whether Assumption 1 or 2 applies, our earlier analysis raises six challenges:

- The poor and immobile in *non*-adopting areas, especially to the extent that they trade, will lose heavily if their competitors acquire transgenic seeds. It will be harder to escape uncompetitiveness by shifting the crop-mix than with GR, partly because transgenics is less host-crop-specific. Transgenics-free zones catering to non-transgenics markets are unpromising.
- Paradoxically, company monopolisation of seeds (given the inappropriateness to poor farmers of many privately researched traits) can hand seed control to poor farmers, as they are induced to obtain transgenic seed legally or otherwise and creolise it to suit their trait preferences (see Herring; Roy et al., 2007).
- Though such farmer power over transgenics is in many ways welcome, the external and environmental risks argue – even if Assumption 2 is correct – for more public-sector, publicly accountable transgenics research (and more incentives to reorient private-sector research) that seeks *sustainable* crop traits sought by the *local* rural poor. For instance, while private transgenics may brilliantly create single-gene resistance to a pest, only public action is likely to address the intersection of the poor's (private-good) and the environmental (public-good) needs for sustainable, horizontal (polygene), moderate resistance or tolerance.
- Widespread small-farm adoption of transgenic crops in some developing countries – and the likely spread of small-farm adoption to more crops, traits and countries – underlines *the need to see bio-safety management as participatory extension with teeth, not as legalistic regulation.* The latter may make sense for rich countries with many technicians and a few hundred giant farms, but it is elsewhere doomed to failure.
- Regulatory restriction on transgenics seeks to cut (sometimes real) environmental and (largely bogus) health hazards. Regulations would be fewer, simpler and quicker if legislators considered, as they should, the (almost unassessed) hazards

of non-transgenic varieties, including those bred conventionally – and the hazards of delaying transgenic varieties: more VAD, greater pest risk, and loss of competitiveness, especially for the poor.
- Finally, developed countries' transgenics strategies matter to the poor of the developing world. If EU regulations stop transgenic imports, developing countries are deterred from adoption.

A cross-cutting influence on how transgenics follows up GR as an engine of poverty reduction is how agricultural-resource, demographic and nutritional situations, in the years of probable transgenics spread in developing countries, *differ* from the situations during the GR.

How a big technical change such as the GR or transgenics affects poverty depends on both social formations and initial situations. The GR changed those radically. In 1960, perhaps three-quarters of the world's dollar-poor lived (overwhelmingly from staples production) in Bangladesh, China, India, Indonesia, Mexico, Pakistan, the Philippines and Bangladesh. Today, these countries have better than halved dollar poverty (in GR regions, much better). They still contain over half the world's dollar-poor, mainly rural,[51] but often depending for income mainly on cash crops, rural non-farm sector activity, and urban work. Once the Green Revolution has slashed poverty and labour is shifting from the land, much poverty reduction can come mainly from such non-farm sources. Most of the world's dollar-poor, however, still mainly depend on agriculture – in Africa and some of Asia, still largely staples farming. They are unlikely to escape poverty unless nearby farm employment grows and becomes more productive, creating the mass demand to permit later shifts to cash crops, rural non-farm activity, and urban work. The *initial* escape route from dollar poverty is likely to be through cereals-based, labour-intensive farm production and employment. In steering through that escape route, transgenics appears well placed to improve productivity of crops adapted to agro-ecologies recalcitrant to seed-fertiliser-irrigation-based development. Transgenics is also well adapted to address the intensifying water crisis (Nuffield, 2004: 3.42), though it will need help from new water science. Though it is imperative to avoid hype, seed science is pointing in the right direction. But its increasingly private owners do not face sufficient incentives to seek the right goals. Changing that depends on human agency.

Notes

1. (i) 'Green Revolution' (GR) and 'genetic modification' (GM) are 'code': GR, for bred changes in plant architecture (including rice and wheat dwarfing) and in stress resistance, applied in developing countries in 1963–85; GM, for transgenics. The codings are inaccurate and unfortunate – see Herring's Introduction to this issue – but save space and semantics. (ii) 'Poor' in this paper means dollar-poor, admittedly a very restricted meaning (Lipton and Ravallion, 1995). (iii) Some primary sources are in Nuffield Council on Bioethics 2004 (Working Group: D. Burke, M. Gale, M. Lipton, S. Thomas and A. Weale). Nothing in this paper is necessarily the view of, or should be attributed to, the Working Party or Council.
2. Except for nutrition, there is little guidance to impact of agrotechnical change specific to the well-being of the *non*-working poor (mainly small children, women in late pregnancy, the aged, and the ill). In most developing countries their incomes normally depend almost wholly on, and hence rise and fall with, those of the working poor.

3. The rural poor comprised over 70 per cent of the world's dollar-a-day poor around 2000, and are projected (despite urbanisation) to comprise over half until 2035 (Ravallion, 2000).
4. In the 1970s and 1980s, plants with (more or less unknown) chemical- or radiation-induced mutations were widely planted, without regulation or containment, in an effort to improve GR outcomes. Such chance experiments carried massively greater dangers of unwanted crossing than do today's contained trials of plants with known gene insertions, yet escaped largely without complaint from environmentalists or NGOs.
5. Despite the recent award of a World Food Prize for 'quality protein maize', no persuasive new evidence threatens this 30-year-old 'Gopalan consensus' (Lipton with Longhurst, 1989: 244–6). However, the consensus is rejected by one of the world's most distinguished nutritionists (Dr Nevin Scrimshaw).
6. For discussion and references see Strauss and Thomas 1995. In the 1970s surveys reported calorie-income elasticity (CIE) as low as 0.1 *at the mean* in some poor groups. However, later work suggests CIE closer to 0.4 in the poorest deciles. CIE apart, nutritional status also rises as extra income leads to health-seeking behaviour. Also, GR seeds make staples cheaper and – if they are on one's own farm, paid as wages, or sold nearby – easier to obtain. The debate about CIE continues; the problem is partly how to increase it! Some people within a community, and some communities and cultures (for example, with less feeding bias against girls), show 'positive deviance' and thus have far less protein-energy (and perhaps micronutrient) malnutrition even when poor; GR or transgenics impact may vary among households or communities. More knowledge here could improve transgenics impact on nutrition for those at risk. GR success with staples induced farmers to divert land from other protein sources such as pulses in South Asia. However, protein loss from pulses was outweighed by protein gain from more, cheaper rice and wheat. So, if transgenics digresses in search of higher protein content, the nutrition of the poor may well suffer.
7. Sundaram and Tendulkar (2002: 43). Many households get income from more than one source.
8. Lipton with Longhurst (1989: 178–88), and sources cited there. It is wrong to read such falls (for example, from 40 per cent to 15 per cent) as 'declining elasticity of employment to yield growth'; most of the falls (due to tractors, reaper-binders and weedicides) would have happened whatever the yield growth, and were therefore softened by it. The falls also affect the poor in non-GR areas, limiting their migrant labour income. On the large scale of such income, see David and Otsuka (1994). Migrant labour also cuts labour supply, and so bids up real wages, in the non-GR source areas.
9. Even then, research planners should ask: would the poor gain *more* if the same research resources were applied to, say, water-economising rather than herbicide-resistant varieties?
10. That is, allowing for effects, if any, of the new variety on crop-mix; on urban-rural, intersectoral and interregional flows of inputs, outputs and spending; and on prices.
11. Because of concentration by area, an increasingly large part of national staples output in, say, India fell when there was a bad year for wheat in the Punjab. Higher inter-farm covariance outweighed lower within-farm variance; the coefficient of variation of national staples output rose. See Hazell (1984: 302–11).
12. As happened twice in 1972, after tungro virus attacked SE Asia's highly popular IR22 rice, while *H. Maydis*, Southern corn blight, attacked the currently most widely planted maize hybrid in the US.
13. It is fashionable to *claim* that 'African' soil-water regimes are inherently more variable, unstable or fragile than 'Asian', and to *infer* that 'African' plant breeding should not seek a widely usable variety, but many varieties for many agro-ecological niches. The claim is baseless. Even if it were not, the inference would remain false.
14. As with the wheat-dwarfing gene, this rice newcomer expresses gibberellins, raising yield additionally to the effect of dwarfing.
15. However, to *maximise* biodiversity, the preservation of older non-transgenic varieties is increasingly important, depending mainly on comprehensive, well-maintained formal collections. It is doubtful whether long-run risk-minimisation is practicable mainly via in situ conservation and/or non-transgenics refugia, especially in land-scarce countries with hundreds of thousands of tiny and inevitably loosely regulated farms.
16. The alternative of environment-saving yield growth through a more varied crop-mix, while viable locally, is infeasible globally due to the shift of world demand towards animal products, which raises demand for feed cereals.
17. 'If world crop yields had not been tripled [in 1960–19]92, we would have ploughed 10–12 million square miles of additional uncultivated land for low-yield crops. To avoid this happening [from 1992], we must . . . triple the yields from the world's existing farmland again' (Avery 1997).

18. FAOSTAT (updated February 2004). For Sub-Saharan Africa 1.5 per cent excludes, and 3.8 per cent includes, Madagascar, South Africa and Sudan.
19. See Drought-Resistant Soy in Pipeline (2004).
20. See Fitzgerald and Perkins (2005): 'Major seed companies are racing to be first to market with the drought trait. Pioneer Hi-Bred International Inc. and Monsanto Co., among others, are investing heavily in the effort. . . . More arid grain-producing regions in other parts of the world may have more trouble with drought than the Midwest, but seed companies are focused on delivering drought tolerance first to this market because it is the world's biggest seed corn market'.
21. 'Excluding South Africa, [SSA] accounts for less than 1 per cent of world fertilizer consumption' in 2000–01 (International Fertiliser Industry Association 2004). In 2000, SSA had about 10 per cent of the world's arable land (FAOSTAT, April 2005.)
22. This is proved by the co-existence of non-declining rates of return to staple seed research *for* the developing world from the 1960s to the 1990s with sharply declining staples yield growth *in* the developing world (see, respectively, Alston et al., 2000; Lipton 2000). Seed research must have been achieving its non-falling returns by shifting from yield-enhancement goals towards prevention of (pest-induced) *falls* in yield. This is especially clear as returns to research were maintained while staples prices fell and input costs rose, rendering a given return to research harder to achieve.
23. C. Delgado (1999). If 'filtered' through livestock feed, 5–7 times as many staples are needed to supply enough human energy to humans.
24. Notably with repeated enlargements of EU to include more agriculturally orientated countries (while retaining heavy protection and/or producer and export subsidies for its main products), and with the abandonment by the US (in the 2002 Farm Bill) of the remnants of agricultural free trade.
25. (Make your minds up 2004). General-equilibrium work shows big costs (income and poverty reduction foregone) to developing countries seeking a transgenics-free export market niche (Anderson et al., 2002, Abdalla et al., 2003).
26. 'Transgenics-free zones' that defy the preferences of farmers will prove unpoliceable (Herring, this issue). Soy seeds 'smuggled' into South Brazil from Argentina confirm the infeasibility of regulating, respectively, inter- or intra-national seed movements.
27. This requires area-specific incorporation into breeding goals. In much of Asia, but not in Africa (Svedberg, 1990), the worst-nourished are small girls, prioritising absorbable calories in staples-based weaning gruels.
28. Through more: infections aged 40–60; diabetes; heart disease; and weight gain for urbanisers (Lipton, 2001).
29. On HIV/AIDS and transgenics incentives, see case discussion in this paper of glyphosate resistance.
30. Implications include joint strategising for biotechnology, led by the international public sector but involving all interested, including small farmers, labourers, and poor consumers; public will and capacity to 'buy in' significant private R & D skills in transgenics; and a big rise in public R & D for transgenics with poverty-reducing priorities.
31. Agricultural research tends to produce results (for example, seeds) saving mostly the factor (labour, land or capital) scarcest for the 'customer' farmers (Hayami and Ruttan, 1985). But this tendency does not explain much of the variance in factor-saving bias, which also responds to (a) public research and policy; (b) inherent features of scientific discovery.
32. Some developing-country public sectors (China, India, Brazil, Mexico) have large capacity in biotechnology. The CGIAR has an annual budget of over $300 million: long static in real terms (though perhaps being revived by the challenge grants), diffused over many institutions and purposes, and with below 10 per cent aimed at biotechnology.
33. In S Africa '*Bt*... increases yields by 40 per cent and is pesticide-saving, halving spraying. It is not labour-saving as the extra harvest labour compensates for less spraying labour. The results for small farmers in Argentina are similar, whereas in China there was zero yield gain and a five-fold reduction in pesticide. This is because the Chinese farmers over-used pesticides prior to Bt, whereas these South African smallholders have difficulty obtaining chemical inputs... yield gains should be greater for most smallholders in Sub-Saharan Africa and *Bt* cotton should not reduce employment' (Shankar and Thirtle, 2005; compare Shankar and Thirtle, 2003; Thirtle et al., 2003).
34. Huang et al. (2003). Comparing surveys for 1999, 2000 and 2001, they show that 'adoption of *Bt* cotton continues to increase [yield] in 2000 and 2001... [in] all provinces in our sample... *Bt* cotton

farmers also increased their incomes by reducing use of pesticides *and labour*...[and] have less health problems [due to] reduced pesticide use'. Typically, yield gains were 10%, but vary greatly; see below and Roy et al. (this issue).

35. The paper (in *Science*) is cited in Qaim and Zilberman (2003).
36. Depending on place and time, control of cotton bollworm might leave more cotton to feed other pests, not affected by *Bt*. This would show up in profitablility or income analyses such as that of Qaim and Zilberman (2003).
37. As with (legal) creolisation of CIMMYT maize HYVs by smallholders in Chiapas, Mexico (Bellon and Risopoulos, 1999).
38. Similar differentiation explains why Indian NGOs commenting on early drafts of Nuffield (2004) were hostile, but their African counterparts much less so, to herbicide-resistant crops.
39. Extra land planted to cassava does little to offset falling employment per hectare (Pachico et al., 2002: 356–67).
40. I am grateful to Ron Herring and Ravi Kanbur for the computerised-banking analogy and for this observation.
41. Sen (1981) shows that most famine-affected people, and by extension many chronically undernourished, suffer from lack of entitlements (claims) to food, not of food availability. Some – not Sen – wrongly infer that more food production, for example via transgenics, cannot make the poor less hungry. India's huge public grain stock, alongside mass hunger, is taken to show that (extra) national food production cannot improve food security. This is fallacious. The most cost-effective *feasible* way to raise the poor's food entitlement is to raise their earnings, relative to food prices. The most cost-effective way to do that, in rural Africa and Asia where two thirds of the world's of the dollar-poor live, is usually to raise *nearby, small-farm, labour-intensive* staples production. This is consistent with the fact that more cereals from non-poor Punjabi (or EU) farmers, grown capital-intensively and sold to public stock, do (at best) nothing to reduce poverty or, therefore, hunger.
42. The share of land in farms below 1ha has been rising in all these areas, in 1970–90 and in 1990–2000 (FAOSTAT, 2005). Small farms show higher value-added per hectare (but less per unit of labour) than larger farms (Lipton, 2005).
43. Dr Ingo Potrykus (personal communication, 2003) has emphasised that regulatory delays impose a career disincentive, especially on young scientists, against public-purpose, non-company work on transgenic crops aimed to benefit the poor.
44. Work on transgenic virus-resistant sweet potato – 'in Kenya...estimated to raise yields 15 per cent' (Nuffield, 1999), '[to] prevent dramatic...reductions in yield of [a] major food crop of many poor people in Africa' (Nuffield 2004) – failed, stopping in early 2004. de Grassi contrasts this programme with cheaper, less publicised, more successful conventional breeding in Uganda. http://www.ahbfi.org/sweetstatement.htm 2004 objects to this as hindsight, citing the transgenics work's *ex ante* plausibility, and yield in experience and training.
45. Transgenic soy and yellow maize are almost entirely used for animal feed.
46. This aggravates the poor's difficulty in avoiding VAD through varied diets. The poor need cheap calories to survive; 'let them eat carrots' is not a useful message, especially if non-staple Vitamin A sources do not grow locally.
47. Even these studies do not assess bio-absorption by under-fives with VAD, for example, through breast-milk and rice gruels.
48. Assumptions: very low bio-absorbability; all vitamin A from Golden Rice; uselessness to those with VAD unless it contributes 100 per cent of RDA. (Greenpeace, 2000 [published before the improved new generations of enriched Golden Rice]). Informally, eating requirements are magnified further. I heard 9 kg confidently asserted, without evidence, by a voluble opponent of transgenics at the IDS Sussex conference, 'Biotechnology and the poor', on 1 October 2003.
49. (Beyer and Potrykus, 2003). Requirements would be much lower given the claimed content of new releases to June 2005, but these involve several new transgenic events, each requiring long clearance procedures.
50. Applied in-country research is usually needed, for good returns from seeds researched abroad (Evenson and Kislev, 1976).
51. Seventy per cent of the world's dollar-poor were rural in 2000; more than half will be rural in 2035 (Ravallion, 2000).

References

Abdalla, A., Berry, P., Connell, P., Tran, Q. and Buetre, B. (2003) Agricultural biotechnology: potential for use in developing countries, ABARE *e*Report 03.17, Canberra.

ACC/SCN with IFPRI (2000) *Fourth Report on the World Nutrition Situation* (Geneva: ACC/SCN).

Alston, J., Chan-Kang, C., Marra, M., Pardey, P. and Wyatt, T. (2000) A meta-analysis of rates of return to agricultural R & D: *ex pede Herculem?*, Research Report No. 113, International Food Policy Research Institute, Washington, DC.

Anderson, K., Nielsen, C. and Robinson, S. (2002) Estimating the economic effects of transgenics: the importance of policy choices and preferences, in R. Evenson, V. Santaniello and D. Zilberman (eds), *Economic and Social Issues in Agricultural Biotechnology*, pp. 359–91 (Wallingford: CABI International).

Avery, D. (1997) Saving the planet with pesticides, biotechnology and European farm reform, Bawden Lecture, British Crop Protection Council, Brighton.

Behrman, J. and Deolalikar, A. (1988) Health and nutrition, in H. Chenery and T. N. Srinivasan (eds), *Handbook of Development Economics. Vol. I*, pp. 631–711 (North Holland: Amsterdam).

Bellon, M. and Risopoulos, J. (1999) Expanding the benefits: farmers' transformation of a CIMMYT technology to suit the poor, mimeo, San Jose workshop on agricultural research and poverty, CIAT, Cali.

Beyer, P., Al-Babili, S., Ye, X., Lucca, P., Schaub, P., Welsch, R. and Potrykus, I. (2002) Golden Rice: introducing the β-carotene biosynthesis pathway into rice endosperm by genetic engineering to defeat vitamin A deficiency, *The Journal of Nutrition*, 132, pp. 506S–10S.

Beyer, P. and Potrykus, I. (2003) How much vitamin A rice must one eat?, accessed at: http://www.agbioworld.org/biotech-info/topics/goldenrice/how_much.html

Bouis, H. E. (2007) The potential of genetically modified food crops to improve human nutrition in developing countries, *Journal of Development Studies*, 43(1), pp. 79–96.

David, C. and Otsuka, K. (eds) (1994) *Modern Rice Technology and Income Distribution in Asia* (Boulder: Lynne Rienner).

Delgado, C. L. and International Food Policy Research Institute (1999) *Livestock to 2020: The Next Food Revolution* (Washington, DC: International Food Policy Research Institute).

Evenson, R. E. and Kislev, Y. (1975) *Agricultural Research and Productivity* (New Haven: Yale University Press).

Fan, S., Hazell, P. and Haque, T. (2000) *Targeting Public Investments by Agro-ecological Zone to Achieve Growth and Poverty Alleviation Goals in Rural India* (Washington, DC: International Food Policy Research Institute).

Fan, S., Linxiu, Z. and Haque, T. (2000a) *Growth and Poverty in Rural China: The Role of Public Investments* (Washington, DC: International Food Policy Research Institute).

FAO (2000) Food and nutrition security: why food production matters, in *The State of Food and Agriculture 2000*, (Rome), pp. 199–242.

Fitzgerald, A. and Perkins, J. (2005) Weather worries: hot, dry or wet, DesMoines Register.com, March 27, accessed at: http://www.dtn.com/news.cfm?content=05news/n_032705&sidenav=sn_innews

Frisvold, G., Reeves, J. and Tronstad, R. (2006) Bt cotton adoption in the United States and China: international trade and welfare effects, *AgBioForum*, 9(2), pp. 69–78.

de Grassi, A. (2004) accessed at: http://www.ahbfi.org/sweetstatement.htm

Haas, J., Beard, J., Murray-Kolb, L., del Mundo, A., Felix, A. and Gregorio, G. (2005) Iron-biofortified rice improves the iron stores of non-anemic Filipino women, *Journal of Nutrition*, 135, p. 2823–30.

Greenpeace (2001), Vitamin A: natural sources vs 'golden rice', accessed at: http://archive.greenpeace.org/~geneng/reports/food/VitaAvs.PDF

Hayami, Y. and Ruttan, V. (1985) *Agricultural Development: An International Perspective* (Baltimore: Johns Hopkins).

Hazell, P. (1984) Sources of increased instability in Indian and US foodgrain production, *American Journal of Agricultural Economics*, 60(3), pp. 302–11.

Hazell, P., Jagger, P. and Knox, A. (2001) Technology, natural resource management and the poor (2000), thematic paper for (IFAD 2001).

Hazell, P. B. R., Ramasamy, C., Aiyasamy, P. K. and International Food Policy Research Institute (1991) *The Green Revolution Reconsidered: The Impact of High-yielding Rice Varieties in South India* (Baltimore: Johns Hopkins University Press).

Herring, R. J. (2007) Stealth seeds: bioproperty, biosafety, biopolitics, *Journal of Development Studies*, 43(1), pp. 130–57.

Hossain, M. (1988) *Nature and Impact of the Green Revolution in Bangladesh* (Washington, DC: International Food Policy Research Institute).

Drought-Resistant Soy In Pipeline: Monsanto scientists zero in on drought-resistant soybeans (2004) *St Louis Post-Dispatch—Knight Ridder/Tribune Business News*, 2 June, accessed at: http://www.afaa.com.au/news/news-1453.asp

Huang, J., Ruifa, H., Fan, C., Pray, C. and Rozelle, S. (2003) *Bt* cotton benefits, costs and impacts in China, Working Paper 202, Institute of Development Studies, Brighton.

International Foundation for Agricultural Development (IFAD) (2001) *Rural Poverty Report 2001* (Rome: Oxford University Press).

International Fertilizer Industry Association (2004) Fertilizer consumption, by region, 1970/71 to 2000/01, Africa (excluding Egypt and Libya), accessed at: http://www.fertilizer.org/ifa/statistics/indicators/ind_cn_afr.asp

India's foodgrain crisis and steps to meet it (1959) Agric. Production Team/Ford Fdn, Min of Agriculture, Delhi.

James, C. (2004) *Global Review of Commercialised Transgenic Crops: 2003* (Ithaca: International Service for the Acquisition of Agri-Biotech Applications).

Kerr, J. and Kohlavalli, S. (1999) Impact of agricultural research on poverty alleviation, Environment and Technology Discussion Paper No. 56, International Food Policy Research Institute, Washington, DC.

Lipton, M. (2001) Challenges to meet: food and nutrition security in the new millennium, *Proc. Nutrition Society*, 60, pp. 203–14.

Lipton, M. (2005) Crop science, poverty and the family farm in a globalising world, IFPRI 2020 Discussion Paper, Washington, DC.

Lipton, M. (1994) Food production and poverty, in B. Harriss-White and J. Hoffenberg (eds), *Food: Multidisciplinary Perspectives* (Oxford: Blackwell).

Lipton, M. (2000) Reviving global poverty reduction: what role for transgenic plants?, 19th Sir John Crawford Memorial Lecture, Washington, DC, CGIAR.

Lipton, M. with Longhurst, R. (1989) *New Seeds and Poor People* (London: Unwin Hyman).

Lipton, M. and Ravallion, M. (1995) Poverty and policy, in J. Behrman and T. N. Srinivasan (eds), *Handbook of Development Economics. Vol. 3B*, pp. 2551–657 (Amsterdam: North Holland).

Make your minds up (2004) Editorial, *Farmers' Guardian*, 2.4.

Maxwell, S. (2004) New directions for agriculture in reducing poverty, DfID Consultation Launch Document.

Nuffield Council on Bioethics (1999) *Genetically Modified Crops: The Ethical and Social Issues* (London: Nuffield Council on Bioethics).

Nuffield Council on Bioethics (2004) *The Use of Genetically Modified Crops in Developing Countries* (London: Nuffield Council on Bioethics).

Pachico, D., Escobar, Z., Rivas, L., Gottret, V. and Perez, S. (2002) Income and employment effects of transgenic herbicide-resistant cassava in Colombia: a preliminary simulation, in R. Evenson, V. Santaniello and D. Zilberman (eds), *Economic and Social Issues in Agricultural Biotechnology*, pp. 356–7 (Wallingford: CABI International).

Paine, J., Shipton, C., Chaggar, S., Howells, R., Kennedy, M., Vernon, G., Wright, S., Hinchliffe, E., Adams, J., Silverstone, A. and Drake, R. (2005) Improving the nutritional value of Golden Rice through increased pro-vitamin A content, *Nature Biotechnology*, 23, pp. 482–7 (2005) (published online: 27 March 2005, DOI: 10.1038/nbt1082).

Psacharopoulos, G. and Patrinos, H. (1993) Indigenous people and poverty in Latin America: an empirical analysis, Report No. 30, World Bank, LAC Technical Department, Regional Studies Program.

Pinstrup-Andersen, P., de Londono, N. and Hoover, E. (1976) The impact of increasing food supply on human nutrition, *American Journal of Agricultural Economics*, 58, pp. 131–42.

Qaim, M. and Zilberman, D. (2003) *Bt* cotton in India: Field-trial results and economic projections, *World Development*, 31(12), pp. 2115–27.

Ravallion, M. (2000) On the urbanisation of poverty, mimeo, World Bank, Washington, DC.

Reardon, T., Timmer, C. P. and Berdegue, J. A. (2003) The rapid rise of supermarkets in Latin America and East/Southeast Asia: implications for domestic and international markets for fruits and vegetables, in A. Regmi and M. Gehlhar (eds), *Global Markets for High Value Food Products*, Agriculture Information Bulletin, USDA–ERS.

Roy, D., Herring, R. J. and Geister, C. C. (2007) Naturalising transgenics: official seeds, loose seeds and risk in the decision matrix of Gujarati cotton farmers, *Journal of Development Studies*, 43(1), pp. 158–76.

Rosegrant, M. and Hazell, P. (1999) *Transforming the Rural Asian Economy: The Unfinished Revolution* (Hong Kong: Oxford University Press, for Asian Development Bank).

Sen, A. K. (1981) *Poverty and Famines: An Essay on Entitlement and Deprivation* (Oxford: Clarendon Press).

Sen, A. K. (2001) *Poverty and Famines: An Essay on Entitlement and Deprivation* (Oxford: Oxford University Press).

Shankar, B. and Thirtle, C. (2005) Pesticide productivity and transgenic cotton technology: the South African smallholder case, *Journal of Agricultural Economics*, 56, pp. 97–116.

Shankar, B. and Thirtle, C. (2003) Pesticide productivity with smallholder *Bt* cotton in Makathini Flats, KwaZulu-Natal, symposium paper for 25th conference of the International Association of Agricultural Economists, Durban.

Smith, L. and Urey, I. (2002) Agricultural growth and poverty reduction: a review of lessons from the post-Independence and green revolution experience in India, mimeo, Imperial College at Wye.

Strauss, J. and Thomas, D. (1995) Human resources: empirical modelling of household and family decisions, in J. Behrman and T. N. Srinivasan (eds), *Handbook of Development Economics. Vol. 3B*, pp. 1893–908 (Amsterdam: North Holland).

Sundaram, K. and Tendulkar, S. (2002) The working poor in India: employment-poverty linkages and employment policy options, Issues in Employment and Poverty series: Discussion Paper 4, International Labour Office, Geneva.

Svedberg, P. (1990) Undernutrition in sub-Saharan Africa: is there a gender bias?, *Journal of Development Studies*, 26, pp. 469–86.

Thirtle, C., Piesse, J. and Jenkins, L. (2003) Can TRANSGENICS technologies help the poor?, *World Development*, (31), pp. 717–32.

UN, ECOSOC (2005) *World Population Projections: the 2004 revision*, population database, accessed at: http://esa.un.org/unpp

World Water Council (2000) *A Water-Secure World: Vision For Water, Life and the Environment* (Marseille: World Water Vision).

Zimmermann, R. and Qaim, M. (2002) Projecting the benefits of golden rice in the Philippines, Discussion Paper on Development Policy No. 51, Bonn, Centre for Development Research ZEF.

The Impact of Agricultural Biotechnology on Yields, Risks, and Biodiversity in Low-income Countries

DAVID ZILBERMAN, HOLLY AMEDEN, & MATIN QAIM

I. Introduction

Agricultural biotechnology has been introduced primarily to meet market needs of farmers in rich countries; there are concerns that this technology will not be appropriate for, or serve the needs of, farmers in the South. One concern lies with the yield-enhancement potential of agricultural biotechnology, a major priority for new crop varieties in the South, due to the modest yield effects from use of transgenic varieties (TGVs) thus far in the North (Altieri, 2001). Other concerns are that TGVs are introduced by private-sector organisations that tend to offer a limited range of choices and that the introduction of TGVs may accelerate loss of crop biodiversity.

This paper relies on economic logic to analyse the impacts of TGVs on yields and crop biodiversity. To develop our argument, we introduce a framework to analyse adoption of TGVs. Our analysis suggests that the introduction of TGVs may have different yield effects in different countries, reflecting differences in economic and climatic conditions, and that *TGVs may actually have significant yield effects in low-income countries*. We use empirical results from China, India, and other countries to support our theoretical predictions. We also show that TGVs play a risk-reducing role, and develop a framework to assess the impacts of pest infestation risks on adoption of TGVs. Finally, we argue that under a range of plausible circumstances, introduction of TGVs may not lead to a reduction in crop biodiversity and, in some cases, may actually lead to its enhancement. Thus, some of the concerns about the limited value of TGVs for low-income countries are misplaced. Moreover, the real challenge is to develop the institutional setup that will allow low-income countries to take advantage of the potential of TGVs.

II. The Yield Effects of TGVs: A Conceptual Approach

Pest-resistant TGVs are the primary types of TGVs that have been introduced and used by farmers. These varieties have been modified primarily through the insertion of *Bt* genes into seeds and hybrids of cotton, corn, soybeans, and so forth. Herbicide-resistant varieties have also been developed, most notably, varieties that tolerate Roundup. Our conceptual analysis considers the various impacts of 'pest reducing' TGVs. This generalisation fits Roundup Ready (RR) varieties, but it is more appropriate when analysing *Bt* varieties. This is a simple version of a model introduced in Ameden et al. (2005); it will show that the same technology may have different impacts in different locations depending on prevailing economic and environmental conditions.

We consider a farm with a fixed amount of land where the farmer can vary the quantity of pesticides used and choose a crop variety to produce an agricultural crop. We assume that the farmer can grow three varieties – a traditional variety, denoted as variety L; a TGV version of this local variety denoted as variety M; and a generic TGV denoted as variety G. The output of the traditional variety is denoted by $Y(L)$, the local TGV by $Y(M)$, and the generic TGV by $Y(G)$. Using the damage-control specification of Lichtenberg and Zilberman (1986), the output of each variety is equal to the potential output ($POTY$) net of pest damage, $Y = POTY * (1 - DAMAGE)$, where potential output is the maximum output produced without pest damage. While the potential output of the traditional and local TGV is assumed to be the same, the potential yield of the generic TGV is assumed to be lower. The variable $DAMAGE$ is the fraction of output damaged by pests. Pest damage depends on the original size of the pest infestation ($NoPest$), the quantity of pesticides applied ($Pesticide$), and whether the crop variety was a TGV. The damage can be expressed as a function (F) of these factors:

$$DAMAGE = F(NoPest, Pesticide, TGV),$$

where TGV is an indicator that assumes a value of 0 for the traditional technology and 1 for TGVs.

The TGVs reduce pest damage substantially. Given initial pest infestation and pesticide application, only a small fraction of the damage that occurs with a traditional variety occurs with a TGV. Figure 1 depicts the relationships in the production process.

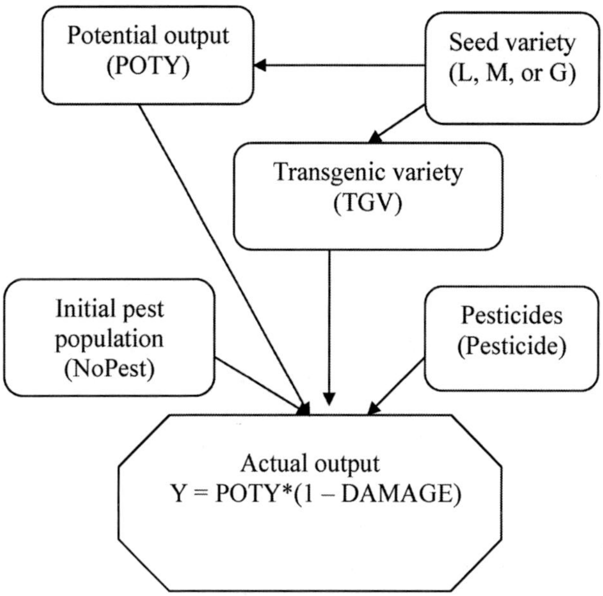

Figure 1. The damage control mechanism

Farmers are assumed to pursue profits, and their gain from application of pesticides is different across varieties. Pesticide application levels are also different across varieties. Therefore, to determine the impact of TGVs on yield given a specific pest infestation level, we must compute the amount of pesticides that would be applied with each variety and assess the yield effect accordingly. The revenue from production is a product of output price (*Yprice*) and output (*Y*).[1] The short-term profits the farmer is assumed to maximise are equal to revenue minus the fixed and variable costs of pesticides. The fixed cost of pesticides, denoted by *Fixpest*, consists of the costs of application equipment and the number of times needed to apply the pesticides. The variable cost of pesticides, denoted by *Pprice*, is the product of the amount of pesticides applied and its price. Thus, operational profits (*OProfits*), which exclude the costs of production and seeds that are used earlier in the season before the arrival of the pests, is the difference between revenues and pesticide costs,

$$Oprofits = Y * Yprice - Pesticide * Pprice - Fixpest.$$

For each variety, the farmer determines whether or not pesticides would be used and, if pesticides are used, how much should be applied. To answer this question, the farmer has to compare maximum short-term profits with no pesticide use and maximum short-term profits with pesticide use. The short-term profits without pesticides is equal to the output price times the yield with untreated pest infestation. The short-term profits, assuming pesticides are used, is the revenues when the pest infestation is treated minus the cost of the pesticides and the fixed costs of application. When pesticides are applied, the optimal level of pesticide use is determined where the incremental gain from increased pesticide use (because of higher yield) is equal to the incremental costs of pesticides.[2] In determining whether or not to apply pesticides, the farmer compares the trade-offs between the extra revenues associated with the damage reduction and the additional cost of the pesticides and their application. Pesticides will be applied if the return from higher yield covers both fixed and variable costs of pesticide applications.

Ameden et al. (2005) show that when pesticides are applied, the quantity applied is likely to increase as the output price or potential output are higher, the damage avoided is more valuable, and the price of the pesticides is lower. Also, pesticide use increases with the size of the initial pest infestation. The likelihood that pesticides are applied (nonzero pesticide) is likely to decline as the fixed costs or the variable costs of pesticides increase. Moreover, the use of pesticides is likely to decrease as TGVs are adopted because the genetic modification reduces pest damage substantially and serves as a substitute for chemical pesticides, reducing their value and productivity.

A very likely outcome is that it is optimal to apply pesticides with the traditional variety and not with the TGVs. Even when pesticides are applied, both with and without TGVs, the adoption of TGVs will reduce pesticide use. The pesticide-saving effect associated with adoption of TGVs is higher the greater is the application level with the traditional variety, and the greater is the efficacy of the TGV in pest control. However, because the potential output of the generic TGV is smaller that the local TGV, fewer chemicals are likely to be applied with the generic TGV, that is,

$$Pesticide\ (L) > Pesticide\ (M) > Pesticide\ (G).$$

Ameden et al. (2005) show analytically that, when a TGV is used instead of the traditional variety, the genetic modification along with the pesticide use will reduce pest damage, resulting in higher yields, and that this effect is likely to be greater in low-income countries. This yield effect depends on the effectiveness of the TGVs and the fixed and variable costs of pesticides. If the fixed and variable costs of pesticide use are low and productivity of pesticides is limited, then the yield effect of TGVs is low. This is likely to be the case in rich countries where pesticides are relatively inexpensive and pesticide-use levels are high, eliminating most pest damage. On the other hand, if pesticide prices and fixed costs of applications are very high relative to output price, as they tend to be in low-income countries, and availability of effective pesticides is severely limited, then pesticide application rates will be low and may even be zero. This, coupled with the humid climates of poor countries (as opposed to the temperate climates of rich countries), is likely to result in much higher initial levels of pest infestations and associated pest damage. Thus, in low-income countries, the introduction of TGVs has significant potential to reduce pest damage and increase yields.

The yield effect of TGVs depends on the specific variety adopted. When a local TGV is adopted, the lower pest damage will result in higher yield ($Y(M) > Y(L)$). If farmers adopt the generic TGV rather than the local TGV, the damage-reduction effect is countered by the lower potential yield of the generic variety. Thus, the net yield effect of the generic TGV is not clear. The yield effect of generic TGV is positive, $Y(G) > Y(L)$, only if the damage reduction effect (that is, $POTY(L) * DAMAGE(L) - \mathrm{POTY}(G) * DAMAGE(G)$) is greater than the yield loss effect (that is, $POTY(L) - POTY(G)$).

This analysis is supported by several studies on the impacts of *Bt* cotton. Frisvold and Tronstad (2003) document relatively low yield effects (5 per cent and below) in the United States where pesticide use is widespread. Huang et al. (2002) and Pray et al. (2002) analyse the impacts of adoption of *Bt* cotton in China, where pesticides are highly subsidised. They find only modest increases in yield but drastic reductions in pesticide use associated with improved worker safety. On the other hand, Qaim and Zilberman (2003) find dramatic yield increases in India in 2002. They analyse the results from 157 farms that planted each of the three varieties (a traditional variety, a TGV of the traditional variety, and a generic TGV) on different plots. Qaim and Zilberman (2003) find that the TGVs reduced pesticide use by 67 per cent, the local TGV increased yield on average by 87 per cent, and the generic TGV by 80 per cent. They suggest that the high yield effects were the result of high pest infestations levels in the year of the study and that impacts were smaller in other years. Qaim et al. (2005) show that during the season 2002–03 the estimated yield effect of *Bt* cotton in India was 34 per cent. There was a 50 per cent reduction on insecticide use on *Bt* cotton plots, and it generated $100 of extra profit per hectare (Ha.) relative to conventional cotton. Roy (2003) reports on the adoption of *Bt* cotton in some areas of India and finds that in certain areas, adopters had low yields and suffered losses. She argues that in most of these cases a generic, imported TGV replaced local varieties. Her findings highlight the importance of variety choice on the success of TGVs. Herring's (2003) analysis of TGVs in India strongly demonstrates that there are significant variations in yield effects of TGVs between seasons, and that suggests that current analyses of the impacts of TGVs should be modified to accommodate this randomness.

III. Random Infestation and the Adoption Decision

The literature on adoption and impacts of TGVs has relied heavily on deterministic models, though empirical evidence suggests that pest damage and levels of infestation vary between periods. This is especially apparent in studies of the impacts of biotechnology in India, where the yield effect of TGVs may vary from 30 per cent to 80 per cent due to varying levels of pest infestation. Thus, the pest damage or yield effect of a TGV is not a single number – it is a statistical distribution, and the modeling of adoption choices should be stochastic. In this section, we introduce this modeling approach.

The initial pest infestation (*NoPest*) is a random variable that may assume several values, each value with a certain probability. For simplicity, assume that there are two values: a low infestation level *Ns*, and a high infestation level *Nh*. The probability of the low level is *Ps*, the high level, *Ph* and, since these are probabilities, $Ps + Ph = 1$. The farmer has to make decisions concerning pesticide use (how much to apply) and choice of seeds (or hybrids). The pesticide decision is determined once the pest infestation level is known. However, the seed decision is made earlier in the season, before the pest infestation level is known. Our earlier analysis of the pesticide choice applies to the case with random infestation, but the choice of seed is modified to accommodate the randomness of infestation.

We first assume that the farmer is risk neutral, and the farmer's objective is to maximise expected profits. Each variety has its own expected operational profits, and the farmer selects the variety with the highest level of expected profits. Expected operational profits are:

$$EOProfits = OProfits\ (POTY,\ Ns)\ Ps + OProfits\ (POTY, Nh)\ Ph.$$

Recall that the operational profits for a given state of nature, denoted by *OProfits*, is the difference between revenue and pesticide costs. The revenue depends on *DAMAGE* as determined by initial pest infestation levels.[3] The operational profits are the overall profits minus the fixed cost of production and genetic materials that are used early in the season.

Ameden et al. (2005) show that, for every state of nature, the local TGV has higher operational profits than the other two varieties and, thus, it has the highest expected operational profits. The generic TGV can have higher expected operational profits than the traditional variety in situations where the savings in pesticide costs and damage reduction using the generic TGV are more valuable than the loss of potential output that the use of this variety entails. Under these conditions, the following holds:

$$EOProfits\ (M) > EOProfits\ (G) > EOProfits\ (L).$$

On the other hand, the seeds of the local TGV are likely to be more expensive than the generic TGV seeds which, in turn, cost more than the seeds of the traditional variety, thus:

$$Seedcost\ (M) > Seedcost\ (G) > Seedcost\ (L).$$

The choice of which variety to use involves weighing the trade-offs between expected operational profits and seed costs. High pest damage, high pesticide costs, and relatively low TGV costs will lead to adoption of the TGV technology. The generic TGV is selected instead of the local TGV when the marginal benefit of the pesticide cost savings and lower seed costs from switching to the generic is greater than the value of the marginal yield losses, thus expected profits are higher. For each variety, expected profits are the average operational profits across pest infestations (denoted by *EOProfits*) minus the nonpesticide-related fixed costs of production (denoted by *Prcost*) and seeds (denoted by *Seedcost*):

$$ExProfit = EOProfits - Prcost - Seedcost.$$

The expected profits of the traditional variety are denoted by *ExProfit* (*L*), and the expected profits of two TGVs are denoted by *ExProfit* (*M*) and *ExProfit* (*G*), respectively.

To understand the implications for the analysis, when pest infestation and damage are randomly distributed, consider the following simple example. Assume that no pesticides are available to fight a particular crop pest, but it can be controlled by a TGV. Suppose that there are two states of nature – no infestation that occurs with probability of 50 per cent and infestation that leads to loss of 50 per cent of the potential output with 50 per cent probability. The yield distribution and expected yield are presented in Table 1 below, for the traditional variety and two TGVs.

Table 1. Yield distributions for various varieties

State of nature	Probability	Yield with traditional variety	Yield with local TGV	Yield with generic TGV
No crop loss	0.5	POTY	POTY	0.95 * POTY
Crop loss	0.5	0.5 * POTY	0.9 * POTY	0.855 * POTY
Expected yield		**0.75 POTY**	**0.95 POTY**	**0.9025 POTY**

Numeral example.

We assume that the TGVs reduce the pest damage by 80 per cent, so that with the local TGVs there is a 50 per cent probability of no damage and 50 per cent probability to have 10 per cent damage ($50\% * 0.2$). We further assume that the generic TGV reduces potential output by 5 per cent. Then with use of the generic TGV, there is a 50 per cent probability of having a 5 per cent yield loss and 50 per cent chance of 14.5 per cent yield loss ($1-0.95*0.9=0.145$).

Let's now assume there are no other costs but seed costs and that seeds of the traditional variety are free. Expected profits for the traditional variety are equal to 75 per cent of the revenue generated by selling the potential output ($0.75 * POTY$), while the expected profits of the local TGV are based on producing 95 per cent of potential output ($0.95 * POTY$). If the price of the TGV seeds is less than $0.2 * POTY$, then the TGV is preferable to the traditional variety even though 50 per cent of the time there is no apparent yield effect of adoption.

The expected profit of the generic variety is derived from $0.9025 * POTY$ minus the seed costs. If the TGV seeds cost more than $0.0475 * POTY$ and the generic seeds cost less than $0.1625 * POYY$, they will be adopted. In this case there will be 50 per cent probability of 5 per cent yield loss relative to the traditional variety, but another 50 per cent probability of yield gain of 71 per cent ($(85.5\% - 50\%)/50\%$). Both local and generic TGVs increase expected yield, but their yield effects relative to the traditional technology vary between seasons. The adoption of the generic TGV may reduce yield in some of the periods, thus judging the technology based on its performance in one period may be misleading.

Impacts of risks are more significant in cases where farmers are risk averse and are concerned not only about the average profits but also about its variability and 'downside events'. Feder et al. (1985) survey a large body of literature showing that adoption behaviour of farmers in low-income countries suggests they are risk averse; this risk aversion is likely to decline with the size of the operation. Risk-averse individuals are ready to pay a premium to reduce the risk they face. The variance of profits is a convenient measure of risk. In our examples above, both TGV technologies increase expected profits and also significantly reduce the variance of profits. For example, the variance of the traditional variety is $0.0625 * POTY^2$ and the variance of the local TGV is $0.0025 * POTY^2$. The adoption of the TGVs divides the variance of profits by 25. Thus, farmers may adopt TGVs because of their positive impact on average profitability and risk-reducing effect.

Other factors may contribute to high yield effects from adoption of TGVs in low-income countries. If lack of access to credit restricts the ability of farmers, especially small farmers, to purchase expensive inputs such as pesticides, then the associated low levels of pesticide use increase the potential yield effect from adopting TGVs. Of course, if credit constraints make it more likely that a generic TGV is adopted rather than a more expensive local TGV, the yield effect will be smaller. Moreover, if farmers' abilities to borrow are severely restricted and TGVs are relatively costly, use of TGVs would be similarly restricted.

Credit availability varies greatly across regions, with poorer farmers located far away from financial centres having a greater difficulty to obtain credit. Thus, while adoption of TGVs is likely to be slow in these areas, our analysis suggests that these areas with severe credit limitations would experience greater yield-increasing effects from adoption.

Combined risk and credit considerations are likely to have a significant impact on farmers' choices. High levels of pesticide use increase farmers' debt-to-income ratio, while low pesticide use could lead to crop failures. For years where pest infestations are suddenly high, or price of output is low, farmers may fail to repay a loan, leading to bankruptcy. Poor farmers without access to credit or institutions for storing wealth cannot address risk effectively by saving. Thus, if TGVs provide a significant risk-reducing effect and are relatively lower in cost compared to pesticides, they are likely to be adopted and be particularly valuable to poorer farmers who are more vulnerable to risk.

Our analysis thus far argues that the concern about potentially low yield effects of TGVs in low-income countries is misplaced. While TGVs may have relatively low yield effects in high-income countries with high use levels of effective pesticides, their yield-increasing potential is high in low-income countries that are subject to high pest infestation levels and do not have effective pest-control strategies. We also saw that yield effects vary randomly, and adoption of TGVs can serve as an insurance mechanism, the impacts of which vary over time. Our analysis suggests that the impact of the TGVs depends on whether a local or a generic TGV is adopted. Adoption of a local TGV will have a higher yield effect, and will help to address another concern – the impact of TGVs on biodiversity, as discussed below.

IV. Biodiversity and the Adoption of TGVs

The practice of using principles of genetics to guide plant breeding is about 100 years old. Selective breeding techniques have been very successful in generating new varieties including the 'Green Revolution' varieties that significantly enhanced agricultural productivity. However, the varieties generated by selective breeding were a distinct genetic departure from traditional varieties. The newer techniques of biotechnology generate new varieties by slightly altering existing varieties, modifying one or two genes, while maintaining the bulk of the original genetic materials. Thus, the impact of biotechnology on crop biodiversity will be slight when the TGVs that are introduced are mainly modified local varieties. However, if a single generic TGV is widely adopted, replacing many varieties, then losses in crop diversity may be significant. Although we do not develop a formal measure of biodiversity, we assume that an increase in the adoption of the generic TGV and, in particular, the replacement of traditional varieties with the generic TGV, represents a loss of crop diversity.

The model and results of Qaim et al. (2005), combined with the results we derived earlier, will be used to identify factors that will lead to wide-scale adoption of generic versus local TGVs. Qaim et al. consider the case of a country where farmers use a large number of local varieties. These are not necessarily landraces, but varieties that were developed through selective breeding to fit local conditions. Here we will investigate analytically the conditions that will lead to adoption of local and generic TGVs in various regions, where adoption is measured by the relative acreage grown with the traditional variety and the local and generic TGVs within each region. Threshold models (see Feder et al., 1985) provide a useful framework to analyse adoption choices and are used here. They assume that farmers in each region are diverse in terms of their human capital, quality of land, and so forth. Farmers are assumed to pursue profits, sometimes adjusted for dynamics and uncertainty considerations. The heterogeneity of farmers may result in differences in their adoption choices. The threshold models reveal the critical (threshold) value of parameters of heterogeneity that separate farmers according to their adoption decisions. These threshold levels are dynamic, changing over time in response to such processes as learning by doing.

As we argued earlier, a farmer will compare the expected gain resulting from the yield effect and pesticide cost reduction of TGVs with the extra costs of the seeds when choosing a crop variety for a specific field. Qaim et al. (2005) identify several patterns of adoption when the heterogeneity of land quality within a region results in differences in potential output. The gains from adoption of TGVs are likely to increase with higher land qualities, and the relative gain from adoption of local TGVs relative to generic TGVs increases with land quality. The local TGVs will be fully adopted and may even lead to expansion of utilised lands to lands that were not farmed with the traditional technology. If at the worst quality land the gain from adoption of the technology is higher than the extra cost needed to pay for the local TGV seeds, then the more it will be adopted. When the local TGV is relatively expensive and the generic TGV is more affordable, partial or full adoption of the generic TGV is more likely.

Thus, the crop biodiversity losses due to the introduction of TGVs are mild when the cost of seeds of local TGVs is low in a large number of regions. This will lead to full or partial adoption of local TGVs in those regions. However, if the price of local TGV seeds is relatively high in most regions, while the generic TGV seeds are relatively cheap and are fully adopted in many regions, the crop biodiversity losses are likely to be substantial.

Several factors will determine the prices of local and generic TGVs and, thus, affect adoption and biodiversity:

1. *The strength of the crop-breeding sector.* In countries with the appropriate scientific infrastructure, such as the United States and China, transgenic modification of crop varieties is done routinely. In some low-income countries, the capacity of the plant-breeding sector is limited, which restricts the number of local varieties that can be modified. In these situations the introduction of TGVs may require reliance on the importation of generic TGV seeds.
2. *Regulatory process and requirements for TGVs.* The cost of developing and introducing new TGV seeds is likely to increase substantially the more restrictive and the less clearly defined the registration requirements are for new TGVs. The introduction of TGVs can be curtailed when the regulatory process is very restrictive. If the regulators are very strict with new local varieties, it may lead to

the importation of generic varieties, which may lead to significant reduction of crop biodiversity.

3. *Institutional arrangements for TGV seed development and marketing*. Qaim et al. (2005) consider two types of institutional arrangements to establish seed prices: Under the first, the public sector develops TGVs while a competitive seed company distributes them and, under the second, monopolies develop and sell TGVs. The first arrangement has been used to introduce new varieties developed using selective breeding in many low-income countries. The Consultative Group on International Agricultural Research (CGIAR) and the national agricultural research services (NARs) develop new genetic materials, and small seed companies sell them to farmers. The second arrangement is more consistent with the reality of agricultural biotechnology in the North, where large multinational corporations with significant monopolistic power have developed and introduced TGVs. The seed price or technology fee charged by monopolies is larger than the price under competition. Thus, Qaim et al. (2005) predict that there will be greater adoption of TGVs when the seed industry is competitive and that more local TGVs will be introduced. Thus, development and introduction of new TGVs by the public sector and marketing through competitive sellers will benefit biodiversity. Note, however, this requires that the public sector has the expertise and skills to produce new TGVs. In most cases the public sector does not have the know-how and capacity to introduce TGVs, and reliance on the private sector may lead to less costly TGVs.
4. *Transaction cost in seed markets and intellectual property rights (IPRs)*. When the private companies consider introduction of TGVs to a country, their strategy may vary according to cost of transaction and the ability to enforce IPRs. One approach they may pursue is to either import TGVs from other countries or to own or control companies that sell a subset of the local seed varieties and offer a small number of TGVs. An alternative approach is to sell the rights to transgenetically modify local varieties to existing seed companies and charge a royalty (technology fee). The second strategy, which is likely to result in more crop biodiversity preservation, is more likely to be pursued as the cost of negotiation decreases and as IPRs are easier to enforce. A broad range of local varieties may be genetically modified and sold cheaply if the local public sector has the required capacity and is not concerned about IPRs.

Our analysis suggests the emergence of several adoption patterns of agricultural biotechnology that vary in their impact on crop biodiversity. In a country like the United States, with a well-developed seed sector and relatively low transaction costs, local TGVs will be widely introduced, even with private sector distribution. Monopolistic pricing, however, will result in partial adoption of biotechnology in many regions and continued plantings of traditional varieties. The restrictive regulatory environment will result in minimal adoption of TGVs in Europe. The significant capacity of the seed sector, the competitive nature of the seed industry, and relatively low regulatory burden is likely to result in a high number of local TGVs in a country like China. However, in Africa, where the local capacity of transgenic modification is very limited, there may be a higher likelihood of importing generic TGVs, and crop biodiversity will suffer. One of the policy challenges is to develop infrastructure at the local level in Africa so that modification of TGVs will not be so difficult and expensive.

The data presented in Qaim et al. (2005) support our general results. They show that thousands of varieties have been modified in the United States, and the acreage per modified variety in some crops is smaller than in other countries. This may reflect the relative low transaction and modification costs in the United States. Low acreage per variety may reflect a high technology fee that results in partial adoption. In the 2001–02 season, more than 1100 varieties of RR soybeans were planted in the US, with about 20,000 ha. on average for each variety. More than 700 varieties of *Bt* corn were planted, each variety on 10,000 ha., on average. In Argentina, 45 varieties of RR soybeans were planted on 10 million ha., with 200,000 ha. per variety. A total of 700,000 ha. of *Bt* corn were planted, with 15 varieties, so 45,000 ha. were planted per variety. In the case of *Bt* cotton, the United States has 19 varieties grown on 2 million ha., while China, with its public sector development of TGVs and subsidised seed sector, has 22 varieties on 1.5 million ha. Our conceptual analysis and empirical examples are consistent with the results in Ammann (2004), who reviews studies on the impacts of the introduction of TGVs in soybeans and cotton in the United States, and finds that genetic uniformity of crop varieties has decreased by 28 per cent during the period of introduction of TGVs in the United States. The evidence suggests that introduction of TGVs around the millennium did not have significant effect on genetic uniformity.

Both our conceptual analysis and data suggest that the introduction of TGVs will not necessarily lead to loss of biodiversity and significant reduction in the number of varieties grown. The introduction of TGVs may actually increase the number of distinct varieties, narrowly defined, when the technological, economic, and regulatory conditions lead to genetic modification of most traditional varieties and when the TGVs are not fully adopted. On the other hand, limited capacity to genetically modify local varieties may affect biodiversity negatively, as it may lead to the adoption of a small number of TGVs (sometimes imported) on lands that were grown with a larger number of local varieties. Strengthening the capacity of the seed sectors in low-income countries and introducing simple mechanisms to allow developers of TGVs easy access to local varieties will increase the biodiversity of TGVs. The preservation of crop biodiversity can also be enhanced by financial incentives, since the biotechnology choices of the individual farmers and the private sector companies that affect biodiversity are economic choices. The public sector and donors concerned about the preservation of local varieties may cover some of the costs associated with development or introduction of local TGVs that would not have been introduced otherwise. Environmental services payment may be introduced to subsidise farmers to grow local varieties instead of generic TGVs.

V. Conclusions and Extensions

This paper addresses two of the primary concerns about the value of using the current generation of TGVs in low-income countries. First, we show that adoption of these TGVs might substantially increase crop yields in low-income countries in spite of low yield effects in rich countries. The different impacts of the technology across locations result from differences in the initial pest control strategies. When

TGVs replace effective pesticides, their impact on yields is limited. However, their value is enhanced where pesticides are not used or have limited impacts. There is evidence that in some low-income countries (for example, China), TGVs reduce pesticide use and, thus, improve worker safety and reduce economic and environmental costs. In other countries (South Africa), they improve yields, and in India they do both. Furthermore, the randomness of the pest infestation leads to varying yield effects over time and, thus, TGVs are also a very valuable source of financial insurance for farmers. The risk-reducing effect of TGVs may benefit mostly farmers who are more vulnerable and who have higher aversion to risk. These tend to be smaller farms and, since TGVs do not have economies of scale, they may hold much promise for the poorer farmers in low-income countries, unlike other modern technologies.

We use a conceptual model to show that TGVs have significant yield-increasing and risk-reducing potential in poor nations. Specifically, where pesticide prices and fixed costs of application are very high relative to output prices, reduced pesticide application rates and warm humid climates produce have high pest-infestation levels; the introduction of TGVs then has significant potential to reduce pest damage and increase yields. Moreover, in areas with great variation in pest damage and pest infestation levels between seasons, farmers may adopt TGVs because of their positive impact on average profitability and risk-reducing effect, especially poorer farmers who have limited access to credit and are thus more vulnerable to risk.

The second concern we address is potential loss of crop biodiversity when TGVs are adopted; we show that adoption of TGVs does not necessarily lead to crop biodiversity loss. We argue that the introduction of transgenic crops is less likely to reduce crop biodiversity in regions with strong seed sectors having fewer barriers to access local seed varieties. The preservation of crop biodiversity can be enhanced by incentives that lead to partial adoption of TGVs and by policies and regulations that provide favourable conditions to introduce local TGVs. Thus, the introduction of TGVs to low-income countries in Africa and other regions should be accompanied by expansion of capacity of the local seed sector.

While the paper shows that the two major concerns about introducing TGVs to low-income countries can be addressed, decision-making concerning use of the technology in poor counties should not be based on ability to address all objections raised by critics of the technology. Instead it should be based on weighing benefits, costs, and risks imposed by the technology, taking into account the likely outcomes if the technology is not introduced (see National Research Council, 2000). The analysis in this paper provides a starting point for qualitative assessment of the plusses and minuses associated with introducing TGVs to low-income countries.

Agricultural systems are heterogeneous and subject to ecologic and socioeconomic variability. Adoption is a complex process that requires availability of superior products appropriate for specific locations, a supportive regulatory economic system, and an effective system for dissemination. Therefore, the diffusion of new TGVs will be limited to locations where they can make a positive difference and where the institutional conditions make them acceptable. Nevertheless, whenever they are adopted, the introduction of TGVs is likely to improve the well-being of adopters as a result of reduction in pesticides and risk-bearing cost, and increase in yield. Adoption of the technology will also benefit its providers, as some may be shareholders of agribusiness firms, but others are local dealers and distributors. The increase in yield may benefit consumers of TGV crops, but nonadopters may lose, as the price of their output will decline. Nevertheless, the overall 'market effects' (impacts on consumers and producers) of supply-enhancing technologies are positive (Qaim et al., 2005).

Due to the scale neutrality of TGVs, their basic economics do not favour larger farms. Risk considerations make them especially beneficial to smaller producers who have fewer options to reduce vulnerability to risk than do larger farmers (Just and Zilberman, 1988). There may be credit and other access constraints that have to be removed before small producers can fully benefit from TGVs.

Since food products absorb a relatively larger share of poor families' income, the yield effects of TGVs and the resulting lower prices will be relatively more beneficial to poorer consumers. Our analysis, thus far, suggests that the introduction of TGVs will improve the overall market surplus and will likely have positive distributional impacts, in the sense that relative gain to poorer individuals is likely to be greater. The recent results of Anderson (2005) and Evenson (2005) suggest significant potential of economic welfare gains to low-income countries from the introduction of existing TGVs. The extent that this potential will be realised depends on implementation of policies and on institutions that aim to enhance adoption. Even though the estimated gains are impressive, they only make a small dent in world poverty, and the existing TGVs are only a small part of a broader effort needed to enhance human welfare.

Evidence suggests that under a broad set of circumstances, the adoption of TGVs will likely lead to significant gains in human and environmental health. The work of Huang et al. (2002) for cotton in China and the findings on rice (Huang et al., 2005) suggest that adoption of pest-controlling TGVs will lead to reduction in pesticide use in some cases, and thus reduced environmental contamination and improved worker safety. The adoption of herbicide-resistant varieties will enhance adoption of low- and no-tillage techniques and reduce buildup of carbon storage and soil erosion. The yield effect of TGVs will also allow food demands to be met, utilise smaller amounts of land, and reduce the incentives to deforestation and conversion of wildlands to farming.

However, these environmental benefits need to be compared to presumed environmental costs. We show that crop biodiversity losses may be substantial only when the institutional and economic setups lead to broad-scale adoption of generic varieties; moreover, these losses can be reduced by policy intervention. Other environmental problems that may emerge from the adoption of TGVs include gene flows to wild varieties and damage to nontarget species. While the gene-flow problem has been recognised by scientists for some time, Gilligan et al. (2003) and Ammann (2004) argue that it is not persistent, and Ammann's survey of the literature does not reveal cases where TGVs have caused substantial damage. Of course, lack of damage thus far is no assurance of the future; monitoring of possible side effects of TGVs should be pursued. With this said, it is likely that the economic, social, and environmental benefits of TGVs in low-income countries outweigh their costs, and there is a strong case for their application in the poorer nations.

VI. Policy Challenges

The primary policy challenge is to develop policies and mechanisms that will harness the potential of biotechnology to benefit the poor in low-income countries. These products may include pest-resistant and herbicide-resistant varieties of crops such as cassava, manioc, and varieties of rice and maize that are appropriate for low-income countries. The range of biotechnology applications has to be expanded to address constraints that affect farmers in poor countries: drought-resistant and salt-tolerant varieties as well as varieties that will nutritionally enrich staple foods.

Private companies control much of the development and distribution of biotechnology in rich countries, but it is unlikely that they will sufficiently invest in products that target the poor in low-income countries. Instead, the development of these products should be carried out mainly by the public sector, continuing the pattern that emerged during the Green Revolution (Evenson and Gollin, 2003). However, this will require investment in research infrastructure in low-income countries, as well as creative arrangements that will allow transfer of knowledge and minimise barriers and costs of accessing new knowledge on development of new varieties. New organisations that aim to reduce the cost of IPRs for agricultural biotechnology for poor countries are taking initial steps in the right direction (Atkinson et al., 2003).

Another challenge is to develop a regulatory process that will maintain a high degree of safety, yet would not impose excessive regulations on developers of new technology. Reduction of redundancy in regulation and establishment of regional regulatory organisations that take advantage of economies of scale will increase the resources available for product development.

A further challenge is to address restrictions on credit availability. The literature suggests that limited access to credit may slow or reduce adoption of agricultural biotechnology by poor farmers owning small farms. Thus, institutional policies that both reduce the cost of credit and increase its availability are needed. For example, instalment plans that require payment of biotechnology 'fees' on a season-by-season basis for farmers utilising transgenic seeds, rather than significant up-front fees, will improve adoption substantially for poor farmers. Alternatively, if the price of seeds is sufficiently low and credit channels are expanded, small farms that did not adopt chemical pesticides will adopt biotechnology seeds.

While creative credit solutions may be effective for commercially viable farms, further solutions are needed for the poorest subsistence farmers, such as price discrimination structures specifically aimed at alleviating poverty. Seed companies could be assured access to commercial seed markets in one area in return for offering seeds at or below cost to poor farmers in other areas. Of course, introduction of credit and pricing policies will require monitoring of farmer behaviour and markets as well as an effective enforcement system. Extension services and the public sector will be challenged to cooperate with the private sector to introduce such mechanisms.

While we argue that biotechnology holds significant promise for low-income countries and that our biggest challenge is effectively introducing it to benefit the poor, we have to recognise the ecological uncertainties and support research and monitoring activities that assess the impact of biotechnology on the environment.

We need to trigger mechanisms for adaptive learning to enhance productivity of biotechnology and, thus, detect its limitations through monitoring. Furthermore, continued support for basic research and reduction of IPRs and regulatory barriers will likely lead to improved varieties with beneficial traits. Biotechnology is in its infancy. Thus far, we have seen several applications that have been very effective for a small number of crops. Overall, though, the impact of biotechnology on agriculture has been modest. Yet, the success thus far is an indicator of promise and potential that has to be utilised through both research and institutional design.

While we emphasise the value and potential of biotechnology for agriculture in low-income countries, we also recognise that the value of complementary approaches. Biotechnology can complement strategies based on improved precision in farm management or empowerment of farmers through augmentation of human capital and collective action. Thus, biotechnology is an important element of an emerging farming strategy, but it has to be integrated with other approaches to enhance productivity, alleviate poverty, and improve environmental quality in low-income countries.

Notes

1. For this analysis, we assume output price is net of harvesting costs.
2. At the optimal pesticide use level, the value of the marginal benefits of pest damage reduction is equal to the price of the pesticides.
3. Recall from the first section that *DAMAGE* = F (*NoPest*, *Pesticide*, *TGV*) and in this case, *NoPest* can take on the values *Ns* or *Nh*. Also, *Oprofits* = *Y* * *Yprice* − Pesticide * *Pprice* − *Fixpest* and *Y* = *POTY* * (1 − *DAMAGE*). Thus, operational profits depend on initial pest infestation levels and the potential yield of the variety, holding other variables constant.

References

Altieri, M. A. (2001) *Genetic Engineering in Agriculture: The Myths, Environmental Risks and Alternatives* (Oakland, CA: Food First).

Ameden, H., Qaim, M. and Zilberman, D. (2005) Adoption of biotechnology in developing countries, in J. Cooper, L. Lipper, and D. Zilberman (eds), *Agricultural Biodiversity and Biotechnology in Economic Development*, pp. 329–57 (Norwell, MA: Springer Publishers).

Ammann, K. (2004) *The Impact of Agricultural Biotechnology on Biodiversity: A Review* (Bern, Switzerland: University of Bern).

Anderson, K. (2005) Interactions between trade policies and GM food regulations, paper presented at the Conference on Economics of Regulation of Agricultural Biotechnologies, Washington, DC.

Atkinson, R. C., Beachy, R. N., Conway, G., Cordova, F. A., Fox, M. A., Holbrook, K. A., Klessig, D. F., McCormick, R. L., McPherson, P. M., Rawlings III, H. R., Rapson, R., Vanderhoef, L. N., Wiley, J. D. and Young, C. E. (2003) Policy forum: public sector collaboration for agricultural IP management, *Science*, 301, pp. 174–5.

Evenson, R. E. (2005) Status of agricultural biotechnology: an international perspective, paper presented at the Conference on Economics of Regulation of Agricultural Biotechnologies, Washington, DC.

Evenson, R. F. and Gollin, D. (2003) Assessing the impact of the Green Revolution, 1960 to 2000, *Science*, 300, pp. 758–62.

Feder, G., Just, R. E. and Zilberman, D. (1985) Adoption of agricultural innovations in developing countries: a survey, *Economic Development and Cultural Change*, 32(2), pp. 255–98.

Frisvold, G. and Tronstad, R. (2003) Economic effects of *Bt* cotton adoption and the impact of government programs, in N.G. Kalaitzandonakes (ed.), *Economic and Environmental Impacts of Agbiotech: A Global Perspective*, pp. 261–86 (New York: Kluwer/Plenum Academic Publishers).

Gilligan, C., Claessen, D. and van den Bosch, F. (2003) Spatial and temporal dynamics of gene movements arising from deployment of transgenic crops, paper presented in Environmental Costs and Benefits of Transgenic Crops in Europe, Wageningen University, Netherlands.

Herring, R. (2003) Underground seeds: the lessons of India's *Bt* cotton episode for representations of the poor, property claims and biosafety regimes, paper presented at the Conference on Transgenics and the Poor: Science, Regulation and Development Strategy, Cornell University, Ithaca, New York.

Huang, J., Hu, R., Fan, C., Pray, C. E. and Rozelle, S. (2002) *Bt* cotton benefits, costs, and impacts in China. *AgBioForum*, 5(4), pp. 153–66.

Huang, J., Hu, R., Rozelle, S. and Pray, C. (2005) Insect-resistant GM rice in farmers' fields: assessing productivity and health effects in China, *Science*, 308, pp. 688–90.

Just, R. E. and Zilberman, D. (1988) The effects of agricultural development policies on income distribution and technological change in agriculture, *Journal of Development Economics*, 28, pp. 193–216.

Lichtenberg, E. and Zilberman, D. (1986) The econometrics of damage control: why specification matters, *American Journal of Agricultural Economics*, 68, pp. 262–73.

National Research Council (2000) *The Future Role of Pesticides in US Agriculture*, The Committee on the Future Role of Pesticides in U.S. Agriculture, Board on Agriculture and Natural Resources, Board on Environmental Studies and Toxicology, National Research Council.

Pray, C. E., Huang, J., Hu, R. and Rozelle, S. (2002) Five years of *Bt* cotton in China – the benefits continue, *The Plant Journal*, 31(4), pp. 423–30.

Qaim, M. and Zilberman, D. (2003) Yield effects of genetically modified crops in developing countries, *Science*, 299, pp. 900–2.

Qaim, M., Yarkin, C. and Zilberman, D. (2005) Impact of biotechnology on crop genetic diversity, in J. Cooper, L. Lipper and D. Zilberman (eds), *Agricultural Biodiversity and Biotechnology in Economic Development*, pp. 283–307 (Norwell, MA: Springer Publishers).

Roy, D. (2003) GMOs and poverty in Gujarat: is there a connection?, paper presented at the Conference on Transgenics and the Poor: Science, Regulation and Development Strategy, Cornell University, Ithaca, New York.

The Potential of Genetically Modified Food Crops to Improve Human Nutrition in Developing Countries[1]

HOWARTH E. BOUIS

I. Introduction

This essay discusses the potential of biotechnology to improve the health and nutrition of consumers in developing countries. In the relatively wealthy countries of Europe, North America and elsewhere, consumers spend perhaps 10 per cent of their income on food. For the most part consumers in developed countries are free of classical nutrient deficiencies, although over-consumption is a problem for some. Also in relatively wealthy countries there is, in general, good access to affordable medical care to meet health needs, and most consumers in rich countries have access to a relatively inexpensive supply of safe and healthy food. In these settings, the possibility that biotechnology might reduce the price of food or make food more beneficial to health is a relatively minor concern. Rather, public debate about genetically modified foods (GMFs) appears to have focused on the potential for harm to either the environment or health without a clear definition of benefit to the consumer.

The situation is quite different in poor countries, where malnutrition and ill health are frequent. Poor consumers typically spend 70 per cent of their incomes on food, and diets consist primarily of staple foods, which lack the vitamins, minerals, and very likely other food components necessary to sustain good health and minimise the risk of adult onset diet-related chronic diseases. In addition, low incomes typically exclude the poor from access to adequate health care.

There are three broad ways that biotechnology may benefit consumers in developing countries. First, biotechnology offers a powerful new tool to improve crop productivity, both by making conventional breeding faster and more efficient and, more controversially, by the insertion of novel genes in a crop species, by use of transgenic methods. This includes the ability to bring new lands with unfavourable growing environments, such as those with highly saline soils, under production. Second, pesticide applications may be reduced through adoption of *Bt*-containing crops (that is, crops with insect resistance owing to the introduction of the *Bacillus thuringiensis* [*Bt*] gene encoding an insect-specific toxin). Reducing pesticide use can improve the health of farmers, in addition to lowering input costs. Third, transgenic methods may be used to improve the micronutrient content and/or bioavailability of commonly eaten foods in developing countries.

The first pathway, that is, improving crop productivity, is associated with an often-asked question: Does feeding the world depend on the use of biotechnology? The answer largely depends on perceptions of the current situation. Countries around the world are already using most of the arable land available for agriculture. As populations continue to grow, increasing the supply of food to meet future world food needs will depend on increasing yields. This, in turn, requires investments in agricultural research to increase the yield potentials of specific food crops as well as fish and livestock production.

Conventional breeding of cereals has been especially successful over the past three decades. Through research and extension efforts associated with the 'Green Revolution', the rate of increase of production of rice, wheat, and maize in developing countries has outpaced population growth and other factors associated with increased demand (such as higher household incomes), such that the inflation-adjusted prices of rice and wheat fell by 30–40 per cent between 1970 and 1997 (World Bank, 1990–99).

World supplies and prices of cereals are interconnected through international trade. Advances in conventional breeding in Australia, Europe, and North America have also contributed to falling cereal prices. Continued low prices of cereals are likely as long as investments in agricultural research continue, although a worrisome trend is that funds for public research in agriculture have been declining (Rosegrant et al., 2001).

It is often said that there is enough food in the world to feed everyone. Except for some pockets of extreme poverty and dislocations owing to natural disasters and civil strife, this is largely the case for cereals. Cereals (and roots and tubers) provide the cheapest sources of energy to meet requirements for poor populations. The availability of affordable staple foods prevents explicit hunger, as defined narrowly in terms of energy requirements.

However, a range of non-staple foods, such as animal and fish products, fruits, pulses, and vegetables, are foods rich in bioavailable vitamins and minerals, or micronutrients, and possibly other food components, which are necessary for good health and a productive life. As discussed in more detail below, micronutrient undernutrition is widespread in poor countries, affecting more than one-half of the population in the developing world. This situation is due primarily to diets of low nutritional quality. The poor want to eat larger amounts of non-staple foods, and most often these foods are available in local markets. However, such non-staple foods are simply too expensive to be eaten by the poor in large quantities owing to a combination of high prices and low purchasing power. The inflation-adjusted prices of non-staple foods have not been declining; in some cases they have been increasing over the past 30 years.

Because of poor dietary quality and consequent micronutrient malnutrition, some people, especially children and their mothers, who have higher requirements for vitamins and minerals, have higher mortality, become sick more often, have their cognitive abilities compromised for a lifetime, and are less productive members of the workforce than might otherwise have been the case. Their quality of life and aggregate economic growth are unnecessarily compromised. To make matters worse, the adverse synergy between undernutrition and infection results in nutrition requirements above those that characterise the needs of well-nourished populations.

In developing countries, demand for non-staple foods is growing faster than demand for cereals, as ever-richer households purchase more non-staple foods than cereals, with consequent pressure on non-staple food prices to increase. In addition, increasing the productivity of non-staple foods through agricultural research (and so lowering prices) is much more expensive than increasing the productivity of staple foods, because of the far larger number of non-staple foods involved.

From this perspective, then, the world is already failing to provide an adequate supply of non-staple foods for a large proportion of its population, and prospects are somewhat bleak for short- to medium-term improvements. Biotechnology affords a powerful new tool to help increase the supply of food produced in the world and redress, in part, situations characterised by poor dietary quality.

A third pathway through which biotechnology can improve the nutrition and health of consumers in developing countries is through increasing the vitamin and mineral content and bioavailability of staples and other foods, an approach which is the focus here.

Although it is obviously important to provide an accounting of the range of specific efforts being developed to improve the nutritional content and health-enhancing properties of particular foods, it is also necessary to place these activities in the context of alternative interventions (that is, interventions that do not involve the use of biotechnology) that might solve these malnutrition and health problems in other ways. This is accomplished by (i) examining the relative costs of nutrient supplementation and commercial nutrient fortification programs that are already being implemented to reduce micronutrient malnutrition; and (ii) examining the recent history of attempts to improve the micronutrient content of staple food crops – efforts that involve use of both biotechnology and conventional plant breeding. Recent developments in efforts to breed iron- and beta-carotene dense rices are described in this essay as a concrete example, illustrating several key generic issues associated with the use of biotechnology to breed for characteristics that have direct benefits for consumers.

It is important to keep in mind that the final, sustainable solution to micronutrient malnutrition is a substantial improvement in dietary quality by higher consumption of pulses, fruits, vegetables, fish, and animal products, which the poor need and already desire, but cannot presently afford. Realisation of this goal of high quality diets will require (i) substantial investments by farmers, private businesses and public agencies to build the infrastructure to produce and bring to market the requisite supply of non-staple foods; (ii) sound government policies to stimulate agricultural and economic growth; and (iii) considerable increases in the incomes of the poor during the course of economic development over many decades; or (iv) other strategies (for example, subsidies) designed to enhance the accessibility of these products to wider populations. Meanwhile, breeding staple foods that are dense in minerals and vitamins can provide a low-cost, sustainable, although as yet unproven strategy for reducing existing levels of specific micronutrient undernutrition.

Sections II to V (below) discuss, in some detail, iron- and beta-carotene dense rice, as an example of the development of a GMF crop of potential importance in developing countries.

II. Potential Contribution of GM Technology: A Case Study on Breeding of Iron- and Beta-carotene Dense Rices to Help Alleviate Malnutrition

Background to the Problem of Micronutrient Malnutrition in Poor Countries, its Economic Costs and Costs of Present Interventions to Address the Problem

It is only relatively recently that nutritionists working in developing countries have demonstrated conclusively that, during times of relative economic and political stability, many children and adults, particularly women in their child-bearing years, suffer more from a lack of essential vitamins and minerals in their diets than from a lack of energy. People, for the most part, are not aware that their diets lack such trace nutrients and so do not associate these deficiencies with listlessness, poor eyesight, impaired cognitive development and physical growth, more frequent and severe bouts of illness, and high mortality. For this reason, the general problem of poor dietary quality has been dubbed 'hidden hunger'.

It is useful to summarise very briefly the scale and consequences of micronutrient malnutrition in developing countries. It is estimated that over three billion people in developing countries are iron-deficient (Administrative Committee on Coordination, Subcommittee on Nutrition, and International Food Policy Institute, 2000). The problem for women and children is more severe because of their greater physiological need for iron. In poor countries, more than half of pregnant women and more than 40 per cent of non-pregnant women and preschool children are anaemic. Iron deficiencies during childhood impair physical growth, mental development and learning capacity. In adults, iron deficiency reduces the capacity to do physical labour. Severe iron deficiency is a leading contributor to death among women during childbirth, owing to low haemoglobin levels in the light of physiological blood loss.

Globally, about three million children of preschool age have visible eye damage owing to a vitamin A deficiency. Annually, an estimated 250,000 to 500,000 preschool children go blind from this deficiency, and about two-thirds of these children die within months of going blind. Estimates of the subclinical prevalence of vitamin A deficiency range from 100–250 million. A number of clinical trials in developing countries have shown that vitamin A supplementation, as capsules distributed biannually in high doses, as weekly distributed low doses (consistent with levels attainable through food), or as a fortified product in diets can reduce mortality rates among preschool children on average by 23 per cent (Beaton et al., 1993).

Deficiencies in several other micronutrients, in particular zinc (Hotz and Brown 2004), may also be widespread, with equally serious consequences for health. However, because for some of these micronutrients, specific indicators are lacking to screen for deficiencies, they have not received as much attention as iron, vitamin A and iodine deficiencies.

The World Bank (1994) estimates that at the levels of micronutrient malnutrition existing in South Asia, 5 per cent of gross national product is lost each year due to deficiencies in the intakes of just three nutrients, namely iron, vitamin A and iodine. For each 50 million in population, this translates into an economic loss of US$1 billion per year.

Some general cost calculations can be made for interventions to relieve iron and vitamin A deficiencies by supplementation and fortification in South Asia, which has a total population of roughly 1.25 billion people. The costs of vitamin A pills themselves are low: an often-quoted cost of vitamin A supplementation, which includes the cost of delivery, is US$0.50 per person per year (that is, US$0.25 per capsule; World Bank, 1994). If one in every 12.5 people in South Asia were to receive supplements (that is, 100 million people in total), this would cost US$50 million per year, or US$500 million over 10 years. An often-quoted cost of iron fortification is US$0.12 per person per year (World Bank 1994). If a particular food vehicle fortified with iron were to reach 33 per cent of the total (but untargeted) population in South Asia (that is, 412 million people), the total cost would, again, be US$50 million annually, or US$500 million over 10 years. In absolute terms, these nutrient-specific interventions may seem to be large amounts of money, but they are very worthwhile investments and actually represent quite small percentages of the total economic activity of the South Asian economies.

The Micronutrient Initiative (MI) estimates that 1 billion vitamin A pills have been distributed worldwide since the inception of vitamin A distribution programmes up to the end of 2000 and that 500 million supplements are now given out annually (UNICEF, UNU, WHO and MI, 1999). A conservative estimate of the annual, recurrent cost of global vitamin A supplement programmes is, therefore, US$125 million.

Although benefit–cost ratios are quite high, supplementation and fortification programmes must be sustained at more or less the same level of funding year after year in any given country. If investments are not sustained, benefits disappear.

Investments in plant breeding research and dissemination are far lower (see Section III(ii)) and potentially longer lasting. Benefits of agricultural research at a central location can be leveraged throughout the world and across time. Breeding for staple plants with high bioavailable micronutrient content in their seeds, hereinafter referred to as 'biofortification', treats the underlying cause of micronutrient undernutrition. Although plant breeding can involve relatively long lead times of eight to ten years before nutritious varieties can be developed and their adoption by farmers can be initiated, such a strategy is sustainable once breeding has been completed, and seeds have been disseminated and adopted by farmers. Biofortification has the potential to provide coverage for remote rural populations, which supplementation and fortification programs may not reach, and it inherently targets the poor who consume high levels of staple foods and little else.

However, efficacy and effectiveness trials and other studies have yet to be undertaken to determine both the degree to which nutritionally dense varieties can improve micronutrient status and whether there are unintended negative effects.

According to an expert group convened to consider micronutrient supplementation and fortification with iron, iron overload disorders and haemochromatosis are rare, even in those populations of European origin most susceptible to them; therefore, the use of fortification and supplementation as public health interventions for preventing and controlling iron deficiency should not be constrained (UNICEF, UNU, WHO and MI, 1999). It is safe to assume that the same conclusions would apply to biofortification. If biofortification eventually were to replace supplementation in some areas, there is the possibility that biofortification might lessen risks of overdosing, such as that recently reported in India, where toxicity associated with a mass distribution of a vitamin A concentrate to children was thought, possibly, to have been the result of a changed delivery routine (Sharma, 2001).

Improving the Nutrient Content and Bioavailability of Minerals and Vitamins in Staple Food Crops Through Biofortification

Breeding strategies. There are three breeding sub-strategies that may be applied individually or in various combinations. These are (i) reducing the level of antinutrients in food staples, which inhibit the bioavailability of minerals and vitamins; (ii) increasing the levels of nutrients and compounds that promote the bioavailability of minerals and vitamins; and (iii) increasing the mineral and vitamin content. Experience with increasing the mineral and vitamin content of rice is discussed in some detail in section II(iii).

Reducing antinutrients. A breeding strategy of lowering the level of antinutrients, such as phytic acid, which binds metals in grain, has often been suggested as a way to increase the bioavailability of minerals already consumed.

Phytin is the primary storage form of phosphorus in most mature seeds and grains and is an important compound required for early seed germination and seedling growth (Wise, 1995). Phytin plays an important role in determining mineral reserves of seeds and thus contributes to the viability and vigour of the seedling produced (Welch, 1993).

Graham and Welch (1996) argue that selecting for seed and grain crops with substantially lower phytin content could have an unacceptable effect on production, especially in regions of the world where there are soils of low phosphorus status and/or poor micronutrient fertility. Such attempts to lower the antinutrient content of seeds and grains significantly require a major shift in seed or grain composition. Because most of the antinutrients known to occur in seeds and grains are major organic constituents of these organs, they may play beneficial roles in plant growth and may have other unidentified roles in human health. Therefore, Graham and Welch argue against a breeding strategy that attempts to increase iron bioavailability by reducing antinutrient content. Antinutrients may also reduce absorption of potentially toxic trace minerals such as cadmium and other heavy metals (McLaughlin et al., 1999).

However, Raboy (2002) has developed low phytic acid (or *lpa*) mutant varieties of maize, rice, and barley. The phytic acid content of *lpa* seeds is reduced by 50 per cent to 80 per cent compared with non-mutant seeds. The total amount of phosphorus remains the same – phytic acid is replaced by inorganic phosphorus, which does not bind a range of trace minerals.[2] These mutations typically have little observable effect on other seed or plant characteristics. These mutants are presently being tested for agronomic performance and effects on micronutrient status in humans.

A related strategy is to increase the level of phytase. Phytase can break down phytates, thus increasing bioavailability of minerals. Phytases have been used successfully in animal feeds (Nelson et al., 1971; Rimbach and Pallauf, 1993; Yi et al., 1996). Phytase is present in the endosperm of rice, for example, but is destroyed when rice is boiled for eating. Heat-stable phytase may be introduced in food staples through genetic transformation (Holm et al., 2002).

Increasing levels of compounds that improve absorption and utilisation of nutrients. Certain amino acids, such as cysteine, enhance iron and/or zinc bioavailability (Hallberg, 1981). These amino acids occur in many staple foods but their concentrations are lower than those found in meat products. A modest increase in the concentrations of these amino acids in plant foods may have positive effects on iron and/or zinc bioavailability in humans. Iron and zinc occur only in micromolar amounts in plant foods, so only micromolar increases in the amounts of these amino acids may be required to compensate for the negative effects of antinutrients on iron and zinc bioavailability. These amino acids are normal constituents of plants, so relatively small increases in their concentrations in plant tissues should not have adverse consequences on plant growth (Graham and Welch, 1996).

An additional possibility is that as intakes of beta-carotene (converted to vitamin A after ingestion) are increased, the utilisation of absorbed iron may be promoted and vice-versa. That is, there are possible synergies between increasing intakes of these two nutrients (Garcia-Casal et al., 1998). There is already considerable evidence for synergism between vitamin A and zinc intakes (Smith, 1996).

Feasibility. Will famers adopt nutrient-rich varieties?

Impact on plant nutrition is a particular concern of agriculturalists, whose first priority may be to raise the yield potential of plants rather than to address human nutrition problems. If yield and profitability are compromised by breeding for nutrient content, farmers may not adopt nutritionally enhanced cultivars. However, relatively recent research has shown that mineral-dense seeds assist plants in resisting disease and other environmental stresses (Welch and Graham, 1999; Graham et al., 2001). More seedlings survive and initial growth is more rapid. Ultimately, yields are higher, particularly in trace mineral 'deficient' soils in arid regions.[3]

Will breeding for micronutrient-dense seeds change the processing or consumer characteristics of staple foods? Mineral micronutrients comprise a tiny fraction of the physical mass of a seed; for example, 5–10 mg/kg (ppm) in milled rice, and as much as 100 mg/kg in micronutrient-dense bean seeds. It is not known whether such small amounts will alter the appearance, taste, texture, or cooking quality of foods. The dissemination strategy for trace minerals, then, would be to include the mineral-density trait in as many varietal releases as possible, in order to benefit a high proportion of the population, without having to rely on behavioural change as a condition for success. This would be much the same approach as adding fluoride to drinking water in developed countries. As with the example of fluoride, information on increased nutrient content would be made publicly available.

However, increasing levels of beta-carotene will turn varieties from white or light colours of yellow to dark yellow and orange. Often white varieties are much preferred by consumers (for example, milled rice, wheat flour, maize, cassava). Major nutrition education programs will probably be necessary to encourage consumers to switch to more nutritious varieties. However, if these nutrition education programs are successful, the yellow–orange colour will mark the more nutritious varieties from the less nutritious, and a disadvantage will have been turned into an advantage (Hagenimana and Low, 2000).

Increasing Micronutrient Content in Rice

Conventional breeding. In a survey of traditional and modern varieties of rice, Graham et al. (1999) reported that iron density in unmilled rice varied from 7–24 mg/kg (ppm) and zinc density from 16–58 mg/kg. Nearly all the widely grown 'green revolution' varieties were similar, at about 12 and 22 mg/kg for iron and zinc, respectively. The best lines discovered in the survey of the existing germplasm collection were, therefore, twice as high in iron and 1.5 times as high in zinc as the most widely grown varieties. These ratios, therefore, are benchmarks for iron and zinc density, which seem to be reasonably attainable using conventional breeding.

Expression of micronutrient-density traits has been tested over a wide range of environments, and although the environmental effect itself is strong, the genotype effect is consistent across environments (implying that the genotype–environment interaction is not serious), a finding that is sufficient to encourage a breeding effort. Environmental factors considered by one or more crop programs include acid soils, alkaline soils, saline soils, acid-sulphate soils, iron-deficient soils, time of planting, field site, season, nitrogen fertilisation, phosphorus fertilisation, potassium status, elevation and drought stress (Graham et al., 2001).

Transgenic methods.

Increasing iron. Ferritin is an iron-storage protein, and forms of it are found in animals, plants, and bacteria. Ferritin is used by both plants and animals as the storage form of iron (Theil et al., 1997). The ferritin gene has been isolated and sequenced from a number of plants, including soybean, French bean, pea, and maize. Orally administered ferritin, in non-transgenic foods (Beard et al., 1996) or from transgenically-derived ferritin-rich rice (Murray-Kolb et al., 2002) has been demonstrated to provide a source of iron for treatment of iron deficiency in rats.

Human studies with extrinsically radiolabelled animal ferritin have indicated that, when added to a meal, iron from ferritin is only about half as well absorbed as iron from vegetables (Martinez-Torres et al., 1976; Taylor et al., 1986) and iron as ferrous sulphate (Skikne et al., 1997). However, a recent study has shown that intrinsically radiolabelled iron in soybeans is highly bioavailable (27 per cent) to iron-deficient women (Murray-Kolb et al., 2003). About 50 per cent of the iron in the soybeans was ferritin; the remaining iron, ferric phytate, is not thought to be highly bioavailable.

Goto et al. (1999) report improving the iron content of rice by transferring the entire coding sequence of the soybean ferritin gene into Japonica rice. The introduced ferritin gene was expressed under the control of a rice seed-storage protein glutelin promoter, to mediate the accumulation of iron specifically in the grain. The transgenic seeds stored up to three times more iron than normal seeds. Iron levels in the whole (unmilled) seeds of the transformants varied from 13–38 mg/kg (ppm); pooled mean values were 23 mg/kg for transformants and 11 mg/kg for non-transformants. The average iron content in the endosperm was 3.4 mg/kg in the transformant and 1.6 mg/kg in the non-transformant.

This underscores the importance of determining where in endosperm the iron (and other trace minerals) are deposited and how mineral levels are affected by milling. Trace minerals cannot be synthesised by plants and must be transported from the soil, through the plant and to the seed. Several processes and genes may be involved. The authors speculate that the amount of iron accumulation is restricted by transport of iron to the ferritin molecule, rather than simply by levels of ferritin protein. Thus, it may be possible to store larger amounts of iron in the ferritin molecule by co-integrating the ferritin gene and the iron reductase-like transporter gene. Robinson et al. (1999) report a newly discovered ferric-chelate reductase gene, which allows plants to absorb more iron from the soil, thus introducing the possibility of widening the scope of rice varieties with high iron uptake.

A doubling of the iron content in rice using a ferritin gene derived from *Phaseolus vulgare* has been reported (Gura, 1999). Metallothionine was also expressed in the rice grain, increasing the cysteine content seven-fold.

The ongoing work of exploiting the sequence of the rice genome is also proving invaluable to the acceleration of our understanding of the internal iron transport processes and should open up new options for iron accumulation in the rice endosperm (Gross et al., 2003; Koike et al., 2004).

Introducing a gene which codes for heat-stable phytase. Phytic acid or phytate is present predominantly in the aleurone layers of the rice grain, and while most of these tissues are removed during polishing of the grain, remaining amounts of phytate may have some effects on the bioavailability of iron (and other micronutrients). The level of phytase, which breaks down phytic acid, is normally low in milled rice. Several studies have already demonstrated the usefulness of adding phytase to the rice diets of poultry (for example, Adrizal et al., 1996; Munaro et al., 1996; Farrell and Martin, 1998; Martin et al., 1998; Teichman et al., 1998). The phytase that does exist in rice seeds will hydrolyse the phytic acid present at lower pH levels, if seeds are left to soak until germination in water. However, boiling destroys the phytases that occur naturally in rice.

In one study, an amino acid was changed in the protein sequence to make the phytase heat stable (Pasamontes et al., 1997). The phytase was active under the pH conditions of digestion and degraded phytic acid in a very short time, during model *in vitro* digestion.

Lucca et al. (2001) reported the introduction into rice of a transgene for this particular heat-stable phytase from *Aspergillus fumigatus*; the transgene increased the level of phytase activity 130-fold. Unfortunately, after expressing the gene in the grain, the phytase was no longer stable to heat and lost its activity on boiling.

A new approach is now under investigation in which the *A. fumigatus* phytase gene would break down phytate during seed maturation (when pH levels [6–7] would be conducive to phytase activity). Fusing the globulin promoter to the phytase gene, which tests have shown is expressed only in the endosperm of rice, which is consumed after milling, should prevent seed germination being compromised (after harvesting and replanting) owing to low phosphorus content. The hope is that the phosphorus in the outer layers of the seed (which is milled away for consumption but not for planting) would remain unaffected (Lucca et al., 2002).

Increasing levels of compounds that improve absorption and utilisation of nutrients. Levels of lysine, which is an important, essential and limiting amino acid in rice and which might promote the uptake of trace minerals, can be improved by transgenic methods (Datta and Bouis, 2000). Introduction of two bacterial genes, expressing DHSPS (dihydrodipicolinic acid synthase) and AK (aspartokinase) enzymes, encoded by the *Corynebacterium dapA* gene and a mutant *Escherichia coli lysC* gene, have enhanced lysine levels about five-fold in canola and soybean seeds (Falco et al., 1995). More recent work shows that a single novel storage protein can lead to significant increases in the rice endosperm (Sun and Liu, 2004). The private sector has been successful in increasing the lysine content of maize and this product has entered regulatory review in a number of countries.

Adding beta-carotene. Beta-carotene, a precursor of vitamin A, does not occur naturally in the endosperm of rice. Ye et al. (2000) originally reported generating transgenic plants that produce grain with yellow-coloured endosperm by inserting three genes to complete the biosynthetic pathway for carotenoids, including one gene taken from daffodil. Biochemical analysis confirmed that the colour represented beta-carotene. The reported level of total carotenoids in one gram of the dry, uncooked transformed rice (Golden Rice) was 1.6 μg/g. More recently, Paine et al. (2005) have reported substantial increases in these levels (37 μg/g total carotenoids and 31 μg/g provitamin A carotenoids, 84 per cent of which are beta-carotene), by replacing the gene from daffodil with a gene from maize. In addition, only one other gene, the phytoene reductase, had been shown to be also necessary – the level of endogenous lycopene cyclase (added originally as the 'third gene' in the Ye et al. work) has been shown to be sufficient in the rice endosperm to drive the pathway to the provitamin A carotenoids (Schaub et al., 2005). A discussion on the degree to which vitamin A intakes may be improved by the use of Golden Rice in the rural Bangladesh context is provided below (Section III(i)).

III. Outcomes and Impacts of Breeding for Iron- and Beta-carotene Dense Rice

Impact on Iron and Beta-carotene Intakes

For poor populations, food staple consumption so dominates diets that primary food staples can provide in the range of 40–55 per cent of total iron intakes (Bouis et al., 2000). If a single food staple provides 50 per cent of total iron intakes for a poor population, then a doubling of the iron density in that food staple will result in a 50 per cent increase in total iron intakes. It would seem evident that a 50 per cent increase in intakes of iron, if it were to be as bioavailable as iron already in the diet, would be of considerable benefit to anaemic women with such low iron intakes. With respect to rice, rural Bangladeshi women consume an estimated 400 grams of milled rice per day. If the level of iron in rice can be increased by (say) +10 mg/kg, which seems feasible (Gregorio et al., 2000; Graham et al., 2001), this represents an increased intake of +4 mg of iron a day on a base intake of 6–8 mg iron per day (Bhargava et al., 2001).

With respect to vitamin A intake from Golden Rice for these same Bangladeshi rural women, 400 grams of rice times a density of 31 μg/g beta-carotene gives a daily intake of 12,400 μg/day. Reducing this figure by 25 per cent to take account of loss of beta-carotene due to boiling (presumed to be an overestimate of losses), and converting Retinol Equivalents (RE) using a ratio of 12 μg beta-carotene per RE, gives an intake of 775 RE. The WHO/FAO recommended daily intake (RDI) is only 500 RE. This is over and above present sources of vitamin A in the diet, which already provide an estimated 50 per cent of the RDI for adult women. Thus, adult women would consume twice the RDI each day if they switched exclusively to Golden Rice.

Rice intakes for non-breastfed pre-school children are about half those for adult women, so that vitamin A intakes from Golden Rice may be estimated at 387 RE, as compared with a WHO/FAO recommended daily intake of 450 RE for four to seven year-olds. This is over and above present sources of vitamin A in the diet, which already provide an estimated 25 per cent of the RDI for preschool children. Thus, four to seven year-olds would consume approximately 10 per cent more than their RDI each day if they switched exclusively to Golden Rice.

Costs and Benefits of a Biofortification Strategy

Development of iron- and zinc-dense varieties of rice or wheat using conventional breeding might cost as much as US$10 million each over ten years, including the costs of nutrition safety and efficacy tests, the costs of dissemination in selected regions, and the costs of an evaluation of nutritional and economic impact (Hunt, 2002). As indicated above (Section II(i)), such an investment (of US$20 million in two crops, rice and wheat) is projected to have far-reaching impacts if efficacy and effectiveness are achieved. Under such conditions, it would be reasonable to extrapolate benefits to countries around the world, as countries would need only to invest in adaptive breeding and dissemination costs. Moreover, benefits should be sustainable at low maintenance costs. Benefits from breeding advances typically do not disappear after initial investments and research have been successful, as long as an effective domestic agricultural research infrastructure is maintained.

A simulation model developed for Bangladesh and India based on development of iron- and zinc-dense varieties of rice and wheat, which were assumed to be adopted on only 10 per cent of approximately 83 million hectares planted to rice and wheat, supported the expectation of enormous economic benefits from a biofortification strategy (Hunt, 2002). Total costs were estimated to be US$42 million which, in addition to the US$20 million in central development costs for rice and wheat, included costs of adaptive and maintenance breeding and extension costs. These conservative assumptions suggest that the returns that come on-stream during the second decade of research and development would be about $ 1.2 billion in benefits from better nutrition.

A more formal economic evaluation of biofortification in Bangladesh and India, in which the ratio of the present value of benefits to the present value of costs was discounted at a 3 per cent rate (commonly used for evaluation of social benefits), gave a benefit-cost ratio of 19, for returns, simply for better iron nutrition in humans; a similar ratio was found by Horton and Ross (1998) for fortification in South Asia. A different way of expressing the concept of discounting over time is the internal rate of return, in which the interest rate is calculated at which benefits would equal costs plus interest, if the funds were to be borrowed to make the investment. In this case the internal rate of return would be 29 per cent, measured as benefits to human nutrition (Hunt, 2002).

In the longer term (years 11–25 of the simulation model) it was estimated that a total of 44 million person-years of iron deficiency would be prevented if nutritionally improved varieties were to be adopted on 10 per cent of rice and wheat areas in Bangladesh and India. This was based on a conservative assumption of only a 3 per cent reduction in the prevalence of iron deficiency among those consuming the high-iron rice. Thus, costs would be about US$1 per case of anaemia per year prevented and US$0.03 per person per year for those whose iron intakes would be increased by 50 per cent through consumption of iron-dense rice and wheat.

These estimates have been made for conventional breeding. It is not known whether transgenic methods would be more or less costly than conventional breeding methods, in part because the higher costs of meeting stricter regulatory procedures for transgenic crops are not yet known. These higher regulatory costs for transgenic methods may be offset by the greater costs and time of backcrossing involved with conventional techniques. However, the relative costs of transgenic and conventional methods are of secondary importance, in the sense that, even if total costs for either method doubled the assumed costs for conventional breeding of $10 million over ten years, biofortification would still be a very cost-effective proposition. For example, the very conservative benefit calculations above included only benefits for India and Bangladesh, while the varieties developed under full development cost assumptions could be used in other countries around the world as well. It is very difficult to predict benefits and breeding costs ex ante, so that it is important to make conservative estimates of benefits.[4] Extrapolations are made on the assumption of specified levels of efficacy and effectiveness for interventions characterised by uncertainties that result from incomplete knowledge.

There is no a priori reason to believe that research costs will be higher or lower for a widely grown crop (for example, for a cereal such as rice) than for a 'minor' crop (for example, pulses). Given constant costs, benefits in agricultural research will always be higher for the more widely grown crop. Thus, there is a bias in research investments toward those crops and, consequently 'minor' crops lag behind in terms of productivity increases. If the costs of developing transgenic crops are high owing to the need to meet regulatory procedures, the tendency to undertake research only on the more widely grown crops will increase. On the other hand, if biotechnology can lower the costs of undertaking agricultural research by making the process more efficient, this tendency will be mitigated.

IV. Knowledge Gaps in Breeding for Iron- and Beta-carotene Dense Rice

The Genotype Factors that Optimally Determine Bioavailability

Bioavailability to humans of micronutrients in plant foods depends on a myriad of factors, which interact to determine the ultimate bioavailability of a particular micronutrient to an individual, eating a mixed diet, within a given environment (Fairweather-Tait and Hurrell, 1996; Welch 2002). These factors include the combination of nutrients, antinutrients, and absorption enhancers present in the nutrient enhanced rice and the meal in which it is eaten. Because of this complexity, the data obtained using *in vitro* systems, animal models and test meals are always ambiguous (House, 1999; Van Campen and Glahn, 1999). Only data obtained on reducing the prevalence of micronutrient deficiencies using feeding trials in test populations under free living conditions can delineate the actual efficacy of using micronutrient-enriched varieties of plant foods as an intervention tool. However, it is impractical to test, in this way, the bioavailability of selected micronutrients, in the numerous genotypes of staple plant foods that can be generated in plant breeding programs (Graham and Welch, 1996; Graham et al., 2001). The ultimate proof of the biofortification strategy will lie in effectiveness studies undertaken in areas where nutritionally improved varieties are introduced.

Consumer Acceptance

There are various potential barriers to the adoption, for example, of Golden Rice.

- It is yellow in colour, whereas Asian consumers who stand to benefit the most from its introduction are used to eating white rice.
- There may be consumer acceptance issues, potentially related to safety, ethics and localisation of the decision-making process in agricultural development and associated processes.
- There may be concerns that technological solutions used in isolation will exclude alternative approaches, including for example, failure to harness technology to social policy.

Whether public health education initiatives can address all these concerns is open to question. Examples of successes in programs encouraging consumers to switch from white to yellow or orange varieties of staple food crops provide some encouragement (Hagenimana and Low, 2000).

Relative Efficiency of Transgenic and Conventional Breeding Methodologies

Much experience has been accumulated with conventional breeding and it continues to become more efficient, in part owing to new techniques developed through the use of biotechnology. Yet, clearly the capability to move genes across species holds great potential, as demonstrated by the examples presented above for rice, to accomplish goals that cannot be attained with conventional breeding. As experience is gained with biotechnology, development of transgenic varieties will become more efficient and cost-effective. It is difficult if not impossible to determine how investments should best be allocated between conventional breeding, which will give benefits now, and transgenic methods, which will generate future benefits.

V. Conclusions

The following lessons may be drawn concerning the potential usefulness of biotechnology in helping to provide more nutritious foods in developing countries.

- To be of value, plant breeding has to be demonstrated to be safe (see below) and more cost-effective than alternative interventions already in place to reduce micronutrient undernutrition. Such cost-effectiveness is anticipated, in large measure because of the far-reaching impacts of plant breeding. A relatively small, fixed, initial investment in research could benefit the health of millions of poor people all over the world, and at the same time improve agricultural productivity on lands that are presently among the least productive. Nutrient dense varieties of crops can be developed for a fraction of the recurrent estimated annual costs of supplementation programmes in developing countries (for example US$1.65 billion for vitamin A supplementation programmes) and can reach far more people, as the benefits can be extended across countries and time.
- There should be aspects of the breeding strategy for which biotechnology is superior to conventional breeding techniques. For rice, this is the case for adding beta-carotene-related and heat-stable phytase genes. In the long run, as more is understood about the factors driving translocation of minerals in plants, biotechnology may also be a more efficient way to increase trace mineral density.
- Where biotechnology is superior to conventional plant breeding, it must be recognised that the same conditions that apply to products developed through conventional plant breeding, must also be met, as follows:
  - there should be no serious, negative agronomic consequences of the characteristic being added;
  - any noticeable changes in the colour, taste, texture, cooking qualities, and other features associated with the characteristic being added should be acceptable to the consumer;
  - the characteristic being added should result in a measurable improvement in the nutritional status of the malnourished population; and
  - there should be no unacceptable nutritional effects.

The conditions, relating to safety, acceptability and efficacy, in particular, have yet to be firmly established. Nutritionally improved transgenic foods have not been fully developed and tested for their potential to improve micronutrient status, even using *in vitro* techniques and animal models, let alone in human subjects with micronutrient deficiencies.

Definitive proof of the benefit of nutritionally improved, transgenic foods obviously depends on further development, release and eventual impact evaluation, all of which are many years into the future. Nevertheless, in the shorter term, lessons learned through initial experiences with conventional breeding efforts show just how cost-effective a biofortification strategy can be. Clearly, the nutritional quality of staple foods can be substantially improved using transgenic methods compared with what can be accomplished using conventional breeding. Knowledge in the area of nutritional genomics should expand rapidly in the future, and more efficient techniques for producing transgenic plants will be developed. There are potentially substantial benefits to the poor in relation to costs.

Characteristics of food, such as appearance, taste, and colour, which are important to consumers in both developed and developing countries, can also be improved through biotechnology; however, such research currently probably benefits wealthy consumers in developed countries more than poor consumers in developing countries.

Ultimately, good nutrition depends on adequate intakes of a range of nutrients and other compounds, in combinations and levels that are not yet completely understood. Thus, the best and final solution to eliminating undernutrition as a public health problem is to provide increased consumption of a range of non-staple foods. By reducing the cost of producing food, biotechnology will, perhaps, make its most important contribution to reducing malnutrition. However, this will require several decades to be realised, informed government policies, and a relatively large investment in agricultural research and other public and on-farm infrastructure.

In conceptualising solutions for a range of nutritional deficiencies, interdisciplinary communication between plant scientists and human nutrition scientists holds great potential. Human nutritionists need to be informed, for example, about the extent to which the vitamin and mineral density of specific foods, as well as compounds that promote and inhibit their bioavailability, can be modified through plant breeding. Plant breeders need to be aware of both the major influence that agricultural research may have had on nutrient utilisation in the past (for example, the bioavailability of trace minerals in modern varieties versus bioavailability in traditional varieties), and the potential of plant breeding for future improvements in nutrition and health.

Notes

1. This essay is a modified version of Bouis et al. (2003). Permission granted by Elsevier Press is gratefully acknowledged.
2. When such mutants are used as animal feeds, this also avoids what has become a serious pollution problem – excretion of unutilised phytic acid.
3. Biofortification will not deplete soils of trace minerals. Sufficient trace minerals are present in most soils for thousands of crops. Some soils are labelled 'trace mineral-deficient' because the properties of these soils are such that the trace minerals are chemically bound and so unavailable to the plant. Roots of trace mineral-efficient lines release compounds into the soil, which chemically unbind the trace minerals in the soils and make them available to the plant. Thus, a strategy of breeding for trace mineral efficient lines makes use of an untapped and abundant resource. This is a very different situation from depletion of macronutrients in the soil, such as nitrogen and phosphorus (Graham and Welch, 1996).
4. There are agronomic benefits associated with high zinc seed density which are not discussed here (Graham et al., 2001). Possible higher crop yields are not counted as possible benefits nor benefits to better zinc nutrition. During germplasm screening, a high correlation between iron and zinc density in rice and wheat seeds has been found; therefore breeding for both characteristics simultaneously is feasible.

References

Administrative Committee on Coordination, Subcommittee on Nutrition, and International Food Policy Research Institute (2000) *Fourth Report on the World Nutrition Situation* (Geneva, United Nations).

Adrizal, P. E. P. and Sell, J. L. (1996) Utilization of defatted rice bran by broiler chickens, *Poultry Science*, 75(8), pp. 1012–7.

Beard, J. L., Burton, J. W. and Theil, E. C. (1996) Purified ferritin and soybean meal can be sources of iron for treating iron deficiency in rats, *Journal of Nutrition*, 126, pp. 154–60.

Beaton, G. H., Martorell, R., Aronson, K. J., Edminston, B., McCabe, G., Ross, A. C. and Harvey, B. (1993) *Effectiveness of Vitamin A Supplementation in the Control of Young Child Morbidity and Mortality in Developing Countries* (ACC/SCN State of the Arts Series, Nutrition and Policy Discussion Paper No. 13) (Geneva: Administrative Committee on Coordination, Subcommittee on Nutrition), accessed at: http://www.unsystem.org/scn/

Bhargava, A., Bouis, H. and Scrimshaw, N. (2001) Dietary intakes and socioeconomic factors are associated with the haemoglobin concentration of Bangladeshi women, *Journal of Nutrition*, 131, pp. 758–64.

Bouis, H., Graham, R. and Welch, R. (2000) The Consultaive Group on International Agricultural Research (CGIAR) Micronutrients Project: justification and objectives, *Food and Nutrition Bulletin*, 21(4), pp. 374–81.

Bouis, H., Chassy, B. M. and Ochanda, J. (2003) Genetically modified crops and their contribution to human nutrition and food quality, *Trends in Food Science and Technology*, 14, pp. 191–209.

Datta, S. and Bouis, H. E. (2000) Application of biotechnology to improving the nutritional quality of rice, *Food and Nutrition Bulletin* (United Nations University Press), 21(4), pp. 451–6.

Fairweather-Tait, S. and Hurrell, R. F. (1996) Bioavailability of minerals and trace elements, *Nutrition Research Reviews*, 9, pp. 295–324.

Falco, S. C., Guida, T., Locke, M., Mauvais, J., Sanders, C., Ward, R. T. and Webber, P. (1995) Transgenic canola and soybeans seeds with increased lysine, *Biotechnology*, 13, pp. 577–82.

Farrell, D. J. and Martin, E. A. (1998) Strategies to improve the nutritive value of rice bran in poultry diets. III. The addition of inorganic phosphorus and a phytase to duck diets, *British Poultry Science*, 39(5), pp. 601–11.

Garcia-Casal, M. N., Layrisse, M., Solano, L., Arguello, F., Llovera, D., Ramirez, J., Leets, I. and Tropper, E. (1998) Vitamin A and β-carotene can improve nonheme iron absorption from rice, wheat and corn by humans, *Journal of Nutrition*, 128, pp. 646–50.

Goto, F. T., Yoshihara, T., Shigemoto, N., Toki, S. and Takaiwa, F. (1999) Iron fortification of rice seed by the soybean Ferritin gene, *Nature Biotechnology*, 17, pp. 282–6.

Graham, R. D. and Welch, R. M. (1996) *Breeding for Staple-food Crops with High Micronutrient Density: Long-term Sustainable Agricultural Solutions to Hidden Hunger in Developing Countries* (Agricultural Strategies for Micronutrients Working Paper 3) (Washington, DC: International Food Policy Research Institute).

Graham, R. D., Senadhira, D., Beebe, S. E., Iglesias, C. and Ortiz-Monasterio, I. (1999) Breeding for micronutrient density in edible portions of staple food crops: Conventional approaches, *Field Crop Research*, 60, pp. 57–80.

Graham, R. D., Welch, R. M. and Bouis, H. E. (2001) Addressing micronutrient malnutrition through enhancing the nutritional quality of staple foods: principles, perspectives and knowledge gaps, *Advances in Agronomy*, 70, pp. 77–142.

Gregorio, G., Senadhira, D., Htut, H. and Graham, R. D. (2000) Breeding for trace mineral density in rice, *Food and Nutrition Bulletin*, 21(4), pp. 382–6.

Gross, J., Stein, R. J., Fett-Neto, A. G. and Fett, J. P. (2003) Iron homeostasis related genes in rice, *Genetics and Molecular Biology*, 26, pp. 477–97.

Gura, T. (1999) New genes boost rice nutrients, *Science*, 285, pp. 994–5.

Hagenimana, V. and Low, J. (2000) Potential of orange-fleshed sweet potatoes for raising vitamin A intake in Africa, *Food and Nutrition Bulletin*, 21(4), pp. 414–8.

Hallberg, L. (1981) Bioavailability of dietary iron in man, *Annual Review of Nutrition*, 1, pp. 123–7.

Holm, P. B., Kristiansen, K. N. and Pedersen, H. B. (2002) Transgenic approaches in commonly consumed cereals to improve iron and zinc content and bioavailability, *Journal of Nutrition*, 132(Supplement), pp. 514–6.

Horton, S. and Ross, J. (1998) *Economic Consequences of Iron Deficiency* (Ottawa, Canada: The Micronutrient Initiative).

Hotz, C. and Brown, K. (eds) (2004) Assessment of the risk of zinc deficiency in populations and options for its control, *Food Nutrition Bulletin*, 25, pp. S189–95, accessed at: http://www.izincg.ucdavis.edu/publications/FNBv25n1supp2zinc.pdf

House, W. A. (1999) Trace element bioavailability as exemplified by iron and zinc, *Field Crop Research*, 60, pp. 115–41.

Hunt, J. M. (2002) Reversing productivity losses from iron deficiency: the economic case, *Journal of Nutrition*, 132(4, Supplement), pp. 794–801.

Koike, S., Inoue, H., Mizuno, D., Takahashi, M., Nakanishi, M., Mori, S. and Nishizawa, N. K. (2004) OsYSL2 is a rice metal-nicotianamine transporter that is regulated by iron and expressed in the phloem, *Plant Journal*, 39(3), pp. 415–24.

Lucca, P., Hurrell, R. and Potrykus, I. (2001) Genetic engineering approaches to improve the bioavailability and the level of iron in rice grains, *Theoretical and Applied Genetics*, 102, pp. 392–7.

Lucca, P., Hurell, R. and Potrykus, I. (2002) Fighting iron deficiency anemia with iron-rich rice, *Journal of the American College of Nutrition*, 21(3), pp. 184S–90S.

Martin, E. A., Nolan, J. C., Nitsan, Z. and Farrell, D. J. (1998) Strategies to improve the nutritive value of rice bran in poultry diets. IV. Effects of addition of fish meal and a microbial phytase to duckling diets on bird performance and amino acid digestibility, *British Poultry Science*, 39, pp. 612–21.

Martinez-Torres, C., Renzi, M. and Layrisse, M. (1976) Iron absorption by humans from hemosiderin and ferritin, further studies, *Journal of Nutrition*, 106(1), pp. 128–35.

McLaughlin, M. J., Parker, D. R. and Clarke, J. M. (1999) Metals and macronutrients – food safety issues. *Field Crop Research*, 60, pp. 143–63.

Munaro, F. A., Lopez, J., Lopez, S. E. and Rutz, F. (1996) Effect of phytase on boiavailability of phosphorus in diets with defatted rice bran for broiler chickens, *Revista da Sociedade Brasileira de Zootecnia*, 25(5), pp. 932–43.

Murray-Kolb, L. E., Welch, R. M., Theil, E. C. and Beard, J. L. (2003) Women with low iron stores absorb iron from soybeans, *American Journal of Clinical Nutrition*, 77, pp. 180–4.

Murray-Kolb, L. E., Takaiwa, F., Goto, F., Yoshihara, T., Theil, E. C. and Beard, J. L. (2002) Transgenic rice as a source of iron for iron-depleted rats, *Journal of Nutrition*, 132(5), pp. 957–60.

Nelson, T. S., Shieh, T. R., Wodzinski, R. J. and Ware, J. H. (1971) Effects of supplemental phtase on the utilization of phytate phosphorus by chicks, *Journal of Nutrition*, 101, pp. 1289–94.

Paine, J. A., Shipton, C. A., Chaggar, S., Howells, R. M., Kennedy, M. J., Vernon, G., Wright, S. Y., Hinchliffe, E., Adams, J. L., Silverstone, A. L. and Drake, R. (2005) A new version of Golden Rice with increased pro-vitamin A content, *Nature Biotechnology*, 23, pp. 482–7.

Raboy, V. (2002) Progress in breeding low-phytate crops. *UNKNOWN*.

Rimbach, G. and Pallauf, J. (1993) Enhancment of zinc utilization from phytate-rich soy protein isolate by microbial phytase, *Zeitschrift für Ernahrungswissenschaft*, 32, pp. 308–15.

Robinson, N. J., Procter, C. M., Connolly, M.L. and Guerinot, M. L. (1999) A ferric-chelate reductase for iron uptake from soils, *Nature*, 397, pp. 694–7.

Rosegrant, M. W., Paisner, M., Meijer, S. and Witcover, J. (2001) *Global Food Projections to 2020: Emerging Trends and Alternative Futures* (Washington, DC: International Food Policy Research Institute).

Schaub, P., Al-Babili, S., Drake, R. and Beyer, P. (2005) Why is Golden Rice golden (yellow) instead of red?, *Plant Physiology*, 138(May), pp. 441–50.

Sharma, D. C. (2001). UN vitamin A campaign in India under fire, *Lancet*, 358(9295), p. 1791.

Skikne, B., Fonzo, D., Lynch, S. R. and Cook, J. D. (1997) Bovine ferritin iron bioavailability in man, *European Joural of Clinical Investigation*, 27(3), pp. 228–33.

Smith, J. C. (1996) The vitamin A-Zinc connection, in H. E. Bouis (ed) *Micronutrients and Agriculture* (Washington DC: Federation of American Scientists).

Sun, S. M. and Liu, Q. (2004) Transgenic approaches to improve the nutritional quality of plant proteins, *In Vitro Cellular and Development Biology – Plant*, 40(2), pp. 155–62.

Taylor, P. G., Martinez-Torres, C., Romano, E. L. and Layrisse, M. (1986) The effect of cysteine-containing peptides released during meat digestion on iron absorption in humans, *American Journal of Clinical Nutrition*, 43, pp. 68–71.

Teichmann, H. F., Lopez, J. and Lopez, S. E. (1998) Effect of phytase on the bioavailability of phosphorus in diets with rice bran for broilers, *Revista Braasileira de Zootechnia*, 27(2), pp. 338–44.

Theil, E. C., Burton, J. W. and Beard, J. L. (1997) A sustainable solution for dietary iron deficiency through plant biotechnology and breeding to increase seed ferritin control, *European Journal of Clinical Nutrition*, 21(4, Supplement), pp. S28–31.

UNICEF, UNU, WHO, and MI Technical Workshop (1999) *Preventing Iron Deficiency in Women and Children: Background and Consensus on Key Technical Issues and Resources for Advocacy, Planning and Implementing National Programmes* (Boston, MA: International Nutrition Foundation).

van Campen, D. R. and Glahn, R. P. (1999) Micronutrient bioavailability techniques: accuracy, problems and limitations, *Field Crop Research*, 60, pp. 93–113.

Welch, R. and Graham, R. (1999) A new paradigm for world agriculture: meeting human needs, productive, sustainable nutritious, *Field Crop Research*, 60, pp. 1–10.

Welch, R. M. (2002) Breeding strategies for biofortified staple plant foods to reduce micronutrient malnutrition globally, *Journal of Nutrition*, 132(Supplement), pp. 495–9.

Wise, A. (1995) Phytate and zinc bioavailability, *International Journal of Food Science and Nutrition*, 46, pp. 53–63.

World Bank (1990–99) *Commodity Markets and the Developing Countries. A World Bank Quarterly* (Washington, DC: World Bank), accessed at: www.worldbank.org/prospects/gcmonline/index.htm

World Bank (1990–99) *Global Commodity Markets* (Washington, DC: World Bank), accessed at: www.worldbank.org/prospects/gcmonline/index.htm

World Bank (1994) *Enriching Lives: Overcoming Vitamin and Mineral Malnutrition in Developing Countries* (Development in Practice Series) (Washington, DC: The World Bank).

Ye, X., Al Babili, S., Kloeti, A., Zhang, J., Lucca, P., Beyer, P. and Potrykus, I. (2000) Engineering the provitamin A (β-carotene) biosynthetic pathway into (carotenoid-free) rice endosperm, *Science*, 287, pp. 303–5.

Yi, Z., Kornegay, E. T. and Veit, H. P. (1996) Supplemental microbial phytase improves zinc utilization in broilers, *Poultry Science*, 75, pp. 540–6.

An Ecological Assessment of Transgenic Crops

JANICE E. THIES & MEDHA H. DEVARE

I. Introduction

Food production will have to triple to feed the anticipated world population of approximately 12 billion by 2050 (Population Reference Bureau, 2002). Much of the worldwide arable land and freshwater resources are already strained, leaving only one viable option for increasing food production: to increase yields on available land, despite diminishing water resources and increasing environmental degradation (Serageldin, 1999; Chrispeels, 2000; Vasil, 2003). Transgenic crop technology is among the more promising means of meeting world food needs using stagnant or declining land and water resources, while preventing soil and water degradation.

Yet, potential environmental impacts need to be rigorously tested under a range of environmental conditions to avoid environmental damage, like that caused by inadequate understanding and monitoring during the Green Revolution, such as soil erosion and water contamination. The potential benefits of these crops include: (a) improving crop productivity on existing land and water resources; (b) increasing production stability, thereby reducing the impact on farmers of biotic and abiotic stresses; (c) preventing or even reversing soil and water degradation through reduced tillage and pesticide applications by use of herbicide and pest-resistant varieties; and (d) improving human health through reduced pesticide use and crops engineered for improved nutritional content or vaccine delivery (Ervin et al., 2001; James, 2003). Anticipating these possible benefits, the Nuffield Council on Bioethics (1999) in the UK concluded that it was imperative to enable emerging economies to evaluate their own use of these new biotechnologies to combat hunger and poverty.

The first transgenic crop approved for human consumption in the US was the Flavrsavr® tomato developed by Calgene and released in 1994. This was followed rapidly by commercial release of herbicide-tolerant cotton, soybean and maize in 1995 and crops expressing insecticidal proteins derived from the soil bacterium *Bacillus thuringiensis* (*Bt*) in 1996. In 2005, approximately 8.5 million farmers in 21 countries grew transgenic crops on over 90 million hectares, with 38 per cent of this area located in developing countries. The US, Argentina, Brazil, Canada, and China account for 95 per cent of all transgenic crops grown. Transgenic crops have been most rapidly deployed in industrialised countries, but the growth in transgenic crop area between 2004 and 2005 was 23 per cent in developing countries, about four times higher than in the industrialised countries (James, 2005). The most common traits introduced into transgenic crops are herbicide tolerance, pest resistance, and physiological enhancements such as delayed ripening, improvements in nutrient and/or vitamin content, and drought resistance, in decreasing order of importance between 1996 and 2003 (Hunter, 2000; James, 2003). The global area planted with these crops increased more than 50-fold between 1996 and 2005, to about 90 million ha, with plants engineered for herbicide tolerance constituting about 71 per cent of this acreage, while those expressing genes coding for the Cry insecticidal proteins obtained from the soil bacterium *Bacillus thuringiensis* (*Bt*) represented about 18 per cent and 11 per cent of biotech crops planted containing stacked genes (James, 2005). Krattiger (1997) speculates that *Bt* technology products could save farmers about 2.7 billion of the 8 billion or so USD spent annually on insecticides worldwide. Provitamin A, engineered into rice seed, has the potential to reduce vitamin A deficiency which is common in at least 26 countries where the diet is dominated by rice. Vitamin A deficiency is responsible for over 1 million deaths and about 0.25 million cases of blindness per annum (Hunter, 2000). Papaya is a key natural source of vitamin A. The Hawaiian papaya industry, almost destroyed by papaya ring spot virus, was saved by development of transgenic resistant varieties (see Gonsalves et al., this issue).

Thus far, the development and deployment of transgenic plants has occurred, for the most part, outside any consistent environmental risk assessment framework (Ervin et al., 2003; Hancock, 2003). Ervin et al. (2003) state that:

> ...an ideal public research program should capture the interconnectedness of ecological systems, the essential roles of ecosystem services, nonlinear and

threshold responses to accumulating stresses, and global expansion of the technology. Basic elements of such a program include long-term studies of cumulative and synergistic pesticide resistance effects, potential gene flow problems for those transgenic crop trait-weed complexes with high probabilities of out-crossing and ecological disruption, and scientific protocols for assessing deleterious effects on non-target organisms.

Commercialisation of transgenic crops has proceeded rapidly, essentially discounting the precautionary principle – that whenever reasonable suspicion of harm exists, the absence of scientific certainty or consensus should not postpone preventative action (Raffensperger and Tickner, 1999). The rapid commercialisation of transgenic technology has resulted in public suspicion or outright aversion, fuelled by sensational and often unfounded or inaccurate claims of health risks. For example, a prominent farmer leader in Karnataka, India, cautioned women that *Bt* cotton contains an anti-fertility factor, which caused many women to refuse to pick the cotton. While most cotton varieties produce gossypol, a plant alkaloid used in medicine to bring on menstruation, the risk to women in this regard is no different in *Bt* than in non-*Bt*, traditional cotton varieties. Moreover, in the centuries in which cotton has been cultivated, the production of gossypol by cotton has presented no problems for pregnant women or for fertility of women working in the fields (Rao, 2005).

The main scientific concerns regarding deployment of transgenic crops include: (a) potential adverse effects on non-target organisms; (b) gene flow into wild plant communities or soil organisms; (c) persistence of gene products or crop residues in the environment; (d) changing land use as farmers shift to less ecologically stable monocultures and, in the case of crops with plant incorporated protectants (PIPs); (e) development of resistance by target pest populations. These concerns are nicely captured in a recent publication by the National Academy of Sciences (2002):

> There is substantial evidence that ecological effects of farming practices exert simplifying and destabilizing effects on neighboring natural ecosystems. These effects are of concern because they appear to weaken or destroy ecosystems' capacity for resilience – that is, an ecosystem's ability to return to its initial state despite disturbance. Potential ecological effects of transgenic crops, and other crops bearing novel traits, may be heightened in this destabilized ecological milieu. This argues for a cautious approach to the release of any crop that bears a novel trait. *Equally, it is an argument for a cautious approach to any extensive change in agricultural practices.* (NAS, 2002: 4, emphasis added)

Nevertheless, the enthusiasm of farmers in high- as well as low-income countries for transgenic crops is evidently not tempered by concern for potential long-term ecological effects, just as most farmers resort to pesticides and heavy fertiliser inputs, when available, to protect their livelihoods. While we recognise the complexity of this issue, this essay deals solely with the scientific bases for concern regarding environmental effects of transgenic crops, and the risk assessment framework in which they have been developed, particularly in the context of agriculture in poor countries.

II. Potential Benefits of Transgenic Crops

The possibility of using transgenic technology to alleviate pressing agricultural problems in poor countries has only recently begun to be examined; ever newer uses of technology make analysis a moving target. How can genetic engineering be applied to specific challenges to agricultural productivity in poor countries? We begin with Smale and DeGroote's (2003) premise that biotechnology innovations with the greatest potential for the poor of Sub-Saharan Africa are those that: (a) tackle economically important, biotic or abiotic problems that are not easily addressed through conventional plant breeding or pest control methods; (b) pose little risk of endangering trade through exports to countries that do not accept transgenic products; and (c) can make a difference in the welfare of small-holder farmers as sources of both food and cash. These conditions could be extended to include the poor in other countries as well, with the added caveat that the transgenic innovation in question can; (d) benefit the health of the poor, through improved nutritional content, vaccine delivery, or reduced pesticide use.

Transgenic crops have already proven to be important for increasing agricultural productivity, either by improving the yield potential of plants with existing or lower input levels (Peng et al., 1999; Taylor et al., 2001; Regierer et al., 2002) or by their control of biotic and abiotic challenges to agronomic productivity. In 2003, over 90 per cent of the global transgenic crop area planted was to herbicide and insect-resistant crops (James, 2003). Studies by Huang et al. (2002), Hossain et al. (2004), Wu and Guo (2005), Morse et al. (2004), Pray et al. (2002) and Qaim and Zilberman (2003) on small-holder farms in China, South Africa and India have demonstrated that *Bt* cotton yielded significantly higher while reducing pesticide use. From personal observations in India and Nepal, and Krattiger's (1998) discussion, weeding in most poor countries is an arduous and time-consuming endeavour, generally accomplished by women and children. Although herbicide-resistant crops presuppose the availability of affordable herbicide, their potential is to not only improve yield, but to enable women to pursue income-generating activities and children to attend school. Benbrook (2003) suggested that transgenic crops with enhanced herbicide tolerance may encourage increased herbicide use, which can be both costly and environmentally unfriendly. Will transfer of this technology to resource-poor farmers create dependency on purchase of imported herbicides? For those areas where pervasive weeds make farming non-viable, herbicide-resistant crops could play a valuable role. They may also reduce labour costs and allow farmers to retain residues, a resource-conserving practice, without paying a high penalty in increased weediness. However, where workers depend on weeding for income, use of herbicide-resistant crops could have adverse community impacts. Insect-resistant crops can decrease losses due to pests and positively affect farmers' health through reduced exposure to insecticides. Several disease-resistant transgenic crops have been or are being developed to address pest losses, including peppers (Rowell et al., 2001), sorghum (Krishnaveni et al., 2001), potatoes (Murray et al., 2002; Sharp et al., 2002), oilseed crops (Lu, 2003), tomatoes (Lee et al., 2002), and wheat (Gerhardt et al., 2002).

Transgenic technology is also being used to create crops resistant to abiotic stresses such as drought, soil acidity, and salinity. Possibilities for modifications such as improved storage stability, delayed ripening, and other changes to increase flexibility in distribution and/or facilitate juicing or other processing for greater economic benefit are being explored. Advances in resistance to abiotic pressures will allow crops to be grown in marginal, low productivity areas most commonly the domain of the poor, whereas increased storage stability and delayed ripening will benefit those with few resources to invest in refrigeration and other equipment to increase the shelf-life of agricultural produce.

Drought is the factor most likely to limit agronomic potential on a global scale; salinity affects about 33 per cent of irrigated land worldwide (Kelemu et al., 2003). Although improvements in the resistance of crops to abiotic stresses through conventional breeding have been limited, genetic manipulation of this complex trait has proven more successful. For instance, the ability of rice root systems to provide sufficient water to balance evapotranspirational demand and capacity for osmotic adjustment are important drought resistance traits (Nguyen et al., 1997). However, the conventional breeding approach of selecting for productivity and stability over locations and years is expensive and slow. The development of molecular linkage maps of rice and other advances in molecular technology offer new, more efficient opportunities for creation of drought-resistant plants. Using these technologies, drought-resistant soybean varieties (De Ronde et al., 2001) have been created, as has rice that is both drought and salt tolerant (Garg et al., 2002). Salt-tolerant maize, tomatoes, oilseed rape, and tobacco have been engineered by Liu et al. (1999), Zhang and Blumwald (2001), Zhang et al. (2001) and Singla-Pareek et al. (2003), respectively. Plants genetically manipulated to withstand the high levels of aluminium typical of increasing soil acidity due to improper land management have also been created (de la Fuente-Martinez et al., 1997; de la Fuente-Martinez and Herrera-Estrella, 1999). However, the value of these modifications in the field has yet to be established. Much of the work on improving storage stability and manipulating ripening and processing-related factors paved the way for developing herbicide and insect-resistant transgenic crops but has been limited, for the most part, to transgenic tomatoes (Fromm et al., 1993; Grierson and Fray, 1994; Picton et al., 1995; Kalamaki et al., 2003a, b; Powell et al., 2003) and potatoes (Greiner et al., 1999).

Over three billion people, mostly in the world's poorest countries, are currently afflicted with micronutrient malnutrition (Mason and Garcia, 1993; Welch et al., 1997; WHO, 1999), while almost two-thirds of childhood mortality is associated with nutritional deficiencies (Caballero, 2002). Micronutrient-enriched staple crops developed through genetic modification target the most vulnerable – resource-poor women and children (Combs Jr. et al., 1996; Bouis, 2000; Welch and Graham, 2004). For instance, vitamin A deficiency causes partial or complete blindness in half a million of the poorest children each year (Conway and Toennissen, 1999). By engineering rice, a staple food for half the world's population (Lampe, 1995), to overproduce beta-carotene – a precursor to vitamin A (Ye et al., 2000) – biotechnologists expect to reduce these numbers significantly. Rice is also an excellent choice of cereal grain to target iron deficiency, which causes anemia, and has been identified as a contributing factor in over 20 per cent of post-partum maternal deaths in Asia and Africa (Conway, 1999). Transgenic rice has been produced containing two to four times the levels of iron found in non-transgenic rice, but the bioavailability of this iron has yet to be ascertained (Goto et al., 1999; Lucca,

1999). Cassava is an important source of calories for Sub-Saharan subsistence farmers, but must be processed to remove toxic cyanogens. Transgenic cassava root contains about 1 per cent of the toxic cyanogenic glucoside, linamarin, compared to the non-transgenic isoline, obviating the need for processing to eliminate toxicity (Siritunga and Sayre, 2003). The starch content of potatoes and the content of potent antioxidants known as flavonols in tomatoes have been increased through genetic manipulation (Regierer et al., 2002; Bovy et al., 2002).

By replacing chemical sprays that farmers the world over generally rely on to control pests, insect-resistant (IR) crops can reduce or eliminate adverse effects of such insecticides on human and environmental health (Gray et al., 1993; Gray, 2000; Huang et al., 2002). Phipps and Park (2002) estimate that pesticide use in the European Union would decrease by 14.5 million kg of formulated product, resulting in a savings of about 21 million litres of diesel and a reduction of about 73,000 tons of carbon dioxide if just 50 per cent of the corn, cotton, oil seed rape, and sugar beet grown in the EU were genetically modified varieties. Insecticide use in rice costs over one billion USD worldwide, much of it in the poorest countries of Asia (Krattiger, 1997). The target is the stem borer, even though this pest can avoid contact with the insecticide by penetrating plant tissue. A large portion of these insecticides are hazardous Category I and moderately hazardous Category II chemicals, many of which have been banned for agricultural use in most developed countries (Warburton et al., 1995). A survey of rice farmers in the Philippines indicated that health problems, including eye abnormalities, skin problems, and respiratory ailments, seemed to be correlated with insecticide use (IRRI, 1994). The negative impacts of insecticide use on the environment and on the control of pests by beneficial insects have also been documented (Way and Heong, 1994; Pingali and Roger, 1995). Transgenic *Bt* rice resistant to the stem borer is one alternative to insecticides, and is currently undergoing greenhouse and field testing in Asia (IRRI, 1997; Datta, 1999; Skerritt, 2000).

On the horizon is the potential for using transgenic plants to produce pharmaceuticals and vaccines ('pharming'). Vaccines can be used to prevent or combat many of the diseases that cause illness and death in low-income countries, but are often expensive to produce and use, must be refrigerated and administered by trained personnel, and require sterile delivery through needles that are often difficult and expensive to obtain. The possibility of engineering plants to produce pharmaceuticals and vaccines is attractive because it provides the possibility of cheaper production and easier delivery and use for segments of the world's poorest population in greatest need. Although research in this area is preliminary, vaccines against some infectious gastro-intestinal diseases have been produced in potatoes and bananas (Thanavala et al., 1995), and an anti-cancer antibody recognising lung, breast, and colon cancer cells and potentially useful in diagnosis and therapy has been expressed in rice and wheat grain (Stoger et al., 2000). Fischer et al. (2003) note that five antibodies produced in transgenic plants ('plantibodies') are undergoing clinical evaluation, three of which can be used prophylactically.

From an agro-ecosystem perspective, use of herbicide-tolerant crops allows reduced and zero-till practices to work more effectively. The resulting decrease in soil disturbance is beneficial for retaining soil organic matter, improving soil structure, reducing soil compaction and improving soil water relations. Insect-resistant crops reduce insect damage, not only improving yield, but also demonstrably lowering contamination of crops by mycotoxins produced by opportunistic fungi that infect crops in wounds made by grazing insects (Munkvold et al., 1999). Mycotoxin contamination can be a serious issue in poor countries where contaminated food is often consumed, with particularly high levels of the *Fusarium* mycotoxin, moniliformin, measured in maize samples from Gambia and South Africa (Sharman et al., 1991).

Thus, both realised and potential benefits of transgenic crops to subsistence farmers appear to be many, varied and likely to accrue over time. Yet, potential risks also exist, and public suspicion abounds. The scientific data available so far suggest that non-target insects are not likely to be at great risk. Gene flow and resistance development, however, represent real risks, particularly in resource-poor countries without strong regulatory and monitoring capabilities.

What are the consequences of the use of transgenic crops on the broad scale for small-scale subsistence farmers? Do the benefits outweigh the risks; how might risks be identified and managed to address the needs of the poor? While we cannot definitively answer these questions, we hope to provide an analysis of the available scientific data from field studies and commercial releases of transgenic crops to inform this debate. In an ideal world, these questions would have been addressed before commercialisation, in the context of rigorous and lengthy risk assessments conducted in the field. The rapid commercialisation new transgenic technologies has rendered this assessment impossible.

III. Risk Assessment and Regulatory Frameworks

In the US, three separate government agencies are responsible for oversight of transgenic crop technologies:

(a) The United States Department of Agriculture (USDA) – specifically the Biotechnology Regulatory Service (BRS) of the Animal and Plant Health Inspection Service (APHIS), that is responsible for regulating the introduction (importation, interstate movement, and field release) of genetically engineered organisms such as plants, insects, microorganisms and any other organism that is known to be, or could be, a plant pest.
(b) The Environmental Protection Agency (EPA), that is responsible for assessing the environmental impact of and regulating primarily insect- and herbicide-resistant plants through the Federal Insecticide, Fungicide and Rodenticide Act (FIFRA).
(c) The Food and Drug Administration (FDA), that monitors mammalian toxicity and allergenicity of transgenic crops under the Federal Food, Drug and Cosmetic Act (FFDCA).

Each of these agencies reviews the scientific evidence related to the key areas of concern: resistance management, gene flow, unintended effects and mammalian toxicity. The USDA–BRS requirements for obtaining a permit for field testing, notification procedures and eventual deregulation of a given crop are published on the Biotechnology Regulatory Services web site at http://www.aphis.usda.gov/brs/index.html.

For each new technology, public comment is sought through listings in the Federal Register. Public hearings and independent evaluations of risk by advisory panels precede the regulatory framework for controlling use and monitoring impacts of each technology. The framework is reviewed regularly in the light of new evidence and adjusted accordingly. The USDA–BRS is currently updating and strengthening its biotechnology regulations for the importation, interstate movement and environmental release of genetically engineered organisms. Significant stakeholder input to this process is actively sought.

Most low-income nation regulatory oversight does not approach this level of technical evaluation, nor is substantial biocontainment or monitoring for target pest resistance or potential ecological impacts of a given technology likely. Neither the risk assessment-regulatory framework, nor the requisite financial backing exist in most instances (see Pray and Naseem, 2007), placing resource-poor farmers, and the agroecosystems in which they farm, at particular risk. Even in the US, 100 times more money is spent in developing a technology than is invested in its risk assessment and post-release monitoring.

The regulatory framework developed in the US to address ecological issues surrounding the deployment of transgenic crops is instructive for, but not replicable in, low-income country agriculture. For example, insect resistance management (IRM) strategies for crops containing plant-incorporated protectants (PIPs), such as *Bt*, require growers to establish large 'refugia' or areas of land planted to crops unprotected by the technology in an effort to prevent pests from developing resistance by maintaining a nearby susceptible breeding population. Poorer farmers will find compliance especially difficult. A middle-ground approach to environmental risk assessment was put forward in the Nuffield Council on Bioethics report (2003): which suggested that methods and data from environmental risk assessments could be shared between countries that have similar agricultural environments, thus reducing the burden for any one country. This and similar logical extensions of prior findings could substantially reduce the need for costly local impact assessments and improve the benefits of these technologies for resource-poor farmers.

IV. Transgenic Crops Approved for Commercial Use in the US and Deployed Worldwide

The following categories of phenotypic change of transgenic crops are recognised within the APHIS regulatory structure: altered agronomic properties (AP), such as delayed ripening; altered resistance to viruses (VR), insects (IR), bacterial (BR) and fungal pathogens (FR), and nematodes (NR); enhanced herbicide tolerance (HT) or product quality (PQ); genetic containment technologies (GC) and marker genes (MG); plant-made pharmaceuticals (PMP) and plant-made industrial proteins (PMIP). To date, APHIS has ensured the safe field testing of more than 10,000 genetically engineered organisms and overseen the deregulation of over 70 transgenic plant products.

In the US, 34 HT crops have been granted deregulated status, largely for glyphosate or phosphinothricin tolerance in beet, corn, cotton, rapeseed and soybean, with additional approvals of phosphinothricin tolerance in rice and bromoxynil or sulfonylurea tolerance in cotton. Eighteen PQ crops have been deregulated: altered fruit ripening in tomato; increased polygalacturonase content in tomato; altered oil profiles in rapeseed and soybean; male sterility in chicory, corn and rapeseed; and reduced nicotine levels in tobacco. Deregulated IR technologies number 26 and include corn, cotton and tomato resistant to lepidopteron pests; potato resistant to the Colorado potato beetle; and corn resistant to the European corn borer (Lepidoptera) and more recently, the corn root worm (Coleoptera). There are nine deregulated VR transgenics that include papaya resistant to the papaya ring spot virus (see Gonsalves et al., this collection) and squash resistant to mosaic virus. Several combined or 'stacked' gene transgenic crops have been deregulated, including glyphosate and corn borer tolerance in corn; bromoxynil and lepidopteron tolerance in cotton; phosphinothricin and lepidopteron tolerance in corn; and Colorado potato beetle and potato leaf roll virus resistance (or potato virus Y) in potato. As yet, no petitions for deregulation of PMP or PMIP technologies have been made. However, permits for field-testing these crops have been issued by APHIS, as notified at http://www.isb.vt.edu/cfdocs/biocharts1.cfm. Crops that have received regulatory approval and those that have been planted worldwide are listed in Table 1. A new database listing transgenic crop releases, those currently in regulated field trials and those under development in different regions of the world is now available from the FAO at www.fao.org/biotech/inventory_admin/dep/default.asp?lang=en. As the technology rapidly spreads, how can we assess new environmental risks?

Table 1. Herbicide tolerant (HT), insect resistant (IR), and other transgenic crops that have received regulatory approval (○) and have been planted (●) worldwide (compiled from the AGBIOS GM database at http://www.agbios.com)

	Maize			Soya	Cotton			Canola		
Countries	HT	IR	HT+ IR	HT	HT	IR	HT+ IR	HT	HT+ fertility	Other
Argentina	●	●	●	●	●	●				
Australia	○	○	○	○	●	●		●	●	
Brazil				●						
Canada	●	●	●	●	●	●		●	●	Numerous
China	○	○	○	○	○	●		○		
Colombia		●								
European Union	○	●	●	○				○	○	
Honduras		●								
India						●				
Iran										Bt rice
Japan	●	●	●	●	●	●	●	●	●	
Korea	○	○	○	○						
Mexico				●		●				
Paraguay				●						
Philippines	○	●	○	○	○	○		○		
Romania				●						
South Africa	●	●	●	●	●	●				
Spain		●								
Taiwan	○	○	○	○						
Uruguay			●	●						
United States	●	●	●	●	●	●	●	●	●	In text

V. What Constitutes Environmental Harm?

A major limitation in the present regulatory framework is lack of any clear sense of what constitutes environmental harm. Despite considerable debate, there are no guidelines that establish the magnitude of change that should trigger concern over ecosystem impacts. There are also practical questions of how to assess long-term risk(s), and how much risk citizens are willing to accept (Arriola, 1998). This ambiguity makes it even more difficult to develop an appropriate environmental risk assessment framework for subsistence agriculture. The Ecological Risk Assessment Group at Oxford University suggests that instead of asking, 'Will the product have an impact on the environment?' – as any introduced non-native species is likely to have – the more relevant question is, 'Will it result in more or less harm than the current conventional alternatives?' (Wilkinson, 1998). Chemical control of insects and weeds is the most common 'conventional' alternative. The primary ecological concerns with the use of transgenic crops – pest resistance, transmission of resistance genes in a plant population and non-target or unintended toxic effects – are also present with the use of insecticides and herbicides. Insecticides have caused rapid development of insect resistance in many cases and have well-known and widespread non-target effects. Recent reports from South Africa and elsewhere (FAO, 2004; Morse et al., 2004) are confirming reduced environmental harm and increased yields with the use of insect-resistant crops that are targeted against specific insect pests. In many farming systems, an increase in beneficial insects and a return of song birds has been noted. Potential risks compared to observed benefits of the deployment of transgenic crops in subsistence agriculture are considered in more detail in following sections.

The US regulatory framework has taken the stance that transgenic crops do not pose any *novel* risks. The classic risk evaluation equation defines 'risk' as the product of 'exposure' and 'hazard'. For transgenic crops, 'hazard' is represented by the impact of the transgene(s) on the ecology of the system – be that the biology and population dynamics of a wild plant relative or food-web effects above- and below-ground wrought by changes in target and non-target populations. At present, it is impossible to define how much risk is tolerable for any given system, let alone provide a definition that has relevance across systems. More tangible are the ecological benefits that are beginning to accrue with commercial release and widespread adoption of transgenic crops. Benefit accrual has led the USDA to recently recertify several *Bt* crops. In US practice, there is acute awareness that new transgenic crops must be evaluated on a case-by-case basis and that older releases must continue to be monitored for long-term ecological impacts. In a recent USDA Stakeholder Workshop (2003) needs for: (a) more baseline data; (b) improved pre-release assessment methods that are robust and scale-relevant; (c) useful indicator species to monitor key ecological processes; and (e) effective post-release monitoring were explicitly recognised.

Where does this put subsistence farmers who are keen to adopt IR and HT crop technologies? With the lack of specific testing in a given agro-ecozone coupled with the absence of any long-term data in comparable zones, how can risk possibly be determined? Critics of transgenic crops continue to raise this question; there is no satisfactory answer. The following section addresses this question by examining *cry* genes encoding the *Bt* Cry proteins into crop plants as a case.

VI. The Case of *Bt* Transgenic Crops

Bacillus thuringiensis (*Bt*) is a common, spore-forming, soil bacterium. Thousands of strains of *Bt* have been isolated and unique insecticidal toxins, both 'Cry' or parasporal crystal proteins and 'Cyt' or cytotoxic proteins, continue to be described (Kumar et al., 1996). The activity of the Cry proteins differs between different *Bt* strains; some are active against the Coleoptera (beetles), some against the Lepidoptera (moths and butterflies), and others target the Diptera (flies and mosquitoes) (Gould and Keeton, 1996).

The Cry protein (Figure 1), or delta-endotoxin, is encoded by a single gene carried on a plasmid in *B. thuringiensis* (McGaughey and Whalon, 1992). The gene is expressed during the stationary phase of growth and is sporulation-dependent. *Bt* is an opportunistic insect pathogen that survives as spores in soil at average concentrations of 1000 spores g^{-1} soil in the top 12.7 cm of the profile. The naturally occurring concentration of Cry protein has been estimated at 0.10 g $acre^{-1}$ in agricultural soils. Spray formulations comprised of specific *Bt* strains have been used in agriculture and forestry for over 70 years. *Bt* sprays have also enjoyed widespread use in organic farming because there is no known toxicity to most beneficial insects or to wildlife, including mammals, fish, or birds. *Bt* sprays contain viable bacteria and their associated insecticidal proteins; however, the proteins are rapidly deactivated by sunlight and readily washed off plant leaves and into the soil by rainfall. The fate of the bacterial cells themselves is not known. Uneven application to plants results in leaf areas not well protected by the *Bt* spray. These low-dose regions facilitate resistance development; subsequent breeding of surviving insects perpetuates resistance within a population. To be effective, *Bt* sprays need to

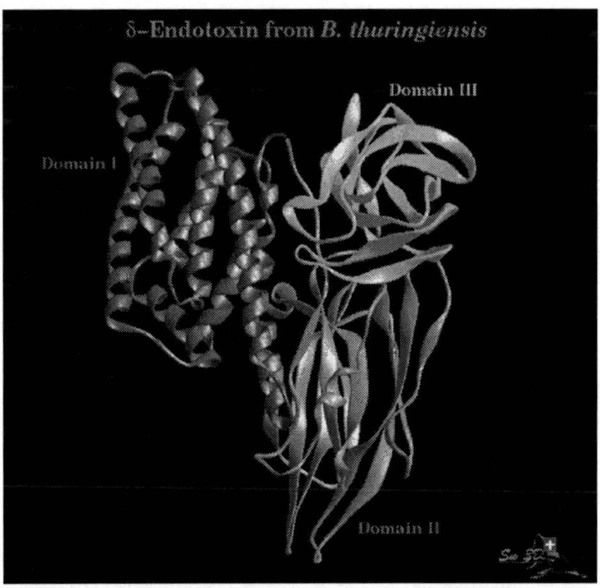

Figure 1. Configuration of the delta-endotoxin from *Bt*

be applied repeatedly, increasing labour costs. By introducing and expressing these genes in plants, farmers obtain the benefit of targeted insect control thus reducing pesticide use in the farming system.

The first report of successful insertion of *cry* genes encoding the *Bt* insecticidal protein into plants was published in 1987 (Vaeck et al., 1987). The first plants to express these traits were tobacco and tomato. It took another eight years to evaluate potential risks and field test crop varieties before the first approval for commercial release was awarded for transgenic *Bt* cotton (Monsanto Corp., St Louis, MO), expressing the Cry1Ac protein active against lepidoptera pests. This was followed closely by *Bt* field corn (Ciba-Geigy and Monsanto) expressing the same protein (USDA, 2005). Introduction of *Bt* genes into corn for the control of European corn borer (*Ostrinia nubilalis*) and more recently corn rootworm (CRW; *Diabrotica* beetles) has been particularly advantageous; chemical control of both types of pests is difficult because they feed either inside the stalk or below ground, locations difficult to reach with insecticides. It is estimated that one in seven insecticide applications in the US is directed toward CRW control (Ely 1993). Different crops, hybrids and genetic events make up the list of currently approved transgenic crops. For most genetic events, the protein is expressed in all tissues of the plant. The protein has been detected in plant root exudates (Saxena and Stotzky, 2000) of most transformed plants and has been shown to persist for varying times in soil, from a few weeks up to 240 days (Stotzky, 2000, 2002). One aim of continuing research is to restrict the expression of the *cry* gene to only green tissue by transforming the plastid DNA, thereby helping to restrict gene flow (see below).

The mode of action of the various Cry proteins is quite similar. The insect ingests either the parasporal crystalline protein (protoxin) or the transgenic *Bt* plant tissue expressing the protein. The protoxin is dissolved in the insect midgut and proteolytically cleaved to form the active protein that binds to receptors on cells of the midgut epithelium (Domains II and III, Figure 1). The activated protein forms pores in the membrane of cells lining the gut (Domain I, Figure 1), which disrupts the osmotic balance of these cells causing them to lyse (Van Rie et al., 1992). The gut becomes paralysed and the insect stops feeding; most insects die within a few hours of ingestion (Marrone and Macintosh, 1993; Neppl, 2000) due to septicemia. Thus, in order for the protein to be toxic, receptors to bind the protein to the midgut epithelial cells must be present. Binding site modification, such that the protein no longer binds, is the most frequent mechanism of resistance to *Bt* proteins (Ferré and Van Rie, 2002).

VII. Non-Target Effects

Much of the scientific literature in this area has focused on the effects of *Bt* crops on non-target arthropods beneficial in controlling natural crop pests. Non-target arthropods may be exposed to *Bt* proteins in several ways: by feeding on plants, by preying on herbivorous arthropods, or by contact with soil receiving *Bt* proteins through root exudation or decomposition of *Bt* plants. Until recently there was no framework in place for assessing specifically the risk of transgenic insect-resistant plants for beneficial arthropods, though protocols have been proposed (Dutton et al., 2003). An optimal protocol would include an evaluation on the effects of the transgenic crop in comparison with the pesticide used if the non-transgenic isoline were grown instead. In the absence of *Bt* crops, farmers typically rely on insecticides to control the target pest, be that *Bt* corn in the US (Gray et al., 1993; Gray, 2000) or *Bt* cotton in Argentina (Qaim et al., 2003) or India (Roy, personal communication, 2005). Farmers use pesticides when available and affordable in the absence of transgenic plants; these chemicals have been proven hazardous to both human and environmental health. Human health hazards are increased in low-income countries where farmers apply pesticides without adequate protection (Jeyaratnam, 1990; IRRI, 1994).

A review of the literature yields several reports on the effects of various transgenic crops on non-target organisms. This controversy entered public consciousness with a correspondence in *Nature* by Losey et al. (1999) entitled ‘Transgenic pollen harms monarch larvae’, which reported possible deleterious effects of transgenic *Bt* corn pollen (event MON 176) on monarch butterflies (*Danaus plexippus* L., Lepidoptera). Media reporting on this study verged on sensationalism. Monarch butterflies figured prominently in anti-transgenic movement literature, despite the fact that Losey et al. (1999) reported results of a feeding assay conducted in the laboratory rather than field-based observations, pollen density on milkweed leaves fed to monarch larvae was not recorded, and the authors urged further data collection to compare environmental risks of transgenic crops with those incurred by using pesticides.

Subsequent studies considered other toxins and other butterflies. Wraight et al. (2000) reported that *Bt* corn expressing Cry1Ab protein had no effect on field mortality of black swallowtail butterflies; Pimentel and Raven (2000) suggested additional testing of non-target effects prior to release, but noted such impacts were likely to be minimal when compared with those of widespread pesticide use in the absence of transgenic crops. Jesse and Obrycki (2000) demonstrated in a monarch butterfly larval feeding experiment that *Bt* corn pollen from two transgenic lines (event 176 and event Bt11) caused significantly higher mortality in larvae feeding for 48 h on milkweed dusted with the pollen. However, no effects on monarch oviposition or larval survival were evident in a second study by the same authors using event MON810 pollen (Jesse and Obrycki, 2003). Shelton and Sears (2001) pointed out that Jesse and Obrycki's conclusion in 2000 was based on some questionable assumptions, including the suppositions that artificial placement of milkweed plants in the cornfield mimicked their natural distribution, and that presence of monarchs and pollen shed occurred simultaneously. Oberhauser et al. (2001) reported that in much of the upper US Midwest there was substantial temporal overlap between monarch populations and corn pollen shed, although this diminished to about 15–20 per cent in Iowa and Maryland. Further, pollen from one of the non-*Bt* controls included in the Jesse and Obrycki (2000) study showed levels of Cry1Ab protein about half that of the *Bt* pollen. Last, pollen samples extracted from plants and used in the feeding assays reportedly contained about 40 per cent plant debris. According to Shelton and Sears (2001), several scientists have shown that inclusion of plant debris in such assays increases larval mortality and significantly reduces weight gain (Hellmich et al., unpublished results).

Critiques of the Jesse and Obrycki (2000) paper notwithstanding, a study by Zangerl et al. (2001) demonstrating adverse sub-lethal effects of event 176 *Bt* corn on black swallowtail butterflies gave credence to their results. Similarly, experiments conducted by Hellmich et al. (2001) and Stanley-Horn et al. (2001) showed that monarch larvae in feeding assays with event 176 *Bt* corn pollen were significantly smaller, and showed reduced weight gain and survivorship compared with those fed event *Bt*11 and MON810 pollen, whose effects were comparable to those of non-*Bt* controls. Although these authors concluded that the overall biological risk posed by corn modified by these events was negligible, the studies indicated that (a) a transgenic crop transformed by different events may behave very differently with respect to non-target effects; and (b) in the case of transgenic corn, event 176 appears to consistently demonstrate high activity against non-target Lepidoptera while events Bt11 and MON810 have proven largely innocuous. Koziel et al. (1993) demonstrated that event 176 corn contains a pollen-specific promoter and consequently exhibits significantly higher levels of the *Bt* protein in pollen than Bt11 or MON810 *Bt* corn. Event 176 *Bt* corn was not re-registered with the US EPA in 2001, primarily due to concerns about resistance management rather than effects on non-target Lepidoptera (Shelton and Sears, 2001; Groot and Dicke, 2002). Thus, it is important to evaluate potential effects of transgenic crops on a case-by-case basis, and with attention to the details of methods used in engineering the transgene and its subsequent expression in plant tissue.

Although some may deem butterfly studies frivolous when people in the low-income world are struggling to feed themselves, the sustainability of the agro-ecosystems they rely on may depend on non-target insects. Although farmers throughout the world often use pesticides, poor farmers generally cannot afford to apply them at the rate and frequency required to effectively control pests. Entomophagous insects and parasitoids in their fields are invaluable in integrated pest management; butterfly larvae are one link of many in the food web that sustains them. A transgenic crop that has deleterious effects on these key predator/prey relationships could set up a domino effect with disastrous consequences for the sustainability of low-input food production systems.

The growing body of literature addressing direct toxicity or indirect food chain-related effects of transgenic crops on non-Lepidopteran arthropods, many of them beneficial, also largely indicates negligible adverse impact. There are laboratory and field studies showing that Cry1Ab *Bt* corn poses little risk to the entomophagous green lacewing *Chrysoperla carnea* (Bourguet et al., 2002; Dutton et al., 2003), aphids *Metopolophium dirhodum, Ropalosiphum padi,* and *Sitobion avenae,* the bug *Orius insidiosus*, the syrphid *Syrphus corollae*, the ladybug *Coccinella septempunctata*, thrips, and hymenopteran parasitoids (Bourguet et al., 2002), the beetle *Coleomegilla maculata* (Zwahlen et al., 2000), predatory arthropods (Orr and Landis, 1997; Pilcher et al., 1997), and soil mites and collembola (Al-Deeb et al., 2003). Field studies with herbicide-tolerant transgenic sugar beets and oilseed rape indicate that the transgenes have no effect on important epigeous arthropod pest predators (Volkmar et al., 2000, 2003). Jorg et al. (2003) observed no effects on the collembolan *Folsomia candida* of two transgenic wheat varieties with enhanced fungal resistance, although a significant effect of variety was detected. Surveys in field plots with transgenic potatoes expressing protease inhibitors have detected no effect on the non-target herbivorous insect *Eupteryx aurata*, or on soil microarthropods (Cowgill et al., 2002; Cowgill and Atkinson, 2003). Potato engineered to express Cry3A had no effect on predators and parasitoids of the target pest, the Colorado potato beetle (Lopez and Ferro, 1995; Riddick and Barbosa, 1998), or the generalist predatory insect *Coleomegilla maculata*, although it reduced the abundance of the specialist predator *Lebia grandis* in comparison with the non-transgenic isoline, likely a result of diminished prey density rather than a toxicity effect (Riddick et al., 1998). There is some concern that *Bt* proteins may affect beneficial insects indirectly by compromising the food quality of their prey. Hilbeck et al. (1998a, b) conducted a study in which green lacewings were fed *Bt*-susceptible and insensitive prey, and concluded that lacewings reared on *Bt*-fed prey had a higher mortality than those fed on insensitive prey. Further, third-instar lacewings fed significantly less from *Bt*-fed insensitive prey than from non-*Bt*-fed insensitive prey (Meier and Hilbeck, 2001). Schuler et al. (1999) also demonstrated an adverse effect of *Bt*-hosts on the development of the parasitoid *Cotesia plutellae*, but this was likely because all hosts died before the parasitoid could eclose from the hosts. Ponsard-Sergine et al. (2002) examined the effect of a relatively *Bt* non-susceptible lepidopteran prey (*Spodoptera exigua* Hubner) that had ingested *Bt* cotton on the survivorship of four important heteropteran predators of cotton pests, and reported reduced longevity for two of the predators. More information on the non-target effects of *Bt* sprays or crops can be found in a review by Groot and Dicke (2002).

None of the studies described above evaluated risks of transgenic crops in comparison with alternative pest controls. A useful example is rice, a staple in many low-income diets. Surveys conducted by scientists at the International Rice Research Institute indicated that frequent and indiscriminate pesticide use was the norm in Asian rice-producing systems, with deleterious effects on beneficial insects instrumental in controlling pest populations as well as on organisms higher in the food chain, human health, and surface and groundwater resources (Heong et al., 1994; Way and Heong, 1994; Pingali and Roger, 1995; Pingali and Gerpacio, 1997). The vast majority of insecticides used in poor Asian countries are banned in most developed nations (Warburton et al., 1995). Transgenic *Bt* rice currently undergoing field-testing could have a very significant positive impact on farmers in low-income countries, by directly or indirectly improving their health, livelihoods, and general quality of life. Nevertheless, it remains imperative for policy-making that there be robust, holistic field trials comparing risks of specific transgenes *vs.* specific insecticides in the same spatial-temporal frame.

Non-target implications of transgenic crops extend to effects on non-arthropod denizens of the soil, especially microorganisms. The literature contains conflicting evidence. Donegan et al. (1995) reported a transient stimulation of soil bacterial and fungal populations by two of three *Bt* cotton lines. These lines also produced changes in the structure of bacterial communities. In experiments to evaluate the persistence of Cry1 proteins from *Bt* cotton leaves incorporated into soil microcosms, Palm et al. (1996) found that degradation of the protein was microbially-mediated as suggested by various reports from the Stotzky group (Koskella and Stotzky, 1997; Crecchio and Stotzky, 1998). However, it is difficult to make conclusions regarding persistence of the Cry protein as the rate of degradation varied greatly between experiments with 0.1 to 85 per cent of original extractable protein remaining after 28 days. Recent field studies with other transgenic crops have also demonstrated changes in microbial and protozoan populations in soil (Di Giovanni et al., 1999; Donegan et al., 1999; Griffiths et al., 2000; Lukow et al., 2000).

Data from ongoing field research in our laboratory comparing effects on the soil microbial community of Cry3Bb *Bt* corn and the insecticide tefluthrin indicate that neither *Bt* corn nor the insecticide had an effect on the soil activity or community diversity of a range of organisms, including bacteria, fungi, protozoa, and nematodes (Devare et al., 2004). Similar greenhouse and laboratory experiments conducted with Cry1Ab *Bt* corn and no insecticide by Blackwood and Buyer (2004) also indicated no effect on the bacterial community. Research conducted by Wu et al. (2004) on *Bt* rice in Hangzhou, China found no negative effect on a range of microbial indicators. Cowgill et al. (2002) compared the effects of a nematicide with cystatin-producing, nematode-resistant transgenic potatoes and observed that the nematicide reduced the bacterial fraction of the microbial community while the transgenic potatoes decreased both the bacterial and fungal components of the community. Although there are few reports in the literature on the effects of *Bt* corn on parasitic or beneficial nematodes, one study demonstrated that the nematode *Caenorhabditis elegans* showed extensive gut damage, decreased fertility, and increased mortality when fed *Bt* protein (Marroquin et al., 2000). Al-Deeb et al. (2003) and Cowgill et al. (2002) however, found no impact of Cry3Bb *Bt* corn or transgenic potatoes expressing cystatins, respectively, on free-living nematodes in the soil.

A number of studies have also shown that Cry proteins from *Bacillus thuringiensis* resist microbial degradation by binding to clay and humic acid fractions in soil, but retain insecticidal activity despite being bound (Venkateswerlu and Stotzky, 1992; Tapp and Stotzky, 1995; Koskella and Stotzky, 1997; Crecchio and Stotzky, 1998; Tapp and Stotzky, 1998; Saxena et al., 1999). Further, Saxena and Stotzky (2000) showed that larvicidal activity increased with the length of time *Bt* corn plants were grown, suggesting that microbial degradation may not have kept pace with the increased root exudation of Cry protein into soil over time. An alternative explanation may be that the protein was unavailable for degradation as it was bound, and concentrated with time on the surface of clays and humic materials. There is very little information on the fate and persistence of *Bt* proteins in the field. One field study reported that while the levels of Cry1Ab toxin in no-till corn decreased to 0.3 per cent of the initial concentration after 200 d, degradation was initially delayed and lower than in a conventionally tilled system (Zwahlen et al., 2003). However, Head et al. (2002) did not detect Cry1Ac protein in soil samples taken about 90 days after the last planting from fields planted with *Bt* cotton for several consecutive years. Hopkins and Gregorich (2003) studied intact and fractionated soil in which *Bt* maize had been grown for 4 years. They found that the Cry1A protein was only found in quantifiable amounts in the $>2000\ \mu m$ fraction, associated with partially decomposed plant material, and was detected only in trace amounts in the smaller size fractions, where more fully decomposed residues would be captured.

Our preliminary results from substrate-induced respiration assays indicate that CO_2 evolved from soil from field plots planted with corn borer-resistant *Bt* corn producing Cry1Ab protein is lower than that from isogenic non-*Bt* corn (Devare and Duxbury unpublished data). The Stotzky group has reported that finely-ground *Bt* corn, rice, potato, and tobacco residues degrade more slowly than their non-*Bt* counterparts in laboratory studies for reasons that are unclear (Stotzky, 1999, 2002; Flores et al., 2005). While the mechanism by which Cry1Ab might depress microbial activity is not known, it is unlikely to be directly toxic to microorganisms. Saxena and Stotzky (2001) suggested that higher lignin content in *Bt* corn might be responsible for a slower rate of residue decomposition, yet, lignin content did not differ between non-*Bt* and *Bt* plants for any of the other crops tested, despite lower observed soil respiration rates between the isolines (Flores et al., 2005). In current field trials, we have found that the in-field decomposition rates of two *Bt* corn hybrids (Pioneer 34N44*Bt* and NC+ 4990Bt) expressing the Cry1Ac protein do not differ from their non-transgenic cultivars, Pioneer 34N43 and NC+ 4880. Bacteria and fungi colonising residues also do not differ between residues from any of the cultivars in a DNA fingerprinting assay. However, decomposition rates and microbial community fingerprints differed by both residue placement and plant part, with residues buried at 5 cm decomposing faster than those placed on the soil surface (Londono et al., in review).

Empirical studies have yielded inconsistent results regarding the persistence of residues from transgenic crops. The absence of a clear picture elicits concern about potential effects on soil microorganisms, which are indispensable in nutrient turnover and maintaining soil structure and fertility. Soil organic carbon plays a pivotal role in nutrient cycling and in improving plant-available water reserves, soil buffering capacity, and structure (Doran et al., 1994). Potential effects on the microbial community or organic matter dynamics could have significant effects on agro-ecosystems, particularly in poor countries where many farmers lack resources for purchased inputs and work marginal soils. The USDA (2003) has acknowledged the need to study plant-mediated effects that are independent of expressed proteins, such as changes in plant composition, morphology or physiology that may affect non-target organisms and the need for improved protocols for testing and elucidating mechanisms responsible for changes in organic matter decomposition.

VIII. Gene Flow

Gene flow between crops and their wild relatives is common and, between plants capable of hybridising, inevitable if grown within the crop's pollen dispersal range (Ellstrand et al., 1999). Although use of border rows containing non-transgenic isolines or trap crops or perimeter fallow zones is strongly suggested, it is generally accepted that complete containment of transgenic pollen in male fertile lines is impossible. Therefore, genes will move (Snow, 2002). How far and how fast will be determined by prevailing weather patterns and conditions at the time pollen is shed, the method of dispersal (for example, wind or insects), the amount of pollen produced and its longevity, the distance between the donor and compatible recipient and whether the crop out-crosses or is self-pollinated. Arriola (1998) suggested that biotechnology companies and researchers should be providing data on crop/wild hybridisation rates and initial hybrid fitness in order to better inform the debate on the real risks of a given transgenic crop. However, anticipating and quantifying the hazards associated with a given plant accruing a transgenic trait in a given environment presents a real challenge. The controversial aspects of gene flow are the cultural and biodiversity implications of 'genetic pollution' of natural landraces (Quist and Chapela, 2001) or neighbouring crops (BBC News, 2000), and the potential ecological consequences should 'superweeds' be inadvertently created by the transfer of genes conferring enhanced survival and fitness to recipient plants (Rissler and Mellon, 1993).

Conventional breeding has already led to wholesale replacement of landraces with elite line monocultures, thereby reducing biodiversity. Such a result is no more, nor less, a concern with transgenic crops. Will transgenic traits, such as drought or pest resistance, lead to an expansion in the niche of transgenic crops that might lead to reduced biodiversity in the surrounding areas? Could the widespread adoption of stress-tolerant plants, for example, increase pressure on fragile, marginal lands, thus potentially destroying valuable natural ecosystems? Will transgenes confer increased fitness so as to enhance the invasiveness of weedy species? What is the likelihood that plants carrying transgenic traits could have other, unintended effects on non-target species or trophic interactions? The answers to these pressing questions are unknown. The influence that a transgene may have on a population is largely determined by the increased fitness that it confers in the natural habitat. Knowing resultant changes in fitness is necessary to predict the extent and potential impact of gene flow events.

Most regulatory frameworks that have been developed dictate that transgenic crops not be grown in areas where wild relatives are endemic, specifically, to limit gene flow and the potential for 'genetic contamination'. This issue was recently highlighted by the Quist and Chapela (2001) report of transgenic DNA introgressing into traditional maize landraces in Mexico. They used a polymerase chain reaction protocol to detect the presence of the 35S promoter (p-35S) from the cauliflower mosaic virus (CMV), a common element in transgenic constructs. They obtained a positive PCR signal in five of the seven pooled maize seed samples they tested. Their results led to a six-year de facto moratorium on the planting of transgenic crops in Mexico. The results of Quist and Chapela (2001) have recently been challenged by the findings of Ortiz-Garcia et al. (2005). They sampled maize seeds from 870 plants in 125 fields in 18 locations in Oaxaca, Mexico in 2003 and 2004. They tested 153,746 seeds for the presence of the 35 S promoter of the CMV and the nopaline synthase gene from *Agrobacterium tumefaciens*; one or both of these transgenic elements are present in all commercial varieties of transgenic maize. They detected no transgenic sequences in any of the samples tested. While not contesting the report of Quist and Chapela (2001), they suggest that the lack of detection of 'genetic contamination' likely stems from reduced GM imports and education of local farmers.

No wild relatives of corn, soybean, or potatoes are present in the US, hence planting of these transgenic crops is not restricted. Transgenic cotton, however, is restricted in Florida and Hawaii due to presence of other *Gossypium* species. Such restrictions presuppose a strong regulatory and monitoring infrastructure. Thus, the risk of out-crossing is likely to be increased in resource-poor countries. The stealth seeds scenario in Gujarat, India (see Herring, forthcoming) may provide an unplanned, but useful, model system in which to monitor the effects of transgene introgression into wild plant communities. No precautionary steps were taken to prevent pollen movement within the system. However, cotton pollen does not travel far from its source and crosses are rare; gene flow to wild relatives of cotton is a very remote possibility. However, this situation may indeed help us understand risks posed by the flow of *Bt* genes in cotton into either wild hybridising species or non-transgenic local cultivars.

Even without hybridising, with non-cultivated relatives, the potential for gene flow to non-transgenic crops of the same cultivated species is of real concern, especially for out-crossing species such as corn, canola, and sunflower (Marshall et al., 2001). Although corn pollen, like cotton, is heavy and generally does not travel far, it has been shown to pollinate nearby, non-*Bt* corn, even if at very low levels. Consequently, a serious issue is the effect on organic growers who could lose organic certification if crops become 'genetically-contaminated' and on neighbouring farmers who may be accused of patent violations if transgenes are found in their fields (BBC News, 2000).

IX. Confinement Issues

In the US, very few permits have been issued for plants engineered to produce pharmaceutical and industrial compounds. However, an increase in requests for and scale of field testing is anticipated. In 2002, APHIS had approved the testing of pharmaceutical-producing plants at 34 sites comprising 130 acres; most test sites were under 5 acres. In 2003, APHIS changed regulations to improve confinement of these transgenic crops. Changes included increasing the perimeter fallow zone from 25 to 50 feet to ensure that test plants were not inadvertently commingled with those to be used for food or feed; restricting planting of food or feed crops in the following season to reduce the possibility of volunteer plants being inadvertently harvested with the following crop; and requiring the use of dedicated planting and harvesting machinery and storage facilities for the regulated crop. Additional conditions for field testing corn engineered to produce pharmaceuticals were established, including an isolation distance of one mile for open-pollinated corn and a half-mile for controlled pollination corn field tests. This increased stringency stemmed from the failure of ProdiGene, Inc, to follow regulations when testing corn designed to produce a pig vaccine. In one trial, nearby crops may have been pollinated by the vaccine-producing transgenic crop, leading to an order for complete destruction of nearby crops and payment by the firms for all damages to local farmers (Ellstrand, 2003).

Public and regulatory concern over potential environmental consequences of plant-made pharmaceuticals (PMP) or plant-made industrial proteins (PMIP) and the inability to contain flow of transgenes in field situations may prevent biotechnology firms from pursuing deregulation of these crops (Ellstrand, 2003). For PMP and PMIP crops, the very real concern of gene-flow requires that these crops be tightly contained and monitored and thus not be widely disseminated to farmers.

Since gene flow from crops to wild populations is not only likely, but inevitable, possible risks and consequences require research. Primary research needs as identified by a USDA Risk Assessment Stakeholder Workshop (2003) are: (a) identifying factors controlling the rate of flow and fate of genes in wild populations as well as the spatial dimensions; (b) developing means to detect transgenes in wild populations; (c) understanding the consequences of gene flow, such as fitness effects, impacts on plant community structure and variations in these for different constructs; (d) developing biological containment strategies and evaluating their stability, such as auto-suicidal mechanisms, tissue-specific excision, gene silencing and chloroplast transformation; (e) new means of physical confinement, such as trap-crop border rows; (f) means to mitigate effects, slow the spread or limit expression; and (g) longer-term monitoring.

Resource-poor countries cannot be expected to address such complex environmental questions on their own, nor to develop the sophisticated tools required for monitoring. Providing solid, science-based data which poor-country regulatory personnel can use to evaluate risk from their own perspective and decide what is culturally and environmentally acceptable requires global cooperation. The FAO recently launched a new database called FAO–BioDeC. The database is designed to help gather, store, organise, and disseminate updated baseline information on the state-of-the-art of crop biotechnology products and techniques, which are in use, or in the pipeline. The database includes about 2000 entries from 70 developing and transitional economies, readily available at www.fao.org/biotech/inventory_admin/dep/default.asp

X. Insect Resistance Management

Concern over resistance development in target insects reflects desire to retain efficacy of the IR crop for the longest possible time. Impacts of insect resistance on the environment *per se* have not been of particular concern. Financial investment in product development, testing and regulatory approval is so high that developers seek to maintain product efficacy for as long as possible. A robust insect-resistance management (IRM) strategy extends the useful life of the technology. Concern over insect resistance to IR crops stems from a wealth of information on insect resistance to insecticides. Insect resistance to *Bt* has been demonstrated in laboratory trials; however, to date, only one major Lepidopteron resistance mechanism has been reported from the field where *Bt* sprays – not transgenic plants – had been used: a reduction in toxin binding on gut brush border target sites in the pest *Plutella xylostella*, the diamondback moth (Tabashnik et al., 2003).

Resistance has been slowed in pest populations mainly by presence of a high dose of *Bt* toxin expressed in plant tissues. A high dose decreases the likelihood of survivorship and ensures that insects are not exposed to sub-lethal doses that might promote development of resistance. To date, despite *Bt* crops being grown on more than 15 million ha worldwide, an increase in the frequency of resistance caused by exposure to *Bt* crops in the field has not yet been reported (Fox, 2003; Tabashnik et al., 2003).

The high-dose strategy has been combined with the use of refugia to reduce the likelihood of target insects developing resistance to *Bt* (Tabashnik, 1994). Refugia serve to maintain susceptible insect populations that will interbreed with potentially resistant insects, thereby diluting the frequency of resistance alleles in the population and delaying resistance (Roush, 1994; Tabashnik, 1994). The rate at which resistance will develop using any chemical control strategy is difficult to predict. The rate depends on many assumptions regarding resistance management strategies, their acceptance and effective implementation by growers, the genetics of resistance in the target pest and its population and behavioural biology (Tabashnik, 1994; Gould et al., 1994). For plants expressing the Cry insecticidal proteins derived from *B. thuringiensis*, the US EPA requires firms to develop and provide an IRM plan.

In turn, firms require that growers agree to and administer an IRM strategy. Biotechnology companies have been charged with ensuring and reporting grower compliance. The IRM strategy for Monsanto's Yieldguard® rootworm-resistant corn, for example, requires growers to plant at least 20 per cent of their total corn acreage to a non-resistant corn variety, preferably one from a similar genetic background (www.monsanto.com/monsanto/us_ag/content/stewardship/irm/2004/yieldgard.pdf). If *Bt* cotton will also be grown in the region, then the refuge must represent 50 per cent of the planted acreage. In contrast, there are no refuge requirements set for any other insect management product or approach. If insecticide or other plant protection is used on the non-resistant variety, the transgenic corn must also be treated. Finally, the acreage on which the non-resistant corn is planted must be owned by the grower (not on a neighbour's land).

Recently, with the introduction of Bollgard and Bollgard II cotton (Monsanto), which contains two 'stacked' Cry genes, alternatives to these stringent requirements have been sought and tested. In addition to the 20 per cent sprayed refuge option, a 5 per cent unsprayed refuge option has been tested over the last several years and so far is proving satisfactory for maintaining efficacy of the crop (US EPA, 2004). In addition, a new 'community refuge' option has been trialled. In this option, the requirement for a farmer to own all the land on which the refuge is grown has been relaxed. Instead, farmers can combine neighbouring fields within a one square mile area that can act as a 20 per cent sprayed refuge or a 5 per cent unsprayed refuge. Still, participants in the community refuge option must have a community refuge coordinator and appropriate documentation is required, including annual resistance monitoring. An annual compliance assurance program, grower education, remedial action plans, and annual reporting are also required. Hence, despite relaxation of some requirements, those for training and proof of compliance have been maintained.

Resource-poor farmers are ill-equipped to comply with these requirements, nor are they likely to do so. First, with limited land area, they can ill afford to plant unprotected crops that are likely to yield little. Farmers are inclined not to plant susceptible seed even when freely provided (W. Green, personal communication, 2005). Hence, area requirements for refugia are not likely to be met, increasing the likelihood that insect resistance may develop. Second, a primary benefit of insect-resistant crops is the decreased use of insecticides. Should insect resistance develop or farmers choose to spray unprotected crops, the resultant need to use insecticide on the entire crop negates the primary benefit of the use of the transgenic. Finally, stringent land ownership requirements limit the ability of farmers to cooperate and use a landscape mosaic approach to resistance management. An organised, enforceable system of oversight is costly and unlikely. Instead, more creative options for providing refugia that will help delay the inevitable onset of pest resistance need to be sought. One such option is the use of alternate hosts as natural refugia (US EPA, 2004).

The success of the refugia approach depends upon several conditions: that the resistance trait is recessive, that there is random mating, that adults will travel sufficiently between toxic plants, that there is complete lack of insecticidal action in the refugia and that farmers can implement and maintain them properly. If adult insects do not travel between the refuge and treated area, resistance will quickly develop in treated areas, while susceptible populations continue to mate amongst themselves in untreated areas. If refugia are subjected to any insecticide, the susceptible population available for mating with *Bt*-exposed individuals will decrease (Tabashnik, 1997). Whether alternate hosts randomly distributed in the landscape can satisfy these requirements is currently unknown. However, in China, where there are no requirements for farmers to plant refuges, resistance to *Bt* has not been detected in the nine years since its release, despite high rates of adoption (Wu and Guo, 2005). Scientific priorities include better understanding the biology and ecology of target pests relative to refuge efficacy, continued testing and validating alternative resistance management strategies and improvement of monitoring methods.

In the end, any resistance management strategy can only be successful if implemented properly. It is crucial that growers – in resource-rich or resource-poor countries – understand resistance management strategies and their importance in postponing insect resistance for as long as possible (Wu and Guo, 2005), thereby improving their own livelihoods and that of their communities.

XI. Genetic Use Restriction Technologies (GURT)

The expression of genes is generally regulated by internal factors such as proteins or other products of metabolism, or external factors such as climate. GURT refers to biotechnology-based methods to regulate gene expression, primarily to restrict plant propagation from a second generation of seed. GURTs potentially have beneficial applications for farmers, consumers, and the environment that should not be overlooked in debates over intellectual property rights. There are three distinct aspects of these technologies: use restriction, environmental containment of transgenic material, and agricultural productivity (FAO, 2001). There are two types of GURT: variety-specific (V-GURT), which results in sterile seeds and has earned the 'terminator gene' or 'terminator technology' labels, and the controversial use restriction of a specific trait (T-GURT), requiring external inducers to activate gene expression (Oliver et al., 1995; UNEP, CBD and SBSTTA, 1999).

Unlike V-GURT, T-GURT would enable farmers to save their own seed, but lack access to the added traits in the absence of chemical activators. For instance, one such technology involves a complex three-gene system with one gene producing a protein that interferes with normal plant embryo development to prevent seed germination. Application of chemicals such as tetracycline triggers the expression of this gene which prevents a recombinase gene from being repressed by another protein. Once the recombinase is expressed after chemical application, the protein lethal to the plant embryo is expressed. Seed sold to farmers would therefore be pre-treated with either tetracycline or other chemicals such as copper and steroids (NAS, 2000).

In addition to their use restriction properties discussed above, GURTs represent an environmental containment strategy through their ability to eliminate the spread of transgenic seed (V-GURT) or transgenes to wild plant relatives (T-GURT). For instance, this technology could potentially eliminate the problems of 'volunteer' plants that appear from seed left in the field after harvest. Volunteer plants must be eliminated before the next crop is planted because they are hosts for pests and pathogens and can nullify the benefits of crop rotation. As with any growth regulator applied to crops, there are possible environmental or human health issues associated with the use of chemical activators (that is, tetracycline, copper, steroids).

Potential productivity advantages from genetic use restriction technologies include T-GURTs that enable a farmer to restrict the expression of a trait when there is an advantage in doing so – for example, at a specific phase of crop development or during periods of biotic or abiotic stress (FAO, 2001). Productivity gains from V-GURTs include the ability to safeguard the integrity of adapted maternal breeds or to prevent pre-harvest sprouting.

A major concern of this technology is that it could hurt small farmers by ending 'brown-bagging', the practice of collecting seed at harvest and bagging it to use as the next year's planting stock. Few US farmers do this; it is much more common in low-income countries where farmers rely on it for seed multiplication. Countries where brown bagging is common practice will still be able to save their traditional seeds and other public varieties, but would have to expend resources each year in the purchase of transgenic seed. However, loss of cost savings from brown-bagging must be weighed against the productivity gains from superior new varieties that could increase yield and quality and reduce input costs and losses such as those due to pests or adverse soils and weather.

The US National Academy of Sciences has suggested that:

> Private corporations and research institutions should make arrangements to share GM technology, now held under strict patents and licensing agreements, with responsible scientists for use for hunger alleviation and to enhance food security in developing countries. In addition, special exemptions should be given to the world's poor farmers to protect them from inappropriate restrictions in propagating their crops. (NAS, 2000: 2)

Raising the economic incentive for aiding in poverty alleviation may be a viable alternative to influencing firm behaviour, rather than leaving open only the option of developing terminator technology to recover investments in transgenic crops.

XII. Pipeline Technologies

A wide variety of crops employing a diversity of transgene products are under development. Crops with resistance to viruses, pathogenic bacteria and fungi, and nematode and insect pests are nearing release (for example, Murray et al., 2002; Sharp et al., 2002). Newer technologies include: improved crop resistance to limiting environmental factors such as drought, salinity and toxic metals (de la Fuente-Martinez et al., 1997; Liu et al., 1999; Garg et al., 2002; Song et al., 2003); enhanced nutrient content (Goto et al., 1999; Lucca, 1999; Bovy et al., 2002), including nutriceuticals; pharmaceutical or 'pharm' crops (Thanavala et al., 1995; Stoger, 2000); crops that bioaccumulate toxins for use in soil remediation (Song et al., 2003); and crops that produce industrially useful products, such as starches and plastics.

How will environmental concerns and monitoring efforts change to address potential risks posed by these new crops? Which are relevant to the needs of the poor? For crops resistant to pathogens and pests, environmental risks would be no different from those discussed above. For crops with stacked genes conferring multiple resistances, the regulatory framework and recommendations for refugia and buffer zones are likely to remain constant or be relaxed due to increased efficacy. For the most part, they are considered no more environmentally risky than those crops already approved for commercial use. However, pharm crops and those expressing industrial proteins present entirely new classes of risk that will require more robust strategies for monitoring and evaluating environmental risk. Any plant that expresses proteins with mammalian bioactivity or with potential to increase the fitness of animal or plant pathogens requires keen scrutiny. Cultivation of pharm crops and PMIP require very strict confinement regulations and will likely be sown on a very small area worldwide. Such crops are not likely to be cultivated by poor farmers because of the confinement issues discussed above.

XIII. Summary and Conclusions

Livelihoods of the poor depend on ecological integrity. While the environmental safety record of transgenic crops is much improved over chemical safety measures, there is sufficient evidence of potential impacts to remain vigilant. Each new transgenic product may present its own potential for unintended environmental impact; the generally accepted recommendation is that each be evaluated on a case by case basis (NAS, 2000). Once the general behavioural traits of a new crop are known, how important is it that potential environmental impact be repeatedly tested in every new region? The Nuffield Council on Bioethics (2003) suggested that methods and results of environmental risk assessments could be shared between countries that have similar agricultural environments, thus reducing costs for any one country. As more information on the environmental benefits of growing transgenic crops accrues (Carpenter et al., 2002; Bennett et al., 2003, 2004; FAO, 2004; Morse et al., 2004), the more likely such data sharing will satisfy regulatory needs. This and similar logical extensions of prior findings could substantially reduce unwarranted restrictions and improve the benefits of these technologies for resource-poor farmers.

There is still no scientific understanding of the level of environmental impact that should trigger concern. Yet, we know that each new crop developed by conventional breeding will elicit environmental changes at the micro-scale, and potentially the meso-scale. Why then are transgenic crops held to substantially higher standards? In part, caution may be a response to the grave environmental consequences of the Green Revolution, consequences no one wishes to repeat. In part, it may simply be fear of the unknown or unfamiliar.

Of the technologies approved for commercial release, those with the most relevance to resource-poor farmers are those that confer insect resistance, herbicide resistance, viral resistance or improved agronomic traits. Informed decision making and wise policies cognisant of the potential for unintended effects could allow the poor to benefit from these new technologies. The recent National Academy of Sciences report (2000) suggests that each country implementing transgenic technology ensure that its infrastructure is sufficient to support the safe development, transfer and application of biotechnology, with special attention to developing relevant policies, information systems and training in risk assessment and biosafety procedures. An overarching international body to maintain and disseminate a public database that includes all newly released varieties and their performance in different environments is recommended. This task has been taken up by the FAO (http://www.fao.org/biotech/inventoryadmin/dep/default.asp?lang=en). Collection and access to such data are critical to inform the debate surrounding benefits and risks to the poor from the use of transgenic crops. Vasil (2003) states that 'transgenic crops rightly deserve to be an integral part of international agriculture in the twenty-first century'. While this is indeed the case for technologies with proven track records for environmental safety, it is still too early for so sanguine a view. Gene flow, resistance management and soil ecology issues, among others, warrant caution and long-term monitoring in the field. We conclude that while there is reason to be optimistic about the potential for transgenic crop technologies to be beneficial in low-income agriculture, there is also an imperative to exert the highest scientific and regulatory standards possible to ensure negligible long-term ecological effects of deploying transgenic crops.

References

Al-Deeb, M. A., Wilde, G. E., Blair, J. M. and Todd, T. C. (2003) Effect of *Bt*corn for corn rootworm control on nontarget soil microarthropods and nematodes, *Environmental Entomology*, 32, pp. 859–65.

Arriola, P. (1998) Benefits and risks of genetic modification in agriculture. Are we too late?, accessed at: http://ezproxy.library.cornell.edu:2195/nature/debates/gmfoods/gmfoods frameset.html

BBC News (2000) GM firm sues Canadian farmer, World, Americas, 6 June.

Benbrook, C. M. (2003) Impacts of genetically engineered crops on pesticide use in the United States: the first eight years, Biotech Infonet, Technical Paper No. 6.

Bennett, R., Buthelezi, T. J., Ismael, Y. and Morse, S. (2003) *Bt*cotton, pesticides, labour and health. A case study of smallholder farmers in the Makhathini Flats, Republic of South Africa, *Outlook on Agriculture*, 32, pp. 123–8.

Bennett, R., Ismael, Y., Morse, S. and Shankar, B. (2004) Reductions in insecticide use from adoption of *Bt* cotton in South Africa: impacts on economic performance and toxic load to the environment, *Journal of Agricultural Science*, 142, pp. 665–74.

Blackwood, C. and Buyer, J. (2004) Soil microbial communities associated with *Bt* and non-*Bt* corn in three soils, *Journal of Environmental Quality*, 33, pp. 832–6.

Bouis, H. E. (2000) Enrichment of food staples through plant breeding: a new strategy for fighting micronutrient malnutrition, *Nutrition*, 16, pp. 701–4.

Bourguet, D., Chaufaux, J., Micoud, A., Delos, M., Naibo, B., Bombarde, F., Marque, G., Eychenne, N. and Pagliari, C. (2002) Ostrina nubilalis parasitism and the field abundance of non-target insects in transgenic *Bacillus thuringiensis* corn (*Zea mays*), *Environmental Biosafety Research*, 1, pp. 49–60.

Bovy, A., deVos, R., Kemper, M., Schijlen, E., Pertejo, M. A., Muir, S., Collins, G., Robinson, S., Verhoeyen, M., Hughes, S., Santos-Buelga, C. and van Tunen, A. (2002) High-flavonol tomatoes resulting from the heterologous expression of the maize transcription factor genes LC and C1, *Plant Cell*, 14, pp. 2509–26.

Caballero, B. (2002) Global patterns of child health: the role of nutrition, *Annals of Nutrition and Metabolism*, 46, pp. 3–7.

Carpenter, J., Felsot, A., Goode, T., Hammig, M., Onstad, D. and Sankula S. (2002) Comparative environmental impacts of biotechnology-derived and traditional soybean, corn, and cotton crops, Council for Agricultural Science and Technology, Ames, IA, accessed at: http://www.cast-science.org

Chrispeels, M. J. (2000) Biotechnology and the poor. *Plant Physiology*, 124, pp. 3–6.

Combs, Jr. G. F., Welch, R. M., Duxbury, J. M., Uphoff, N. T. and Nesheim, M. C. (1996) Food-based approaches to preventing micronutrient malnutrition: an international research agenda. Summary report of an international workshop, Salt Lake City, UT, November 6–9, 1995. Ithaca, New York: Cornell International Institute for Food, Agriculture, and Development.

Conway, G. (1999) *The Doubly-Green Revolution: Food for All in the Twenty-First Century* (Ithaca: Cornell University Press).

Conway, G. and Toenniessen, G. (1999) Feeding the world in the twenty-first century, *Nature*, 402, pp. C55–8.

Cowgill, S. E. and Atkinson, H. J. (2003) A sequential approach to risk assessment of transgenic plants expressing protease inhibitors: effects on nontarget herbivorous insects, *Transgenic Research*, 12, pp. 439–44.

Cowgill, S. E., Bardgett, R. D., Kiezebrink, D. T. and Atkinson, H. J. (2002) The effect of transgenic nematode resistance on non-target organisms in the potato rhizosphere, *Journal of Applied Ecology*, 39, pp. 915–23.

Crecchio, C. and Stotzky, G. (1998) Insecticidal activity and biodegradation on the toxin from *Bacillus thuringiensis* subsp. *kurstaki* bound to humic acids from soil, *Soil Biology and Biochemistry*, 30, pp. 463–70.

Datta, S. K. (1999) Transgenic rice products: roadmap for future impact, in *Abstracts, General Meeting of the International Program on Rice Biotechnology, 20–24 Sep. 1999*, Phuket, Thailand, pp. 12–13.

de la Fuente-Martinez, J., Ramirez-Rodriguez, V., Cabrera-Ponce, L. and Herrera-Estrella, L. (1997) Aluminum tolerance in transgenic plants by alteration of citrate synthesis, *Science*, 276, pp. 1566–8.

de la Fuente-Martinez, J. M. and Herrara-Estrella, L. (1999) Advances in the understanding of aluminum toxicity and the development of aluminum-tolerant transgenic plants, *Advances in Agronomy*, 66, pp. 103–20.

De Ronde, J. A., Cress, W. A., and Van der Mescht, A. (2001) *Agrobacterium* mediated transformation of soybean seed with the GUSINT marker gene, *South African Journal of Science*, 97, pp. 421–4.

Devare, M. H., Jones, C. M. and Thies, J. E. (2004) Effect of Cry3Bb transgenic corn and tefluthrin on the soil microbial community: biomass, activity, and diversity. *Journal of Environmental Quality*, 33, pp. 837–43.

Di Giovanni, G., Watrud, L. S., Seidler, R. J. and Widmer, F. (1999) Comparison of parental and transgenic alfalfa rhizosphere bacterial communities using BIOLOG GN metabolic fingerprinting and enterobacterial repetitive intergenic consensus sequence-PCR (ERIC-PCR), *Microbial Ecology*, 37, pp. 129–39.

Donegan, K. K., Palm, C. J., Fieland, V. J., Porteous, L. A., Ganio, L. M., Schaller, D. L., Bucao, L. Q. and Seidler, R. J. (1995) Changes in levels, species and DNA fingerprints of soil microorganisms associated with cotton expressing the *Bacillus thuringiensis* var. *kurstaki* endotoxin, *Applied Soil Ecology*, 2, pp. 111–24.

Donegan, K. K., Seidler, R. J., Doyle, J. D., Porteous, L. A., di Giovanni, G., Widmer, F. and Watrud, L. S. (1999) A field study with genetically engineered alfalfa inoculated with recombinant *Sinorhizobium meliloti*: effects on the soil ecosystem, *Journal of Applied Ecology*, 36, pp. 920–36.

Doran, J. W., Varvel, G. E. and Culley, J. L. B. (1994) Tillage and residue management effects on soil quality and sustainable land management, in R. C. Wood and J. Dumanski (eds), *Sustainable Land Management for the Twenty-First Century: Vol. 2, Plenary Papers*, pp. 59–74 (Canada: University of Lethbridge).

Dutton, A., Romeis, J. and Bigler, F. (2003) Assessing the risks of insect resistant transgenic plants on entomophagous arthropods: Bt-maize expressing Cry1Ab as a case study, *BioControl*, 48, pp. 611–36.

Ellstrand, N. C., Prentice H. C. and Hancock, J. F. (1999) Gene flow and introgression from domesticated plants into their wild relatives, *Annual Review of Ecology and Systematics*, 30, pp. 539–63.

Ellstrand, N. C. (2003) Current knowledge of gene flow in plants, *Philosophical Transactions of the Royal Society of London B Biological Sciences*, 358, pp. 1163–70.

Ely, S. (1993) The engineering of plants to express *Bacillus thuringiensis* delta-endotoxins, in P. F. Enwistle, J. S. Cory, M. J. Bailey and S. Higgs (eds), *Bacillus thuringiensis, an Environmental Biopesticide: Theory and Practice*, pp. 105–24 (New York: Wiley).

Ervin, D. E., Batie, S. S., Carpentier, C. L. and Welsh, R. (2001) Transgenic crops and the environment: The economics of precaution, *Journal of Agricultural and Resource Economics*, 26, p. 554.

Ervin, D. E., Welsh, R., Batie, S. S. and Carpentier, C. L. (2003) Towards an ecological systems approach in public research for environmental regulation of transgenic crops, *Agriculture Ecosystems and Environment*, 99, pp. 1–14.

FAOBioDeC, Biotechnologies in Developing Countries, accessed at: http://www.fao.org/biotech/inventory_admin/dep/default.asp?lang=en

FAO (2001) Potential impacts of genetic use restriction technologies (GURTs) on agricultural biodiversity and agricultural production systems, Item 6 of the Draft Provisional Agenda of the Working Group on Plant Genetic Resources for Food and Agriculture, Commission on Genetic Resources for Food and Agriculture, July 2–4, Rome, FAO.

FAO (2004) *The State of Food and Agriculture 2003–04* (Rome: FAO).

Federal Register (2003) Field testing of plants engineered to produce pharmaceutical and industrial compounds, *Request for Comments*, 68(6), pp. 11337–40.

Ferré, J. and Van Rie, J. (2002) Biochemistry and genetics of insect resistance to *Bacillus thuringiensis*, *Annual Review of Entomology*, 47, pp. 501–33.

Fischer, R., Twyman, R. M. and Schillberg, S. (2003) Production of antibodies in plants and their use for global health, *Vaccine*, 21, pp. 820–5.

Flores, S., Saxena, D. and Stotzky, G. (2005) Transgenic *Bt* plants decompose less in soil than non-*Bt* plants, *Soil Biology and Biochemistry*, 37, pp. 1073–82.

Fox, J. L. (2003) Resistance to Bt toxin surprisingly absent from pests, *Nature Biology*, 21, pp. 958–9.

Fromm, M. E., Stark, D. M., Austin, G. D. and Perlak, F. J. (1993) Improved agronomic and quality traits in transgenic crops: recent advances, *Philosophical Transactions of the Royal Society of London-B: Biological Sciences*, 339, pp. 233–7.

Garg, A. K., Kim, J., Owens, T. G., Ranwala, A. P., Choi, Y. D., Kochian, L. V. and Wu, R. J. (2002) Trehalose accumulation in rice plants confers high tolerance levels to different abiotic stresses, *Proceedings of the National Academy of Sciences of the United States of America*, 99, pp. 15898–903.

Gerhardt, S. A., Balconi, C. and Sherwood, J. E. (2002) Control of *Fusarium* scab with puroindoline-containing transgenic wheat, *Phytopathology*, 92, pp. S28–9.

Goto, F., Yoshihara, T., Shigemoto, N., Toki, S. and Takaiwa, F. (1999) Iron fortification of rice seed by the soybean ferritin gene, *Nature Biotechnology*, 17, pp. 282–6.

Gould, F., Follett, P., Nault, B. and Kennedy, G. G. (1994) Resistance management strategies for transgenic potato plants, in G. W. Zehnder, M. L. Powelson, R. K. Jansson and K. V. Raman (eds), *Advances in Potato Pest Biology and Management*, pp. 255–77 (St Paul: American Phytopathological Society Press).

Gould, J. L. and Keeton, W. T. (1996) *Biological Science* (New York: Norton Press).

Gray, M. E. (2000) Prescriptive use of transgenic hybrids for corn rootworms: an ominous cloud on the horizon? Crop Protection Technology Conference, January 5–6, 2000, University of Illinois, Champaign-Urbana.

Gray, M. E., Steffey, K. L. and Oloumi-Sadeghi, H. (1993) Participatory on-farm research in Illinois cornfields: an evaluation of established soil insecticide rates and prevalence of corn rootworm (Coleoptera: Chrysomelidae), *Journal of Economic Entomology*, 86, pp. 1473–82.

Greiner, S., Rausch, T., Sonnewald, U. and Herbers, K. (1999) Ectopic expression of a tobacco invertase inhibitor homolog prevents cold-induced sweetening of potato tubers, *Nature Biotechnology*, 17, pp. 708–11.

Grierson, D. and Fray, R. (1994) Control of ripening in transgenic tomatoes, *Euphytica*, 79, pp. 251–63.

Griffiths, B. S., Geoghegan, I. E. and Robertson, W. M. (2000) Testing genetically engineered potato, producing the lectins GNA and Con A, on non-target soil organisms and processes, *Journal of Applied Ecology*, 37, pp. 159–70.

Groot, A. T. and Dicke, M. (2002) Insect-resistant transgenic plants in a multi-trophic context, *The Plant Journal*, 31, pp. 387–406.

Hancock, J. F. (2003) A framework for assessing the risk of transgenic crops, *BioScience*, 53, pp. 512–19.

Head, G., Surber, J. B., Watson, J. A., Martin, J. W. and Duan, J. J. (2002) No detection of Cry1Ac protein in soil after multiple years of transgenic *Bt* cotton (Bollgard) use, *Environmental Entomology*, 31, pp. 30–6.

Hellmich, R. L., Siegfried, B. D., Sears, M. K., Stanley-Horn, D. E., Daniels, M. J., Mattila, H. R., Spencer, T., Bidne, K. G. and Lewis, L. C. (2001) Monarch larvae sensitivity to *Bacillus thuringiensis*-purified proteins and pollen, *Proceedings of the National Academy of Sciences of the United States of America*, 98, pp. 11925–30.

Heong, K. L., Escalada, M. M. and Mai, V. (1994) An analysis of insecticide use in rice: case studies in the Philippines and Vietnam, *International Journal of Pest Management*, 40, pp. 173–8.

Hilbeck, A., Baumgartner, M., Fried, P. M. and Bigler, F. (1998a) Effects of transgenic *Bacillus thuringiensis* corn-fed prey on mortality and development time of immature *Chrysoperla carnea* (Neuroptera: Chrysopidae), *Environmental Entomology*, 27, pp. 481–7.

Hilbeck, A., Moar, W. J., Pusztai-Carey, M., Filippini, A. and Bigler, F. (1998b) Toxicity of *Bacillus thuringiensis* Cry1Ab toxin to the predator *Chrysoperla carnea* (Neuroptera: Chrysopidae), *Environmental Entomology*, 27, pp. 1255–63.

Hopkins, D. W. and Gregorich, E. G. (2003) Detection and decay of the *Bt* endotoxin in soil from a field trial with genetically modified maize, *European Journal of Soil Science*, 54, pp. 793–800.

Hossain, F., Pray, C. E., Lu, Y., Huang, J., Fan, C. and Hu, R. (2004) Genetically modified cotton and farmers' health in China, *International Journal of Occupational and Environmental Health*, 10, pp. 296–303.

Huang, J., Rozelle, S., Pray, C. and Wang, Q. (2002) Plant biotechnology in China, *Science*, 295, pp. 674–7.

Hunter, M. D. (2000) Between hyperbole and hysteria. Entomological issues and the deployment of transgenic plants, *Agricultural and Forest Entomology*, 2, pp. 77–84.

IRRI (1994) *Filling the World's Rice Bowl* (Los Baños: International Rice Research Institute).

IRRI (1997) *Bt Rice: Research and Policy Issues. IRRI Information Series No. 5* (Los Baños: International Rice Research Institute).

James, C. (2003) Global status of commercialized transgenic crops: 2003, ISAAA Briefs No. 30, Ithaca, ISAAA.

James, C. (2004) Preview: global status of commercialized biotech/GM crops: 2004, ISAAA Briefs No. 32, Ithaca, ISAAA.

James, C. (2005) Preview: global status of commercialized biotech/GM crops: 2005, ISAAA Briefs No. 34, Ithaca, ISAAA.

Jesse, L. C. H. and Obrycki, J. J. (2000) Field deposition of *Bt* transgenic corn pollen: lethal effects on the monarch butterfly, *Oecologia*, 1125, pp. 241–8.

Jesse, L. C. H. and Obrycki, J. J. (2003) Occurrence of *Danaus plexippus* L. (Lepidoptera: Dandaidae) on milkweeds (*Ascepias syriaca*) in transgenic *Bt* corn agroecosystems, *Agriculture, Ecosystems and Environment*, 98, pp. 225–33.

Jeyaratnam, J. (1990) Acute pesticide poisoning: a major global health problem, *World Health Statistics Quarterly*, 43, pp. 139–44.

Jorg, R., Battini, M. and Bigler, F. (2003) Transgenic wheat with enhanced fungal resistance causes no effects on *Folsomia candida* (Collembola: Isotomidae), *Pedobiolgia*, 47, pp. 141–7.

Kalamaki, M. S., Harpster, M. H., Palys, J. M., Labavitch, J. M., Reid, D. S. and Brummell, D. A. (2003a) Simultaneous transgenic suppression of LePG and LeExp1 influences rheological properties of juice and concentrates from a processing tomato variety, *Journal of Agricultural and Food Chemistry*, 51, pp. 7459–64.

Kalamaki, M. S., Powell, A. L. T., Struijs, K., Labavitch, J. M., Reid, D. S. and Bennett, A. B. (2003b) Transgenic overexpression of expansin influences particle size distribution and improves viscosity of tomato juice and paste, *Journal of Agricultural and Food Chemistry*, 51, pp. 7465–71.

Kelemu, S., Mahuku, G., Fregene, M., Pachico, D., Johnson, N., Calvert, L., Rao, I., Buruchara, R., Amede, T., Kimani, P., Kirkby, R., Kaarla, S. and Ampofo, K. (2003) Harmonizing the agricultural biotechnology debate for the benefit of African farmers, *African Journal of Biotechnology*, 2, pp. 394–416.

Koskella, J. and Stotzky, G. (1997) Microbial utilization of free and clay-bound insecticidal toxins from *Bacillus thuringiensis* and their retention of insecticidal activity after incubation with microbes, *Applied and Environmental Microbiology*, 63, pp. 3561–8.

Koziel, M. G., Beland, G. L., Bowman, C., Carozzi, N. B., Crenshaw, R., Crossland, L., Dawson, J., Desai, N., Hill, M., Kadwell, S., Launis, K., Lewis, K., Maddox, D., McPherson, K., Meghji, M. R., Merlin, E., Rhodes, R., Warren, G. W., Wright, M. and Evola, S. V. (1993) Field performance of elite transgenic maize plants expressing an insecticidal protein derived from *Bacillus thuringiensis*, *Biotechnology*, 11, pp. 194–200.

Krattiger, A. F. (1997) Insect resistance in crops: a case study of Bacillus thuringiensis (Bt) and its transfer to developing countries, ISAAA Briefs No. 2, Ithaca, ISAAA.

Krattiger, A. F (1998) The importance of ag-biotech to global prosperity, ISAAA Briefs No. 6, Ithaca, ISAAA.

Krishnaveni, S., Jeoung, J. M., Muthukrishnan, S. and Liang, G. H. (2001) Transgenic sorghum plants constitutively expressing a rice chitinase gene show improved resistance to stalk rot, *Journal of Genetics and Breeding*, 55, pp. 151–8.

Kumar, P. A., Sharman, R. P. and Malik, V. S. (1996) The insecticidal proteins of *Bacillus thuringiensis*, *Advances in Applied Microbiology*, 42, pp. 1–4.

Lampe, K. (1995) Rice research: food for four billion people, *GeoJournal*, 35, 253–9.

Lee, T. J., Coyne, D. P., Clemente, T. E. and Mitra, A. (2002) Partial resistance to bacterial wilt in transgenic tomato plants expressing antibacterial Lactoferrin gene, *Journal of the American Society for Horticultural Science*, 127, pp. 158–64.

Liu, Y., Wang, G., Liu, J., Peng, X., Xie, Y., Dai, J., Guo, S. and Zhang, F. (1999) Transfer of *E. coli* gutD gene into maize and regeneration of salt-tolerant transgenic plants, *Science in China Series C: Life Sciences*, 42, pp. 90–5.

Londono, L.-M., Tarkalson, D. and Thies, J. E. (in review) In-field rates of residue decomposition and microbial communities colonizing residues do not differ between Cry1Ab *Bt* corn and non-transgenic isolines.

Lopez, R. and Ferro, D. N. (1995) Larviposiion response of Myioharus-Doryphorae (Diptera, Tachirridae) to Colorado Potato Beetle (Coleoptera, Chrysomelidae) larvae treated with lethal and sublethal doses of *Bacillus-thuringiensis* Berliner subsp plants, *Journal of Economic Entomology*, 88, pp. 870–4.

Losey, J. E., Rayor, L. S. and Carter, M. E. (1999) Transgenic pollen harms monarch larvae, *Nature*, 399, p. 214.

Lu, G. (2003) Engineering *Sclerotinia sclerotium* resistance in oilseed crops, *African Journal of Biotechnology*, 2, pp. 509–16.

Lucca, P. (1999) Genetic engineering approaches to improve the bioavailability and the level of iron in rice grains, paper presented at the General Meeting of the International Program on Rice Biotechnology, September 20–24, Phuket, Thailand.

Lukow, T., Dunfield, P. F. and Liesack, W. (2000) Use of the T-RFLP technique to assess spatial and temporal changes in the bacterial community structure within an agricultural soil planted with transgenic and non-transgenic potato plants, *FEMS Microbiology Ecology*, 32, pp. 241–7.

Marrone, P. G. and Macintosh, S. C. (1993) Resistance to Bacillus thuringiensis and resistance management, in P. F. Enwistle, J. S. Cory, M. J. Bailey and S. Higgs (eds), *Bacillus thuringiensis, an Environmental Biopesticide: Theory and Practice* (New York: Wiley), pp. 221–35.

Marroquin, L. D., Elyassnia, D., Griffiths, J. S., Feitelson, J. S. and Aroian, R. V. (2000) *Bacillus thuringiensis* (Bt) toxin susceptibility and isolation of resistance mutants in the nematode Caenorhabditis elegans, *Genetics*, 155, pp. 1693–9.

Marshall, M. W., Al-Khatib, K. and Loughin, T. (2001) Gene flow, growth, and competitiveness of imazethapyr-resistant common sunflower, *Weed Science* 49, pp. 14–21.

Mason, J. B. and Garcia, M. (1993) Micronutrient deficiency – the global situation, *SCN News*, 9, pp. 11–16.

McGaughey, W. H. and Whalon, M. E. (1992) Managing insect resistance to *Bacillus thuringiensis* toxins, *Science*, 258, pp. 1451–5.

Meier, M. S. and Hilbeck, A. (2001) Influence of transgenic *Bacillus thuringiensis* corn-fed prey on prey preference of immature *Chrysoperla carnea* (Neuroptera: Chrysopidae), *Basic and Applied Entomology*, 2, pp. 35–44.

Morse, S., Bennett, R. and Ismael, Y. (2004) Why *Bt* cotton pays for small-scale producers in South Africa, *Nature Biotechnology*, 22, pp. 379–80.

Munkvold, G. P., Hellmich, R. L. and Rice, L. G. (1999) Comparison of fumonisin concentrations in kernels of transgenic *Bt* maize hybrids and non-transgenic hybrids, *Plant Disease*, 83, pp. 130–8.

Murray, S. L., Thompson, G., Visser, A. and Berger, D. K. (2002) Transgenic potatoes (cv. Late Harvest) show increased tolerance to potato leafroll virus in greenhouse and field trials, *South African Journal of Science*, 98, pp. 97–101.

National Academy of Sciences (2002) *Transgenic Plants and World Agriculture* (Washington, DC: National Academy of Sciences).

National Academy of Sciences (2002) *Environmental Effects of Transgenic Crops: The Scope and Adequacy of Regulation* (Washington, DC: National Academy of Sciences).

Neppl, C. C. (2000) Management of Resistance to *Bacillus thuringiensis* Toxins, Environmental Studies Program, The University of Chicago, accessed at: http://camillapede.tripod.com/bapaper.html

Nguyen, H. T., Babu, R. C. and Blum, A. (1997) Breeding for drought resistance in rice: Physiology and molecular genetics considerations, *Crop Science*, 37, pp. 1426–34.

Nuffield Council on Bioethics (1999) *Genetically Modified Crops: The Ethical and Social Issues* (London: Nuffield Council on Bioethics).

Nuffield Council on Bioethics (2003) *The Use of Genetically Modified Crops in Developing Countries* (London: Nuffield Council on Bioethics).

Oberhauser, K. S., Prysby, M. D., Mattila, H. R., Stanley-Horn, D. E., Sears, M. K., Dively, G., Olson, E., Pleasants, J. M., Lam, W. K. F. and Hellmich, R. L. (2001) Temporal and spatial overlap between monarch larvae and corn pollen, *Proceedings of the National Academy of Sciences of the United States of America*, 98, pp. 11913–18.

Oliver, M. J., Ferguson, D. L. and Burke, J. J. (1995) Interspecific Gene Transfer (Implications for Broadening Temperature Characteristics of Plant Metabolic Processes), *Plant Physiology* 107, pp. 429–34.

Orr, D. B. and Landis, D. A. (1997) Oviposition of European corn borer (Lepidoptera: Pyralidae) and impact of natural enemy populations in transgenic versus isogenic corn, *Journal of Economic Entomology*, 90, pp. 905–9.

Ortiz-Garcia, S., Excurra, E., Schoel, B., Acevedo, F., Soberon, J. and Snow, A. A. (2005) Absence of detectable transgenes in local landraces of maize in Oaxaca, Mexico (2003–04), *Proceedings of the National Academy of Sciences*, 102, pp. 12338–43.

Palm, C. J., Schaller, D. J., Donegan, K. K. and Seidler, R. J. (1996) Persistence in soil of transgenic plant produced *Bacillus thuringiensis* var. *kurstaki* δ-endotoxin, *Canadian Journal of Microbiology*, 42, pp. 1258–62.

Peng, J., Richards, E., Hartley, N. M., Murphy, G. P., Devos, K. M., Flintham, J. E., Beales, J., Fish, L. J., Worland, A. J., Pelica, F., Sudhakar, D., Christou, P., Snape, J. W., Gale, M. D. and Harberd, N. P. (1999) 'Green revolution' genes encode mutant gibberellin response modulators, *Nature*, 400, pp. 256–61.

Phipps, R. H. and Park, J. R. (2002) Environmental benefits of genetically modified crops: Global and European perspectives on their ability to reduce pesticide us, *Journal of Animal Feed Science*, 11, pp. 1–18.

Picton, S., Gray, J. E. and Grierson, D. (1995) The manipulation and modification of tomato fruit ripening by expression of antisense RNA in transgenic plants, *Euphytica*, 85, pp. 193–202.

Pilcher, C. D., Obrycki, J. J., Rice, M. E. and Lewis, L. C. (1997) Preimaginal development, survival, and field abundance of insect predators on transgenic *Bacillus thuringiensis* corn, *Environmental Entomology*, 26, pp. 446–54.

Pimentel, D. S. and Raven, P. H. (2000) *Bt* corn pollen impacts on nontarget Lepidoptera: Assessment of effects in nature, *Proceedings of the National Academy of Sciences of the United States of America*, 97, pp. 8198–9.

Pingali, P. L. and Roger, P. A. (1995) *Impact of Pesticides on Farmer Health and the Rice Environment* (Norwell: Kluwer Academic Publishers).

Pingali, P. L. and Gerpacio, R. V. (1997) Living with reduced insecticide use for tropical rice in Asia, *Food Policy*, 22, pp. 107–18.

Ponsard-Sergine, Guitierrez, A. P. and Mills, N. J. (2002) Effect of *Bt* toxin (Cry1Ac) in transgenic cotton on the adult longevity of four heteropteran predators, *Environmental Entomology*, 31, pp. 1197–205.

Population Reference Bureau (2002) Washington, DC.

Powell, A. L. T., Kalamaki, M. S., Kurien, P. A., Gurrieri, S. and Bennett, A. B. (2003) Simultaneous transgenic expression of LePG and LeExp1 influences fruit texture and juice viscosity in a fresh market tomato variety, *Journal of Agricultural and Food Chemistry*, 51, pp. 7450–5.

Pray, C. E. and Naseem, A. (2007) Supplying crop biotechnology to the poor: opportunities and constraints, *Journal of Development Studies*, 43(1), pp. 192–217.

Pray, C. E., Huang, J., Hu, R. and Rozelle, S. (2002) Five years of *Bt* cotton in China – the benefits continue, *The Plant Journal*, 31, pp. 423–30.

Qaim, M., Cap, E. J. and de Janvry, A. (2003) Agronomics and the sustainability of transgenic cotton in Argentina, *AgBioForum*, 6, 41–7.

Qaim, M. and Zilberman, D. (2003) Yield effects of genetically modified crops in developing countries, *Science*, 299, pp. 900–2.

Quist, D., and Chapela, I. H. (2001) Transgenic DNA introgressed into traditional maize landraces in Oaxaca, Mexico, *Nature*, 414, pp. 541–3.

Raffensperger, C. and Tickner, J. (1999) *Protecting Public Health and the Environment, Implementing the Precautionary Principle* (Washington, DC: Island Press).

Rao, C. K. (2005) Misuse of science to spread suspicion and fear of biotechnology, Foundation for Biotechnology Awareness and Education (FBAE), accessed at: http://www.fbae.org/Channels/Views/Misuse_of_science.htm

Regierer, B., Fernie, A. R., Springer, F., Perez-Melis, A., Leisse, A., Koehl, K., Willmitzer, L., Geigenberger, P. and Kossman, J. (2002) Starch content and yield increase as a result of altering adenylate pools in transgenic plants, *Nature Biotechnology*, 20, pp. 1256–60.

Riddick, E. W. and Barbosa, P. (1998) Impact of Cry3A-intoxicated *Leptinotarsa decemlineata*, (Coleoptera: Chrysomelidae) and pollen on consumption, development, and fecundity of *Coleomagilla maculata* (Coleoptera: Coccinellidae), *Annals of the Entomological Society of America*, 91, pp. 303–7.

Riddick, E. W., Dively, G. and Barbosa, P. (1998) Effect of a seed-mix deployment of Cry3A-transgenic and nontransgenic potato on the abundance of *Lebia grandis* (Coleoptera: Carabidae) and *Coleomagilla maculate* (Coleoptera: Coccinellidae), *Annals of the Entomological Society of America*, 91, pp. 647–53.

Rissler, J. and Mellon, M. (1993) *Perils Amidst the Promise: EcologicalRisks of BD–HRCs in a Global Market* (Cambridge, MA: Union of Concerned Scientists).

Roush, R. T. (1994) Managing pests and their resistance to Bacillus-Thuringiensis. Can transgenic crops be better than sprays?, *Biocontrol Science and Technology*, 4, pp. 501–16.

Rowell, B., Jones, R. T., Nesmith, W., Satanek, A. and Snyder, J. C. (2001) Bacterial spot resistance, yield, and quality of bell and specialty peppers, *Hortscience*, 36, p. 563.

Saxena, D., Flores, S. and Stotzky, G. (1999) Insecticidal toxin in root exudates from *Bt* corn, *Nature*, 402, p. 480.

Saxena, D. and Stotzky, G. (2000) Insecticidal toxin from *Bacillus thuringiensis* is released from roots of transgenic *Bt* corn *in vitro* and *in situ*, *FEMS Microbiology Ecology*, 33, pp. 35–9.

Saxena, D. and Stotzky, G. (2001) *Bt* corn has a higher lignin content than non-*Bt* corn, *American Journal of Botany*, 88, pp. 1704–6.

Schuler, T. H., Poppy, G. M., Kerry, B. R. and Denholm, I. (1999) Potential side effects of insect-resistant transgenic plants on arthropod natural enemies, *Trends in Biotechnology*, 17, pp. 210–6.

Serageldin, I. (1999) Biotechnology and food security in the twenty-first century, *Science*, 285, pp. 387–9.

Sharman, M., Gilbert, J. and Chelkowski, J. (1991) A survey of the occurrence of the mycotoxin moniliformin in cereal samples from sources worldwide, *Food Additives and Contaminants*, 8, pp. 459–66.

Sharp, G. L., Martin, J. M., Lanning, S. P., Blake, N. K, Brey, C. W., Sivamani, E., Qu, R. and Talbert, L. E. (2002) Field evaluation of transgenic and classical sources of wheat streak mosaic virus resistance, *Crop Science*, 42, pp. 105–10.

Shelton, A. M. and Sears, M. K. (2001) The monarch butterfly controversy: scientific interpretations of a phenomenon, *The Plant Journal*, 27, pp. 483–8.

Singla-Pareek, S. L., Reddy, M. K. and Sopory, S. K. (2003) Genetic engineering of the glyoxalase pathway in tobacco leads to enhanced salinity tolerance, *Proceedings of the National Academy of Sciences of the United States of America*, 100, pp. 14672–7.

Siritunga, D. and Sayre, R. T. (2003) Generation of cyanogens-free cassava, *Planta*, 217, pp. 367–73.

Skerritt, J. H. (2000) Genetically modified plants: developing countries and the public acceptance debate, *AgBiotechNet*, 2, February, ABN 040, accessed at http://www.agbiotechnet.com/Reviews.asp?action = display&openMenu = relatedItems&ReviewID = 713&year = 2000

Smale, M. and DeGroote, H. (2003) Diagnostic research to enable adoption of transgenic crop varieties by smallholder farmers in sub-Saharan Africa, *African Journal of Biotechnology*, 2, pp. 586–95.

Snow, A. A. (2002) Transgenic crops – why gene flow matters, *Nature Biotechnology*, 20, p. 542.

Song, W. Y., Sohn, E. J., Martinoia, E., Lee, Y. J., Yang, Y.-Y., Jasinski, M., Forestier, C., Hwang, I. and Lee, Y. (2003) Engineering tolerance and accumulation of lead and cadmium in transgenic plants, *Nature Biotechnology*, 21, pp. 914–19.

Stanley-Horn, D. E., Dively, G. P., Hellmich, R. L., Mattila, H. R., Sears, M. K., Rose, R., Jesse, L. C. H., Losey, J. E., Obrycki, J. J. and Lewis, L. (2001) Assessing the impact of Cry1Ab-expressing corn pollen on monarch butterfly larvae in field studies, *Proceedings of the National Academy of Sciences of the United States of America*, 98, pp. 11921–36.

Stoger, E., Vaquero, C., Torres, E., Sack, M., Nicholson, L., Drossard, J., Williams, S., Keen, D., Perrin, Y., Christou, P. and Fischer, R. (2000) Cereal crops as viable production and storage systems for pharmaceutical scFv antibodies, *Plant Molecular Biology*, 42, pp. 583–90.

Stotzky, G. (1999) 1999 progress report. Toxins of *Bacillus thuringiensis* in transgenic organisms: persistence and ecological effects, Science to Achieve Results (STAR) Program, National Center for Environmental Research of the United States Environmental Protection Agency, accessed ar: http://es.epa.gov/ncerqa/progress/grants/97/ envbio/stotzky99.html

Stotzky, G. (2000) Persistence and biological activity in soil of insecticidal proteins from *Bacillus thuringiensis* and of bacterial DNA bound on clays and humic acids, *Journal of Environmental Quality*, 29, pp. 691–705.

Stotzky, G. (2002) Release, persistence, and effects of larvicidal proteins from *Bacillus thuringiensis* in soil, in *Annual Meeting of the American Society of Agronomy, Agronomy Abstracts* (S03-stotzky120831), Indianapolis, USA.

Tabashnik, B. E. (1994) Delaying insect adaptation to transgenic plants: seed mixtures and refugia reconsidered, *Proceedings of the Royal Society of London Series B: Biological Sciences*, 255, pp. 7–12.

Tabashnik, B. E. (1997) Seeking the root of insect resistance to transgenic plants, *Proceedings of the National Academy of Sciences of the United States of America*, 94, pp. 3488–90.

Tabashnik, B. E., Carriere, Y., Dennehy, T. J, Morin, S., Sisterson, M. S., Roush, R. T., Shelton, A. M. and Zhao, J. Z. (2003) Insect resistance to transgenic *Bt* crops: lessons from the laboratory and field, *Journal of Economic Entomology*, 96, pp. 1031–8.

Tapp, H. and Stotzky, G. (1995) Insecticidal activity of the toxins from *Bacillus-thuringiensis* subspecies *kurstaki* and *tenebrionis* adsorbed and bound on pure and soil clays, *Applied and Environmental Microbiology*, 61, pp. 1786–90.

Tapp, H. and Stotzky, G. (1998) Persistence of the insecticidal toxin from *Bacillus thuringiensis* subsp. *kurstaki* in soil, *Soil Biology and Biochemistry*, 30, pp. 471–6.

Taylor, D. C., Katavic, V., Zou, J., MacKenzie, S. L., Keller, W. A., An, J., Friesen, W., Barton, D. L., Pedersen, K. K., Giblin, M. E., Ge, Y., Dauk, M., Sonntag, C., Luciw, T. and Males, D. (2001) Field testing of transgenic rapeseed cv. Hero transformed with a yeast sn-2 acyltransferase results in increased oil content, erucic acid content and seed yield, *Molecular Breeding*, 8, pp. 317–22.

Thanavala, Y., Yang, Y. F., Lyons, P., Mason, H. S. and Arntzen, C. (1995) Immunogenicity of transgenic plant-derived hepatitis B surface antigen, *Proceedings of the National Academy of Sciences of the United States of America*, 92, pp. 3358–61.

United Nations Environment Program/Convention on Biodiversity/Subsidiary Body on Scientific, Technical, and Technological Advice (UNEP/CBD/SBSTTA) (1999) Consequences of the use of the new technology for the control of plant gene expression for the conservation and sustainable use of biological diversity, paper presented at the fourth meeting of the SBSTTA, Montreal, June.

United States Department of Agriculture, Biotechnology Regulatory Services, accessed at: http://www.aphis.usda.gov/brs/index.html

USDA (2003) Stakeholder workshop, Future Directions & Research Priorities for the USDA Biotechnology Risk Assessment Grants Program, Washington, DC, June 9–10, accessed at: http://www.csrees.usda.gov/funding/brag/pdfs/sum_gene.pdf

USDA (2004) Press Release No. 0033.04, USDA announces first steps to update biotechnology regulations, Washington, DC.

USDA (2005) Petitions of non-regulated status granted or pending by APHIS, accessed at: http://www.aphis.usda.gov/brs/not_reg.html

US EPA (2004) FIFRA Scientific Advisory Panel (SAP) Open meeting, June 8–10, accessed at: http://www.epa.gov/scipoly/sap/2004/june/charge.htmhttp://www.epa.gov/scipoly/sap/2004/june/charge.htm

Vaeck, M., Reynaerts, A., Hofte, H., Jansens, S., DeBeuckeleer, M., Dean, C., Zabeau, M., vanMontagu, M. and Leemans, J. (1987) Transgenic plants protected from insect attack, *Nature*, 328, pp. 33–7.

Van Rie, J. H., Van Mellaert, H. and Peferoen, M. (1992) Mechanism of Insect Resistance to *Bacillus thuringiensis* in *Plodia interpunctella* and *Plutella xylostella*, *ACS Symposium Series*, 505, pp. 191–9.

Vasil, I. K. (2003) The science and politics of plant biotechnology – a personal perspective, *Nature Biotechnology*, 21, pp. 849–51.

Venkateswerlu, G. and Stotzky, G. (1992) Binding of the protoxin and toxin proteins of *Bacillus-thuringiensis* subsp *kurstaki* on clay-minerals, *Current Microbiology*, 25, pp. 225–33.

Virginia Polytechnic Institute and State University, Information Systems for Biotechnology, Charts for Field Test Releases in the US, accessed at: http://www.isb.vt.edu/cfdocs/biocharts1.cfm

Volkmar, C., Kreuter, T., Richter, L., Al-Hussein, M. L., Jany, D., Schmutzler, K. and Wetzel, T. (2000) Ecological studies accompanying the cultivation of transgenic and conventional rape plants in the Central German region from 1996 to 1998, *Zeitschrift fuer Pflanzenkrankheiten und Pflanzenschutz-Journal of Plant Diseases and Protection*, 17, pp. 337–45.

Volkmar, C. M., Al-Hussein, M. L., Jany, D., Hunold, I., Richter, L., Kreuter, T. and Wetzel, T. (2003) Ecological studies on epigeous arthropod populations of transgenic sugar beet at Friemar (Thuringia, Germany), *Agriculture Ecosystems & Environment*, 95, pp. 37–47.

Warburton, H., Palis, F. G. and Pingali, P. L. (1995) Farmer perceptions, knowledge, and pesticide use practices, in P. L. Pingali and P. A. Roger (eds), *Impact of Pesticides on Farmer Health and the Rice Environment*, pp. 59–65 (Norwell: Kluwer Academic Publishers and Los Baños: International Rice Research Institute).

Way, M. J. and Heong, K. L. (1994) The role of biodiversity in the dynamics and management of insect pests of tropical rice-a review, *Bulletin of Entomological Research*, 84, pp. 567–87.

Welch, R. M., Combs Jr., J. F. and Duxbury, J. M. (1997) Towards a 'greener' revolution, *Issues in Science and Technology*, 14, pp. 50–8.

Welch, R. M. and Graham, R. D. (2004) Breeding for micronutrients in staple food crops from a human nutrition perspective, *Journal of Experimental Botany*, 55, pp. 353–64.

WHO (1999) *Malnutrition Worldwide* (Geneva: World Health Organization), accessed at: http://www.who.int/nut/malnutrition_worldwide.htm

Wilkinson, M. (1998) Benefits and risks of genetic modification in agriculture, *Nature Debates*, 1 October, accessed at: http://www.nature.com/nature/debates/gmfoods/gmfoods_frameset.html

Wraight, C. L., Zangler, A. R., Carroll, M. J. and Berenbaum, M. R. (2000) Absence of toxicity of *Bacillus thuringiensis* pollen to black swallowtails under field conditions, *Proceedings of the National Academy of Sciences of the United States of America*, 97, pp. 7700–3.

Wu, K. M. and Guo, Y. Y. (2005) The evolution of cotton pest management practices in China, *Annual Review of Entomology*, 50, pp. 31–52.

Wu, W. X., Ye, Q. F. and Min, H. (2004) Effect of straws from Bt-transgenic rice on selected biological activities in water-flooded soil, *European Journal of Soil Biology*, 40, pp. 15–22.

Ye, X., Al-Babili, S., Kloti, A., Zhang, J., Lucca, P., Beyer, P. and Potrykus, I. (2000) Engineering the provitamin A (beta-carotene) biosynthetic pathway into (carotenoid-free) rice endosperm, *Science*, 287, pp. 303–5.

Zangerl, A. R., McKenna, D., Wraight, C. L., Carroll, M., Ficarello, P., Warner, R. and Berenbaum, M. R. (2001) Effects of exposure of exposure to event 176 *Bacillus thuringiensis* corn pollen on monarch and black swallowtail caterpillars under field conditions, *Proceedings of the National Academy of Sciences of the United States of America*, 98, pp. 11908–12.

Zhang, H. X. and Blumwald, E. (2001) Transgenic salt tolerant tomato plants accumulate salt in foliage but not in fruit, *Nature Biotechnology*, 19, pp. 765–8.

Zhang, H. X., Hodson, J. N., Williams, J. P. and Blumwald, E. (2001) Engineering salt-tolerant *Brassica* plants: characterization of yield and seed oil quality in transgenic plants with increased vacuolar sodium accumulation, *Proceedings of the National Academy of Sciences of the United States of America*, 98, pp. 12832–6.

Zwahlen, C., Nentwig, W., Bigler, F. and Hilbeck, A. (2000) Tritrophic interactions of transgenic *Bacillus thuringiensis* corn, *Anaphothrips obscurus* (Thysanoptera: Thripidae), and the predator *Orius majusculus* (Heteroptera: Anthocoridae). *Environmental Entomology*, 29, pp. 846–50.

Zwahlen, C., Hilbeck, A. R., Gugerli, P. and Nentwig, W. (2003) Degradation of Cry1Ab protein within transgenic *Bacillus thuringiensis* corn tissue in the field, *Molecular Ecology*, 12, pp. 765–75.

Stealth Seeds: Bioproperty, Biosafety, Biopolitics

RONALD J. HERRING

I. Monsanto, Terminators, and the Mud

In July 2003, before a meeting in Palakkad district, south India, to memorialise the peasant leader Keraleeyan, agrarian activists told me and each other of threats from 'the terminator'. I explained to colleagues that the 'Monsanto/terminator/suicide-seed' narrative about *Bt* cotton tests was a product of a Canadian web site, NGOs and instrumental political dramaturgy, not reality. Politely, no one corrected me. At the meeting, a prominent public intellectual – P. A. Vasudevan – said that the current stage of historic agrarian struggles for which Kerala is justifiably well-known is only for 'the mud'; world agriculture will be controlled by Monsanto and Cargill, through biotechnology. Popular forces had learned how to struggle against and defeat the landlords, their *goondas* and police, but they did not know how to fight globalisation.[1] The emergence of a new farmer organisation in a district already intensively organised reflects his analysis: the *Deshiya Karshaka Samrakshana Samithi* (National Agriculturalist Protection Committee, DKKS) was formed to protect farmers from globalisation, one prominent manifestation of which was Monsanto and its terminator technology (DKSS, 2003: 2).

In this diagnosis, the DKSS joined a loose national movement linking external threats to agriculture to multinational corporations and biotechnology. Monsanto was (falsely) attributed ownership of a patent on terminator technology, and (more falsely) of unleashing this bio-cultural abomination on India through field trials of *Bt* cotton. Terminator technology would in theory permit engineering of plants that could not produce viable seeds, generating a biological dependence of farmers on firms beyond that of commercial arrangements.[2] Traditional practices of 'self-organising' agriculture would be replaced by dependency and cash nexus. This construction – linking multinational capital and globalisation to the cultural abomination of 'suicide seeds' – created a powerful political narrative. Paired with Dow Chemicals, which 'brought us Bhopal and Vietnam', Monsanto was said to be planning to 'unleash genetic catastrophes' (Asian Social Forum, 2003). Terminator imaginary persisted in Indian public discourse around biotechnology, demonstrating great power, pervasive reach and persistence in the face of disconfirming evidence. Biopolitics centered on this ideational construction cleaved coalitions seeking social justice and betterment of farmers.

Suicides by debt-ridden farmers – most notably in Warangal district, Andhra Pradesh – were linked explicitly by activists to globalisation of agriculture and new seed technologies.[3] Dependence of farmers on hybrid seeds of multinationals – rhetorically branded 'seeds of death' or 'suicide seeds' – linked field trials of transgenic cotton in 1998 to the opening wedge of terminator technology in India. *Seeds of Suicide* was 'dedicated to the farmers of India who committed suicide' (Shiva et al., 2000). Chapter one of Vandana Shiva's *Biopiracy: The Plunder of Nature and Knowledge* is entitled: 'Piracy through patents'. Dr Shiva's over-riding concern with biotechnology is 'the control of agriculture by multinational corporations' (Shiva, 1997: 91). Activists burning field crops of *Bt* cotton trials called their movement 'Operation Cremate Monsanto'. Terminator seeds were specifically banned by the government of India in 1998 in response, but the movement continued (Herring, 2005).

Monsanto's representative in India publicly refuted charges of suicide seeds: 'Since the so-called terminator gene does not exist today in any plant in any country in the world, the question of its involvement in the field trials currently on in India does not arise' (Dow Jones Agnet, 20 November, 1998). The Chairman of Mahyco, Monsanto's Indian business partner, BR Barwale emphasised that the seeds being tested had been approved by the Department of Biotechnology for trials and have 'nothing to do with the so-called terminator genes'. Monsanto's marketing director for India argued that the farmers' suicides had nothing at all to do with Monsanto's *Bt* seeds (which were not even on the market), but ironically might have been prevented by its technology (Mistry, 1998). With transgenic cotton, he said, farmers would have had less debt from pesticide purchase and less crop loss to bollworms – less poverty, fewer suicides. More obviously, since the hybrid transgenics under testing had been back-crossed into local cultivars, there were at least six generations of *Bt* cottons in India at time of the field tests: a clear indication that no terminator genes were in the cotton (Gosh, 2001: 11). If anti-*Bt*, pro-farmer organisers were interested, a countervailing discourse was readily available, and in high profile.[4]

That transgenics arrived in India through Monsanto yoked biotechnology to multinational capital. Had *Bt* cotton come through the Indian private or public sector – as is now unfolding with both Chinese and Indian varieties – biopolitics would have taken a different turn. But in the event, the technology was inseparable from the property, from multinational capital and from pressures on agriculture from globalisation.

II. Transgenic Cotton in India: From Robin Hood to Cottage Industry

The vicissitudes of cotton farming in India are extreme. Yields are among the lowest in the world, the area under cotton the highest (James, 2002). A bitter irony in the farmer suicides is that insecticides unable to protect crops – either because of insect resistance or dilution and adulteration – were sufficiently strong to kill farmers ruined by debts incurred to purchase pesticides. Sharad Joshi, leader of India's largest farmer organisation, *Shetkari Sanghatana*, illustrated the crisis through a single tragedy: 'It [2001] was a year of miseries for the cotton growers of Maharashtra. Neelkanth Mankar, a cotton grower in Yavatmal district, unable to face creditors, committed suicide'.

But in the neighbouring state of Gujarat, Joshi (2001) noted:

> Through a lucky stroke a nondescript seed company managed to play Robin Hood and smuggle into Gujarat one line of anti-bollworm gene. For three years nobody noticed the difference and then came the massive bollworm rampage of 2001.

There was no way to distinguish transgenic lint or seeds from their appearance, but fields indicated the difference:

> Gujarat saw all its traditional hybrid cotton crop standing devastated, side-by-side the Bt-gene crops standing resplendent in their glorious bounty. The Government was upset and ordered destruction and burning of the bountiful crop.[5]

Neither the 'Cremate Monsanto' movement nor the government bio-safety regulators had noticed the transgenic cotton. Monsanto's partner Mahyco did, and complained to the Genetic Engineering Approval Committee in Delhi. The seeds had been sold as Navbharat 151. They were originally detected in Gandhinagar district of Gujarat in six locations. Press reports typically said that the extent of coverage was over '10,000 hectares' [or sometimes '10,000 acres'] in extent; this number was used in parliament, and has crept into academic accounts. Both suspiciously parallel estimates are groundless. Precisely because these were underground seeds, no one knows exactly the extent or location of plantings. The GEAC investigated during the last week of September, and found Mahyco's charges to be true. The cotton contained the Cry1Ac gene in the construct of Monsanto. The head of Navbharat, Dr D.B. Desai, was summoned to Delhi for the 9 October meeting of GEAC to explain evident violation of biosafety regulations: no transgenic crop could be planted without officially sanctioned tests and final approval by the GEAC. Dr Desai did not appear; his counsel argued that Navbharat did not know there was a *Bt* gene in its seeds. It was sold as a hybrid registered by the government of Gujarat. On 12 October the GEAC met again and ordered the Gujarat government to act. Here regulation encountered the light-switch problem: a switch is thrown but it is not connected to anything. Gujarat state had not set up a biosafety committee, as all states were mandated to do; the GEAC itself has no police powers.[6] A case was registered against Dr Desai for violation of the environmental protection act (1986; Rules, 1989) that regulates transgenic organisms.

The GEAC ordered not only burning of the crop, as the farmers' organisation notably scorned in their resistance ('we will burn with our crops in our fields'), but also mandated (1) a public warning in regional newspapers; (2) retrieval and destruction of seeds from farmers' houses and ginning mills; (3) collection of lint, which was to be stored in steel containers and transported to the Central Institute of Cotton Research in Nagpur for testing; (4) procurement of all yet-to-be harvested crops from farmers; (5) uprooting and burning of the standing crop; and (6) measures to sanitise the fields.

These orders were not carried out, marking an unambiguous political victory for farmers over regulators. Appropriately enough, Gujarat's decision to do nothing to enforce the order was announced in Delhi by the Union Minister for Textiles, Kashiram Rana immediately after a meeting with the Chief Minister of Gujarat, Narendra Modi. Delhi has a deep national interest in cotton production. Textile Minister Rana could see nothing wrong with the controversial seeds; he reasoned that since the *Bt* seeds reduced pesticide use and were favoured by farmers, opposition must be coming from the pesticide lobby. Gujarat state's Minister of Agriculture, Purshottam Rutala, made the same argument: who other than the pesticide lobby had an interest in depriving farmers of a beneficial technology? The consensus, across state and national governments, and eventually the GEAC itself, articulated by Secretary of the Department of Biotechnology Manju Sharma, was that the 'interests of farmers' would not be harmed. Union Agriculture Minister Ajit Singh said delay in approval of the *Bt* seeds would be 'inexcusable', given the damage bollworms were doing to Indian cotton. The consequence was that farmers would continue to grow, save and breed transgenic cottons despite clear evidence of violation of India's environmental protection laws and biosafety procedures.[7]

Failure of the GEAC order indicated that the biosafety regime was out of the hands of regulatory authorities and scientists in 2001 and in the hands of politicians and organised farmers. The central government provided little political support for the hard line originally adopted by the GEAC. Regulation in a federal political system necessitates balancing central and provincial powers. Agriculture is constitutionally a state subject in India; Gujarat's position was clear. Likewise, the government of the neighbouring state, Maharashtra, was actively pressing for immediate approval of transgenic cotton; Maharashtra is the geographic base of the *Shetkari Sanghatana*. The Maharashtra Minister of Agriculture Rohidas Patil announced on 11 December 2001 in the State Legislative Council that the state would make *Bt* cotton seeds available to farmers from 1 January onwards – three months before the Genetic Engineering Approval Committee in Delhi approved the technology.[8] On 25 March 2002, farmer representatives led by Sharad Joshi – a member of the *Kisan* [agriculturalist] *Coordination Committee* (KCC) – threatened to launch a civil-disobedience movement if *Bt* cotton were not approved by Delhi. KCC representatives from cotton-growing states across India – Gujarat, Maharashtra, Punjab and Andhra Pradesh – rallied for immediate approval, and threatened to cultivate transgenic varieties whether or not the government approved.[9] The following day, 26 March, the GEAC approved three varieties of the Mahyco–Monsanto *Bt* cotton, making India the 16th nation in the world to certify a genetically engineered plant for commercialisation, albeit provisionally and in the face of fierce opposition.

Though the officially sanctioned varieties – Bollgard-MECH 12, 162 and 184 – are much more expensive than other hybrid seeds, they sold well and the technology has been licensed to other seed firms. But competition from the stealth seeds and their progeny has been vigorous. On 16 August 2004, Union Agriculture Minister Sharad Pawar stated in parliament that the underground seed market was flourishing and alarming. No one knows how many 'Robin Hoods' – in the construction of Sharad Joshi – are active in rural India, but it is clear that a cottage industry of transgenic pocket breeding has grown up around descendants of the original Navbharat 151 seeds.

There is a rough sequence to this evolution. After Navbharat 151 was banned in 2001, it became scarce, though not impossible to find; farmers sought out its parent lines for breeding.[10] Some saved their seeds after ginning and sold or exchanged or replanted the F2 generation of Navbharat 151, which was no longer available legally in the market. There is deep irony in this spread of the vigorous off-spring of the 'suicide seeds'. These seeds are called 'loose seeds', straight from the ginning mill, unpackaged and unbranded.[11] They may express less *Bt* endo-toxin, but, according to farmers in Gujarat, offer reasonable protection at a very low price.[12]

Nevertheless, many farmers worry about hybrid vigour in cotton and distrust F2 seeds. Given the high cost of official seeds and the scarcity of the very effective NB151, farmers themselves began breeding new transgenic hybrid varieties. They use Navbharat 151 seeds for the male contribution and a local variety especially well-suited to their agronomic conditions as female. From this process, a new Gujarati word has been hybridised: 'Navbharat variants'. There are un-counted branded and packaged *Bt* variants in circulation: Luxmi, Kavach, Viraat, Sarathi, Vaman, Agni, Rakshak, Maharakshak, Kranti, the generic Kurnool *Bt* and simply '151' playing on the original Navbharat 151 variety, among many others. These locally back-crossed hybrids made by farmers are sold by local merchants, who sometimes guarantee the seeds, to distinguish them from the many spurious seeds claiming *Bt* status in the market. To indicate transgenic character semi-covertly, some variants have printed 'BesT Cotton Seed' on the package. There are as well farmer-to-farmer transactions of modified and crossed transgenic seeds with no name.[13] The decision matrix of farmers facing this volatile seed market is complex, as there is great agronomic and cost variation, but farmers in Gujarat have largely naturalised stealth *Bt* as part of their time-tested decision matrix (Roy et al., 2007).

The tension between official seeds and stealth seeds is dynamic, in continuous flux as seed firms jockey for position and new varieties are approved – or de-certified for certain areas, as were two MMBL varieties in Andhra Pradesh in 2005. When the GEAC refused approval for Mahyco–Monsanto's new variety MECH-915 for growing in north India, an advertisement soon appeared in a prominent Hindi daily advertising farmer-grown transgenic seeds available by calling the cell phone of one Piyush Patel. Sonu Jain, relating the story in the Indian Express (Delhi) of 20 April 2002, quotes Patel as saying 'If I live in Gujarat and go to Shimla, I will not die, so the same way these seeds developed in Gujarat will grow'. Patel's *Bt* seeds sold at Rs 555 per packet of 500 grams, less than a third the officially approved Mahyco–Monsanto seeds' price. By June of 2005, I found that the range of locally-hybridised transgenic cotton cultivars in Gujarat sold for Rs 250–700 per packet (roughly enough to plant one acre); F2 transgenic seeds were selling for Rs 10 for the same weight packet. Jayaraman (2004) cites 'industry sources' as estimating that more than half of the transgenic cotton in India comes from unapproved varieties; my discussions with Gujarati seed producers suggest a much higher figure for that state. Data from Navbharat Seeds (personal communication, October 2005) indicate that on an all-India basis, about 34 per cent of the cotton seed packets sold are transgenic, of which 9 per cent are legal and 25 per cent stealth. Yet these estimates apply only to packaged and branded stealth seeds, not to loose seeds. The ratio is highest in the North Zone (Punjab, Haryana, Rajasthan): 107,000 packets of legal transgenic seeds to 1,170,000 illegal packets, together accounting for about a third of cotton acreage.

Dr R. P. Sharma of the GEAC believes this to be a temporary phenomenon: farmers will eventually choose *Bt* cultivars from trusted seed companies and abandon the stealth transgenics: the current state of affairs simply reflects the fact that 'scarcity breeds corruption' (personal communication, New Delhi, 27 June 2005). Scarcity was induced by banning the extremely effective Navbharat 151.

Where did stealth *Bt* originate? Standard accounts have emphasised Gujarat and the Robin Hood construction of Sharad Joshi, but the seeds were available in other locations as well. In early November, 2001, 460 acres of seed farms were found to be producing transgenic cotton in the Kurnool and Mahabubnagar districts of Andhra Pradesh. Chairman of the GEAC A. M. Gokhale claimed that Navbharat was selling transgenic cotton under brand names of *Vijay, Digvijay* and *Jay* in the state. It was widely believed that other stands of parent seeds were growing in the Punjab and Maharashtra.[14] As the matter is *sub judice*, it is difficult to come to definitive answers. Dr N. P. Mehta, the breeder who developed NB 151, has stressed serendipity: his breeding strategy was directed toward early flowering for a crop rotation with sugar cane in Surat; one of his scouts came across an unusually bollworm-resistant cultivar, which proved vigorous upon crossing, producing NB 151 (personal communication, Ahmedabad, 20 June 2005; see also Mehta, 2005). Both Mahyco–Monsanto Biotech Limited and the GEAC believe that the transgene was appropriated from MMBL stock, possibly from the field trials mandated by the GEAC. No definitive answer is now possible; sequestering of *Bt* technology in either the regulatory or property sense has not proved administratively or politically feasible.

Farmers in Gujarat have embraced the agrarian anarcho-capitalism of stealth seeds. This outcome is not in the interest of seekers of innovator rents through state protection of intellectual property – in this case both Monsanto and Navbharat. The long development costs and time, mandated by a biosafety regime that is becoming global, are estimated to be about US$8 million for MMBL; these costs put official seeds at a price disadvantage. In Indian law, there is no restriction on farmer-to-farmer exchange of seeds. Farmer-generated transgenic hybrids that I have seen reference on the package the Indian Seed Act of 1966, Section 24, which protects this right explicitly. Lacking capacity to use patent protection on intellectual property, Mahyco–Monsanto favours strict regulation. In these circumstances, regulatory restriction of official seed varieties confers property-like rights unavailable through patents. If only its seeds are legal, licensing of the technology confers short-term monopoly rents on MMBL. Advocates for Navbharat accuse the GEAC of exactly this: market-rigging through biosafety regulation creates de facto bioproperty rights.

Bt cottons have been in the field too short a time for definitive assessment of either biological or economic success across so varied an agro-ecology as India; results vary with seasonal variations of pests, weather and local agronomics (Rao, 2004b). More systematic data are being compiled, much of it rejected by opponents for being tainted by Mahyco–Monsanto sponsorship, and none long-term enough for robust conclusions.[15] But unless one thinks farmers irrational, there is strong evidence of *Bt* technology's effects in India being similar to those in China, where both the Monsanto and public sector versions of *Bt* cotton have been adopted rapidly by small farmers (James, 2002; Pray et al., 2002). Farmers have done this for higher yields, less pesticide application against bollworms, and higher profits.

Opponents continue to proclaim 'the disaster wrought by *Bt* cotton in India'.[16] This construction is empirically groundless but strategically partisan. Opponents somehow fail to distinguish *Bt* technology from specific cultivars. The reason that there are between 200 and 300 cultivars of cotton grown in India is that cultivars have specific agronomic characteristics: no single variety will do well in all places in all seasons. Farmers know this; many who claim to represent them do not.

There exists no credible evidence of the 'failure' of *Bt* cotton technology in India. There is great variation in performance of cotton cultivars, both *Bt* and non-*Bt*. Reasons for variance are not always discernable, either by farmers or researchers, since there are many unmeasured variables in complex interactions – local climate, soil chemistry, pest variance, water timing, nutrients. Second, spurious seeds are pervasive: some varieties sold as *Bt* are not; some farmers honestly but mistakenly believe their *Bt* crop has failed. Third, there are demands for financial compensation from Mahyco–Monsanto and the government for *Bt* crop failure; there is material incentive to claim poor results. Fourth, the MMLB varieties are clearly not the best germplasm for insertion of the *Bt* gene: many farmers seem to prefer Navbharat and other varieties, legal and illegal. New firms are vigorously entering the market as licensees of MMLB's technology, but with different cultivars. Finally, and most important, none of the claims of failure compare two isogenic varieties, one with and one without the *Bt* gene, to assure control of varietal characteristics (Naik et al., 2005). Rather, all disadvantageous variance across over time and space – which will be extreme in India – is attributed to the *Bt* gene, constructing a biological absurdity. The *Bt* gene codes for a single protein, the Cry1Ac; there is no reason for production of that protein – lethal to *Lepidoptera* – to cause staples to shorten or leaves to wilt.[17]

Perhaps the most careful controlled study to date is of MECH-162 (Bambawale et al., 2004) – compared to the isogenic *non-Bt* MECH 162 and a conventional hybrid. This study used a participatory field trial to test meaningfully paired varieties with and without integrated pest management. Consistent with other studies, *Bt* plants required half the sprayings of other plants, experienced less bollworm infestation, and, somewhat surprisingly, reduced attacks of sucking pests and two natural enemies of cotton. With integrated pest management, the *Bt* variety recorded a yield of 7.1 q/ha and a net return of Rs 10,507/ha. Damage to fruiting bodies was much less with *Bt* plants, which would account for the premium some *Bt* farmers receive for their lint in the market. The authors concluded: '*Bt* Mech-162 used in an IPM mode resulted in highest yields and economic gains to the farmers; pesticide consumption was also reduced' (Bambawale et al., 2004: 1633). *Bt* technology and improved agro-ecological practices each contributed to superior outcomes in this controlled study.

Indian farmers are experimenting widely with *Bt* cottons, both official and unofficial; both categories are multiplying rapidly. Neither duped nor passive puppets of multinational monopolists, cotton farmers continue the endless struggle of agriculture against insects, with a new tool. Their techniques continue traditions of seed saving, seed exchange, and seed experimentation (Gupta and Chandak, 2005; Roy et al., forthcoming). But then what of the elaborate bio-safety regime that is to prevent genetic anarchy? A commission on bio-safety regulations headed by eminent agricultural scientist M. S. Swaminathan concluded in under-stated officialese: 'Public regard and satisfaction for the regulatory systems currently in place are, to say the least, low' (Bagla, 2004). Dr Swaminathan himself was quoted in *Nature Biotechnology* (2004, 22: 1334): 'illegal proliferation of GM varieties must cease or else the bio-safety regulations will be rendered meaningless'. Yet A. K. Dixit, Director of Agriculture for Gujarat, said: 'It is impossible to control something at this large a scale. When we go to the fields, we become targets for trying to take away a beneficial technology from farmers' (Seed Quest, 2003a).

III. Representing Farmers: Biopolitics and Credibility

Coalitions for the poor are difficult to conjure, harder to create and sustain. The experience of a single farmer, little noted at the time, proved diagnostic of the representational problems of the suicide-seed coalition.

In the beginning of Operation Cremate Monsanto in 1998, the *Karnataka Rajya Raitha Sangha* (KRRS) organised the burning of *Bt* cotton crops on two test plots in Raichur and Bellary districts of Karnataka state. The KRRS is a farmer organisation specifically dedicated to protecting Indian farmers – and India – from globalisation, personified by Monsanto (Herring, 2005; Omvedt, 2005). Cremated test plots received international attention; failure to cremate a third trial plot as planned, in Adur village, Haveri *taluq*, went virtually unnoticed. Yet this episode proved diagnostic of difficulties facing any coalition that seeks to be both pro-poor and anti-transgenic. The farmer who owned the plot, *Shri* Shankarikoppa Mahalingappa, was a member of the KRRS, but was unconvinced by its leadership.

He stressed in a conversation with me that he could not depend on the movement's explanation of the new technology, but had to see for himself. He asked for and received police protection for his test crop. *Shri* Mahalingappa had sequestered 100 seeds from biosafety trials on his land; he planted them and found germination by 95 of the 100 'suicide seeds'. The terminator construction dissolved before the organic empiricism of a farmer concerned about pesticide costs and farm income. Shankarikoppa Mahalingappa concluded that 'terminator' talk was 'just propaganda'.[18]

Mahalingappa's dissent illustrated not only the political problem of cultural distance of movement elites from farmers, but also a different view of science. His epistemology was inductive and grounded rather than deductive and derivative. Against the notion that farmers were duped by corporate propaganda, he said: 'no one's word can be taken; you have to see for yourself... farmers must be convinced personally that a crop is beneficial; only the farmers can decide'. His perspective is simply prudent farm management. Mahalingappa now buys Bollgard MECH-162 and finds it profitable despite high seed cost; he spends less on chemicals and gets higher yields because of reduced bollworm damage. Why does he not use the cheaper stealth seeds so popular in Gujarat? He trusts the quality of the official seeds and does not trust the F2 generation of hybrids. The additional cost of seeds is more than compensated by additional revenue: 'it makes money for me'. And what of bio-safety? 'The genes cannot be taken back', but he is not worried about any bad effects, because he's seen none. The foliage seems not to harm insects other than bollworms, nor mammals; he could perceive no threat from the new seeds.

How could there be such a disjuncture between farmer experience and the position of public intellectuals and NGOs that claim to represent the farmer? In a typical press account of Operation Cremate Monsanto:

> 'Farmers,' it is said, led by Prof. Nanjundaswamy of the Karnataka Rajya Raitha Sangha, attacked the Monsanto seed farm near Malaldguda in Raichur district and destroyed the cotton crop in order to protect Indian farmers from the dreaded Terminator Gene. The campaign was repeated in Bellary district and other 'farmers' in Andhra Pradesh did the same thing. Speaking in the name of all of them, Prof. Nanjundaswamy has vowed to repeat it in Maharashtra and Punjab.[19]

This press report notes the ambiguity in characterising 'farmers' active against transgenics. It is now clear that the technology was spreading, not terminating, because farmers were finding ways to obtain illegal seeds and eventually ways to breed transgenic hybrids themselves.

Despite persistent reports of catastrophic failure, and movements to de-certify Mahyco–Monsanto varieties (successful in Andhra Pradesh in 2005), demand for official seeds continues to grow, at times exceeding supply, even at prices triple or more those of other hybrids. Sales of the MMBL Bollgard seeds increased six-fold in 2004–05 over the previous year. Moreover, 21 varieties produced by Indian firms that had sublicensed the *Bt* technology had been approved for planting as of June 2005. Approvals of new *Bt* cultivars continue to come from the GEAC, including the

Bollgard II technology. Farmers continue to scramble for *Bt* seeds but are price-conscious, often opting for stealth seeds over official seeds.[20]

Yet 'the failure of *Bt* cotton' – in an agronomic and economic sense – continues to fill press reports. Refusal to believe that farmers might have some valid experience on which to base a preference for transgenics is diagnostic of representational problems in rural movements headed by metropolitan elites. This epistemological fundamentalism – privileging international public intellectuals over farmers' intelligence – creates a wedge in possible coalitions for the poor.[21] If one grants that Indian farmers and seed firms might be commercially rational, the continuing spread of the Cry1Ac technology must indicate biological and micro-economic success. When I presented this argument to activists against biotechnology in Palakkad district, the response was that farmers had been duped or coerced, falling into the trap of the monopolist Monsanto. Vandana Shiva et al. wrote in the conclusion to their 1999 article in the prominent Indian social-science journal *Economic and Political Weekly* that 'the promotion of genetic engineering by corporations like Monsanto can only be based on dictatorial, distorted and coercive methods'.

If biological and economic failure constitute one strand of resistance to a technology many farmers find beneficial, a second strand is nationalism: biotechnology is unacceptable for its foreign origin and implications for neo-colonial control. Responding to a BBC story that portrayed the farmers of Gujarat as clever pirates of Monsanto's intellectual property – as implied by Sharad Joshi's Robin Hood characterisation – the Research Foundation for Science Technology and Environment (headed by Vandana Shiva) rebutted:

> This rumour about piracy is initiated by Monsanto whose *Bt* cotton has totally failed throughout the length and breadth of the country and to divert attention of the public and policy makers from the failure of its genetically engineered seeds, Monsanto is trying to focus on the outstanding success as unjust and illegal of an indigenously bred cotton variety.[22]

'Indigenously bred', perhaps, but equally transgenic – with the Monsanto Cry1Ac gene.[23] Charges of genetic pollution and the saving construction of indigenous breeding provide an exit from the *cul de sac* of the suicide-seed narrative. Just as the 'Monsanto terminator' construction came from a website in Canada, the revisionist response of anti-transgenic forces bears a striking resemblance to Canadian websites defending Percy Schmeiser. The construction of Canada's most famous biopirate – in Monsanto's view – or victim of biological pollution – in the opposition's view – deploys the defence of gene flow. Mr Schmeiser's defence in the case he lost repeatedly, finally in the Supreme Court of Canada in June of 2004, is that he did not illegally appropriate intellectual property for transgenic canola seed production on his farm, but suffered pollution of his fields by Monsanto's plants.[24] The revisionist reconstruction of RFSTE valorises Gujarati farmers' 'indigenous' breeding, but the major trait responsible for success – resistance to bollworms through the Cry1Ac endotoxin – is reconstructed as incidental and accidental, a result of biological pollution from Monsanto.

The *Bt* controversy began with a nationalist (and Gandhian) theme of resisting foreign threats to India (Herring, 2005). Attacks on Monsanto continue, but *Bt* technology itself has been naturalized. Leader of the farmer organisation *Khedut Samaj* in Gujarat, Bipin Desai, charged that the 'failing' – and foreign – *Bt* technology has been approved by the government, but the successful home-grown *Bt* variety (Navbharat 151), has not. Vice President of the organisation, Labshankar Upadhyay, added: 'The BJP talked about *Swadeshi* [self-reliance]. But it promotes a foreign company at the cost of an Indian firm. And we [farmers] stand to lose'. The RFSTE construction enables a nationalist and populist attack on the biosafety regime: Delhi officially allowed the import and testing of, and then certified, a foreign incarnation of *Bt* technology, but filed a court case against an indigenous plant breeder. The stealth seeds are held to be legitimate in a way the government seeds – which are much more expensive – are not (Mehta, 2005: 77–84, 108–15, 137–9).

Stealth seeds were decisive in turning Indian farmers into a pressure group for transgenic cotton. Farmers encountered transgenics through the mediation of rumour, misinformation, and contradictory official signals (Parmar and Vishvanathan, 2003). The suicide-seed narrative was dramaturgically engaging; but external political intermediation loses power over time unless both framing and objectives resonate with a threshold level of farmers' experience. The micro-economic and biological success of *Bt* technology outweighed the more indirect, distal and hypothetical arguments about foreign control and dangerous genes: oppositional discourse outran agricultural interests. Framing is bounded, ultimately, by interests, even if loosely at any particular time.

The triumphalism of the industry relies on the success and acceptance of *Bt* technology, but may be premature. For all the romanticisation of local knowledges and the village *Volk*, it is not clear that sons of the soil always know best (Herring, 2000). Farmers adopted insecticides not after considering the science and social externalities, but rather from an interest calculation: to protect their crops. Farmers understand the dead-end nature of the pesticide alternative to *Bt* cotton; they are little concerned about biosafety. Uncertainties are externalised to society as a whole, and projected onto an elusive 'biosafety regime'.

IV. Stealth Soy in Brazil

India's *Bt* cotton story suggests rethinking core assumptions in the standard narrative of seeds and states in developmentalist biotechnology. Is India an aberrant case?

Brazil parallels India's experience in three ways: first, the developmental state supported biotechnology. Looking outward to competition in the international economy, Brasília pressed for indigenous and collaborative development of biotechnology. In both nations, support for transgenics from the centre was challenged in the periphery, by forces in civil society. Second, despite considerable political conflict around the bio-safety regime, regulations were rendered irrelevant by the capacity of farmers to find and breed transgenic seeds underground. Third, Monsanto provided a symbolic target and rallying point for opposition at the same time that farmers were finding ways to avoid Monsanto's property claims. Rather than monopoly power, Monsanto encountered not only competition from stealth seeds, but the same difficulty in policing underground seeds faced by the regulatory state. Intellectual property rights fiercely debated in political and policy circles proved illusory when seeds went underground.

Brazil's economy depends on agricultural exports, of which soy is a major component. As competing nations approved transgenic soy much sooner, Brazilian farmers faced cost disadvantages vis-à-vis farmers in Argentina and the United States – the other two of the top three soy exporters.[25] But global imperatives were contradictory in the case of soy. Global market segmentation created a niche for 'GMO-free' soy when opposition to transgenics surfaced in Europe and Japan after 1997; the cost-of-production disadvantage Brazilian farmers faced might be off-set by the premium for 'GMO-free' soy (Paarlberg, 2001: 67) – if exports could be reliably segregated and certified.

Introduction of transgenic crops to Brazil proceeded in what David Hathaway (2002) called 'a vacuum in the exercise of authority'. The state was divided against itself. The political struggle for legitimacy to rule on transgenics engaged different levels of the court system, divisions within the federal government and disputes between the states and Brasília.

Brazil's Biosafety Law was passed by the National Congress in 1994 and issued in January 1995. It granted authority over genetically-engineered organisms – both pharmaceutical and agricultural – to a National Technical Biosafety Commission (CTNBio), consisting of academics, a broad range of government ministries – from Health to Agriculture to Education to External Affairs and representatives of industry and civil society. CTNBio was to be the federal agency responsible for the development and implementation of biosafety policies, ethical codes, and evaluation of risk and environmental threat. Brazil's 1990 Consumer Defense Code mandates labelling of all products to inform consumers of characteristics they have a right to know about. Transgenics are also regulated by the Industrial Property Code of 1996, which explicitly responded to new requirements of WTO's Trade-Related Aspects of Intellectual Property Rights (TRIPS), granting legal protection to inventions related to pharmaceuticals, food processes, and biotechnology.[26]

CNTBio approved commercial release of three varieties of Monsanto's Round-Up Ready (RR) soybeans in September 1998; the representative on the Commission from the Institute for Consumer Defense (IDEC) resigned in protest. Both Greenpeace and IDEC filed legal appeals; in 1999, an injunction (*ação cautelar*) was issued by a federal judge, Antonio Prudente. Commercial cultivation of RR soybeans was legally banned, on grounds that they had not been adequately tested for human health and environmental impacts. More fundamentally, the authority of CTNBio was challenged by a suit in 2000, seeking an injunction against decisions on transgenic-crop releases before the government formulated rules for assessing biosafety. A third decision, issued 14 February 2002 in response to a suit brought by the federal Public Ministry, suspended all further field tests of 'biopesticide' transgenics until Brazil's pesticide legislation is enforced. These decisions combined to produce a 'judicial moratorium' on the commercial release of transgenic crops in Brazil.[27]

Opposition to regularisation of transgenic soy was led by Greenpeace Brazil and IDEC, later joined by a section of the Ministry of Environment – the Brazilian Institute for the Environment and Renewable Natural Resources (*Instituto Brasileiro do Meio Ambiente e dos Recursos Naturais Renováveis* or IBAMA) – and the umbrella *Campanha Por um Brasil Livre de Transgênicos*.[28] As in India, opposition targeted corporate globalisation, and Monsanto in particular. At a 'People's Tribunal' – a functional equivalent of the *Beej Panchayat* of India – Monsanto was convicted of producing seeds sold to Argentina in 1995, subsequently smuggled into Brazil, that 'endangered the environment, biodiversity, human health, the country's agricultural genetic wealth, and the Brazilian economy'.[29]

The federal government held that uncertainty was discouraging private investment in biotechnology, both foreign and domestic, as well as development within the sophisticated public sector. In July of 2000, six senior cabinet members of then-President Cardoso's government (including those for environment and health) signed a 'manifesto' supporting both transgenic crops and the authority of the federal CTNBio. Brazil's National Academy of Science as well as the Brazilian Genetics Society expressed approval of transgenic crops (Poddar, 2004: 30). In December of 2000, legislation granted CTNBio more robust authority to circumvent appeals against its legal competence.

Much of the legal manoeuvring in elite circles was irrelevant to what was growing in the ground. A report issued in 2002 by Liberal Party parliamentary Deputy Ronaldo Vasconcellos exposed widespread planting of glyphosate-resistant soy in Brazil despite the legal ban (Osava, 2002). Seeds had been smuggled from Argentina, and perhaps other neighbours, since at least 1997, and were being reproduced and re-crossed by Brazilian farmers. According to his report, transgenic seeds accounted for up to two-thirds of the crop of Rio Grande do Sul – Brazil's southernmost state; there was evidence as well of seeds' spreading north.[30] Vasconcellos was suspicious of his own numbers and underscored the political consequences of stealth seeds for biotechnology data. He suggested that disseminating exaggerated figures had a political rationale: to present transgenic soy as a *situación de facto (fait accompli*), seemingly irreversible. Indeed, some opponents saw stealth seeds generally as a 'contamination strategy'.[31] As in Gujarat, the actual area planted to underground seeds in Brazil was unknown. Officially, Rio Grande do Sul maintained its own anti-transgenic policy, and stopped most federally authorised field trials in the state. The state government claimed incapacity to halt widespread smuggling from Argentina – though by this time most seeds seem to have been indigenously produced.

The illegal-immigrant seeds themselves had not necessarily been purchased through legal channels. Monsanto was refused patent protection for Round-Up Ready (RR) soybeans in Argentina in 1995. Argentina's largest seed firm, NIDERA, purchased a former Monsanto partner, Asgrow Argentina, and thus acquired RR seeds, which were sold to farmers without technology fees. In 1999, president of the National Seed Institute (INASE) of Argentina, Adelaida Harries, estimated that '25–30 per cent of the soybean and wheat crops are sown with illegal seed, which farmers sell to their neighbours'.[32] Nor could Monsanto count on enhanced chemical sales as their technology spread: Round-Up (glyphosate) was out of patent and 20 companies competed with Monsanto for sales (Paarlberg, 2001: 71–2). Opponents of transgenics in Rio Grande do Sul argued that control of the international border was a federal matter; the fault lay in Brasília – which favoured biotechnology. Other states – Santa Catarina, Mato Grosso do Sul, and Pará – passed state laws giving their newly-created State Biosafety Commissions control over transgenics in anticipation of federal deregulation (Hathaway, 2002). As in India, federalism made for a disjointed and indeterminate regulatory environment.

In January 2001, approximately 1200 Brazilian farmers mobilised by the Landless Workers Movement (MST) stormed a Monsanto biotechnology plant in Rio Grande do Sul in protest against genetically modified food (Reuters, 2001). This protest strategically coincided with the World Social Forum that had convened in Porto Alegre, countering the World Economic Forum simultaneously underway in Davos, Switzerland. Aided by anti-globalisation protesters, farmers uprooted genetically engineered corn and soybeans at the experimental station and wrote on walls: 'The seed of death' and 'Monsanto is the end of farmers!' A local MST leader, Solet Campolete, explained, 'these seeds trick farmers and create dependency on seeds produced by a big multinational'.[33]

Though discursive framing of social movements paralleled that of India, there was also interest-driven opposition: stealth soy was a threat to Brazilian exports. Federations of agriculture in Mato Grosso do Sul – the country's leading soy-producing state – and Paraná demanded action against transgenic soy. Paraná began interdicting plantings in 17 areas and ultimately attempted to blockade seeds by declaring the state a 'GMO-free zone'. Paraná had been conducting negotiations to export non-transgenic soy to China (*Actualización,* 24 March 2003, Issue 8). Both states' federations cited preferences of the European Union and Japan for 'GMO-free' food. The Paraná federation of agriculture's Technical Commission on Grains warned of 'an enormous loss of revenues' because the national system is unable to certify 'reliable separation' of transgenic soy (Osava, 2002).

Opponents of Monsanto, globalisation and transgenics expected support from the federal level when the Workers' Party (*Partido dos Trabalhadores* or PT) government took office on 1 January 2003. The party promulgated an anti-globalisation agenda and endorsed the precautionary principle in environmental policy. Yet, on 25 September, a Presidential Decree (MP131) authorised the planting and harvesting of transgenic soy for one year (Conroy, 2003) – a de facto amnesty. Permission was extended in 2004 for one year. Farmers were allowed to plant transgenic seeds saved from the last harvest, but not to sell them (Osavo, 2004; *Actualización*, 25 September 2003, Issue 3). In October 2004, permission for growing transgenics in the 2005 season was granted as well, subject to registration.[34]

The government had first sought to limit transgenic soy to one state, Rio Grande do Sul. When Governor Requião of the neighbouring state of Paraná declared both his state and Brazil's main grain port, Paranaguá, as GM-free zones, officials began stopping soy trucks at the border and turning away those that tested positive for glyphosate-resistant soy. This practice directly contradicted the presidential decree; it also created a problem of international jurisdiction, as countries such as Paraguay that permit transgenic soy found their shipments delayed or seized. On 11 December 2003, Brazil's Supreme Court unanimously ruled suspension of Paraná's law, while delaying a final ruling on constitutionality (Reuters, 2003). The Liberal Front Party had brought the case, arguing that the state legislature had overstepped its jurisdiction. The perhaps unstable outcome is that there can legally be no 'GM-free zone' in Brazil (Seed Quest, 2003b). On 26 March, another presidential provisional measure (later converted to Law 10688/03) established rules for the commercialisation of the 2002–03 soy crop, including transgenic soy. This measure gave amnesty to farmers who had illegally planted transgenic soy when the ban was in place. There are no official statistics, but estimates run up to 90 per cent of soy in Rio Grande do Sul as transgenic (Neto, 2003; *Actualización*, 5 December 2003, Issue 6).

Significant sectors of civil society objected to post-facto regularisation of stealth seeds in an economy that both consumes and exports soybeans and products on a large scale. The Campaign for a Transgenic-Free Brazil went so far as to brand President Lula – whom they saw as a proponent of the precautionary principle before his election – as a 'crusader' for transgenics ('*un cruzado pro de-GM*', *Actualización*, 24 March 2003, Issue 8). As in India, farmers' organisations were divided; spread and replication of seeds underground clearly demonstrates desirability to many farmers. Some deputies in the Workers' Party (PT) supported what opponents – specifically Paulo Pimenta – called the agenda of big landlords and soya producers of Rio Grande Do Sul. Whether poor farmers opposed or supported transgenics is not established, but there is anecdotal evidence that small farmers have bought into the technology, which should in theory be scale-neutral.[35]

The national state is divided as well. The Agriculture Ministry has supported liberalisation of transgenics; the Environment Ministry, increasingly marginalised, has opposed it (*Actualización*, 18 December 2003, Issue 5). A new bio-security law was proposed in October, 2003; in February of 2004, the amended draft bill was approved by the Chamber of Deputies. Delays and compromises reflected differences in the ruling coalition, particularly conflict between the minister of agriculture, Roberto Rodrigues and Marina Silva, minister of environment, who has opposed de-regulation in the absence of further studies (Neto, 2003; Hochstetler, 2004: 17–9). Agricultural interests recognized these deep divisions and were reluctant to plant transgenic soy, fearing actions by courts, the environment minister and state governments might prevent their marketing.[36]

Why were these seeds spreading? Urbashi Poddar (2004: 31) summarises advantages perceived by Brazilian farmers:

> . . . producers are able to control weeds with less tillage and limit their herbicide applications to just one spray over the top of the plants rather than adhering to a more expensive sequence of pre-emergent and post-emergent sprays . . . used with conventional soybeans. . . . The Brazilian Association of Seed Producers estimates that farmers save approximately 20 per cent on their fertiliser costs using the genetically modified crop, 'which also taxes their land less' —And with higher yields. In justifying use of illegal technology, some farmers responded that the seeds 'serve as an instrument of economic leverage against the subsidy regime upheld by the United States and other . . . industrialized nations'.

Public distrust of transgenics targets opaque concentration of information in unaccountable hands. Though public institutions are being crafted to permit trials and supervise biosafety procedures, much research is under private auspices. Brazil, like India and China, has significant public sector resources for biotechnology. Paarlberg (2001: 70) notes that Brazil's intellectual property protections have encouraged not only foreign investment but also scientific entrepreneurs. Brazil's public sector research organisation, EMBRAPA, has announced a new transgenic soy resistant to Imidazolinone-based herbicides, similar to Monsanto's RR soy resistant to herbicides of the glyphosate family. Testing for commercial release will take time, and imidazolineone is more expensive than glyphosate; nevertheless, the assumption of multinational monopoly of biotechnology is unsustainable. Such outcomes of public science undermine critiques of biotechnology that assume a high-technology monopoly of bioproperty (Reuters, 2004b). Paarlberg notes that plant breeders at EMBRAPA have developed local varieties of soybean, maize, and cotton best suited to Brazilian conditions, and have used Brazil's new IP system to establish ownership of this valuable collection of improved germplasm. These scientists see strengthening of research capacity of universities and public institutes as an antidote to monopoly by multinational firms.[37]

The future of stealth seeds in Brazil remains in legal limbo. Farmers who grow them are in theory required by the federal government to sign an agreement to comply with terms for growing transgenic soy: but there are interpretive disputes as to what compliance means, and how it can be enforced. Moreover, Monsanto claims intellectual property in the germplasm and continues to press its property claims, against uneven resistance by Brazil and its neighbors.[38]

How widespread are stealth seeds beyond Brazil and India? Anecdotal evidence is prolific, hard evidence scarce. Pray and Naseem (2007) note that 'a recent survey of scientists in Latin America by ISNAR indicates... that government scientists were using patented techniques extensively without knowing they were patented in some cases or not caring about it in others'. Qaim and de Janvry (2002) find that if 'black market' seeds for *Bt* cotton in Argentina were included in the calculus, most benefits of the technology would end up in farmers' pockets, not Monsanto's. Pray and Naseem also note that though China has patent laws, plant breeders' rights, and trademarks – in theory protecting both Monsanto and public sector *Bt* cotton varieties, and has stringent biosafety regulations – 'most of the *Bt* cottonseed that farmers planted was produced by farmers or small seed companies and government research institutes who did not apply for bio-safety committee approvals or pay royalties to the owners of varieties or genes'.[39] *Bt* rice has reportedly appeared in the market in China, without bio-safety clearance. Bolivia and Paraguay have had the Brazil-Argentine experience with underground soy; Paraguay has explicitly acknowledged the fact. Finally *Bt* cotton and *Bt* corn have reportedly appeared in Brazil.

Under what conditions will the phenomenon of stealth seeds emerge? Underground seeds have appeared in democratic and authoritarian systems, federal and centralized, strong and weak states. Generalization is unwarranted. Certainly some common factors contribute: monopoly prices of official transgenic seeds – enabled by bio-safety regulations more than market power; structural power of agriculture as a sector and local power of farmers as political actors; permeable bureaucracies; regulatory confusion and delay; weak institutions. These conditions, though widespread, tend to disappear in biotechnology discourse – both developmentalist and oppositional – that reifies seed and state.

V. Reification of Seed and State

The ability of seeds to go underground through farmer stealth strategies undermines surveillance of states and firms assumed in two divergent discourses: opponents of genetic engineering fear monopolisation of property rights, proponents assume bio-safety regimes that can control nature and farmers – a Panopticonic state enforcing global protocols. Neither discourse is proving robust on the ground. Surveillance of nature is no mean task, either macro or nano.[40] Strong interests are at work. Farmers pursue stealth seeds because the technology is affordable and divisible; the genetic roulette they enable has not been a powerful deterrent.

Stealth seeds inject a new dimension of dissention into potential coalitions for the poor: farmers seek them out, NGOs seek to stop the technology. Brazil's PT government has been split, as have representatives of farmers and NGOs. In India, the loss of credibility entailed in the Monstanto-terminator-suicide-seeds campaign divided farmers' movements and reduced the political capability of NGOs supporting farmers' rights and environmental protection. Contested science drives different segments of the active body politic in opposite directions. Indeed, a defining characteristic of bio-politics is the ambiguity of interests. Interests in transgenics must be processed through a complex cognitive screen in which much that needs to be known is not known: the great ideational divide follows neither class nor North–South lines (Herring, 2007). Intellectual property rights are reified as a force against the farmer – as if patents were somehow self-enforcing. Opponents fear the same reifications of bio-property that techno-optimists posit as preconditions for progress. 'Monopoly' by seed firms dominates the argument against transgenics even as stealth strategies defy property claims in the fields. Farmer-bred 'Robin-Hood' *Bt* seeds may be spreading faster than officially sanctioned seeds from 'monopolist' Monsanto, as they are cheaper, and often give better results (Roy et al., 2007). Farmers in Brazil and India pursued benefits of transgenic crops for reasons not credible, or even comprehensible, to their nominal allies in politics and NGOs.

Property rights in landscapes – in macro-nature – evince a long history of conflict between states and local communities, easier to claim in international treaties than in forests and fields (Herring, 2002). Stealth seeds underscore the dependence of property rights in micro-nature on close monitoring mediated by high technology. In both Brazil and India, in different crops, technology developed by Monsanto was appropriated, redeployed and developed by small-scale entrepreneurs and farmers themselves, unmindful of TRIPS negotiators or NGO petitions. In India, Delhi could not know that the cotton plants that survived the bollworm infestation of 2001 were transgenic and illegal prior to genetic testing.[41] Commercial interests diligently provided the knowledge state actors lacked; 'seeing like a state' may be more astigmatic than seeing like a firm. Navbharat's successful *Bt* cotton violated no property rights, but clearly expressed the Monsanto *Cry1Ac* gene. In Brazil's transgenic soy, it is clear where the seeds originated, but it is less clear that Monsanto has a valid property claim after generations of field-level modification of varieties by farmers and seed firms operating independently of Monsanto. Bioproperty claims, like property claims generally, are relational; they mean what actors can make them mean. Ironically, the popular and successful Navbharat 151 seeds were banned – and their germplasm went underground – not because of Monsanto's property claim but because they had evaded bio-safety procedures. In effect, as angry farmers and seed producers in Gujarat protested, to the extent Mahyco–Monsanto had privileged market position and extracted rents from farmers, it was a function not of market power but restriction of alternatives by bio-safety regulators.

Whatever else the genomics revolution may bring, it will certainly bring higher levels of surveillance. Surveillance has real costs, and certainly developmental opportunity costs; good policy rests on analysis that is currently absent. The analytics are missing partly because of inevitably incomplete science at the knowledge frontier: there are known unknowns and unknown unknowns. Because the transgenic genie is out of the bottle, the cost and consequences of bio-safety regimes need attention.

VI. The Goldilocks Paradox: Getting Regulation 'Just Right'

To be normatively acceptable, development policy must argue for priorities at the margin: what are opportunity costs of an extra dollar spent on bio-safety regimes? How do benefits compare to alternative uses of that same dollar? The answer depends on how one conceptualises benefits, how one couches alternatives, the normative position one takes on uncertainty and risk, and the projections one makes from an inevitably incomplete science. Figure 1 sketches the analytics of this problem. The first question in Figure 1 is: what additional risk is introduced by

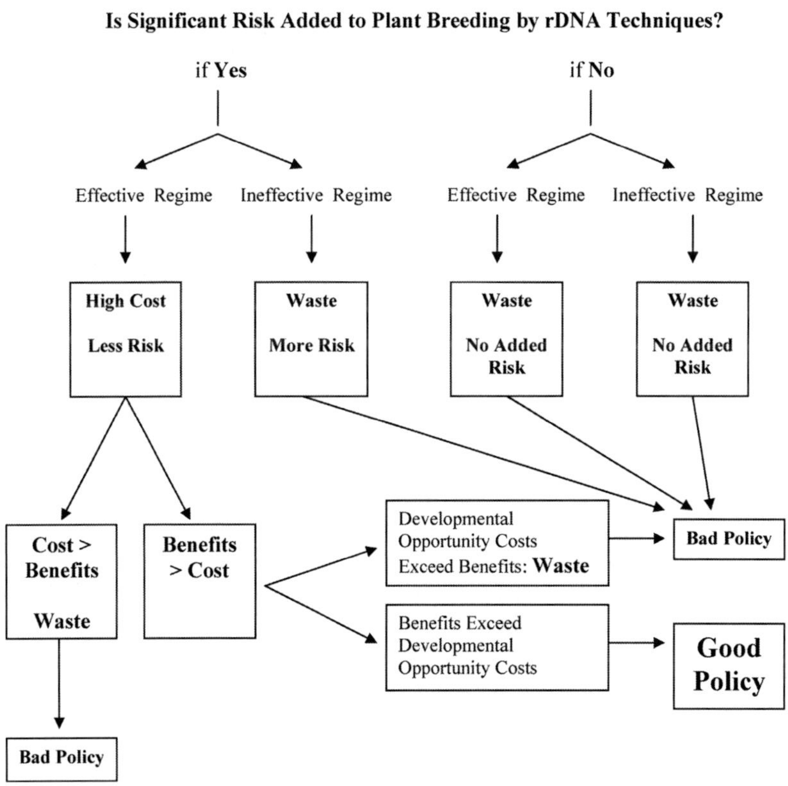

Figure 1. Biosafety regime scenarios and social costs.
Source: Ronald J. Herring, Cornell University

recombinant-DNA technology compared to conventional breeding techniques? Answers in the literature vary from zero to a lot. Gene flow is pervasive in plants; all transfers of genetic materials through any form of breeding constitute some risk, however remote. There may be little if any *incremental* risk in genetic engineering. Moreover, invasive species allow whole genomes to flow through ecologies; these risks may dwarf the risk of single-gene flow, especially since not all transgene flow will improve the fitness of receptive plants: most crosses and mutations are less, not more fit. Yet, control of invasive species seldom receives the level of rhetorical or financial support given to regulation of transgenics, despite enormous costs (Herring, 2007).

If there is no incremental risk in rDNA techniques of plant breeding compared to alternatives, all expenditures – human capital, administrative time, institution building, money – will be wasted, whether the biosafety regime is effective or ineffective. The analogy might be to buying insurance against spells cast by enemies. Alternatively, let us assume that there *is* additional risk – the dominant position. It then matters whether the biosafety regime is effective or ineffective. If institutions are ineffective, incurred costs will go for naught: transgenes will spread and the outcome is again waste. An analogy might be the very expensive 'war on drugs' periodically claimed by American administrations. Third, consider the situation where there *is* additional risk, and the biosafety regime is effective. Here analysis faces a new challenge: should the additional costs be incurred? Who should pay? If costs exceed benefits, and the biosafety regime is nevertheless established and nurtured, there is again waste. If benefits exceed costs, a second level of analysis becomes necessary: if the opportunity costs of the effective regime exceed the costs of the regime, we again are in the world of waste. It is *only* if benefits of a biosafety regime clearly exceeded the value of alternative uses of resources that the policy would be justifiable in developmental terms.

It is hard to avoid the conclusion from the outcomes in Figure 1 that justifiable developmental policy promoting genetic engineering would be restricted to extremely important objectives, or restricted to deployments with very low risk. For low-income countries, opportunity costs are large: microbiologists, lab space, research protocols, administrative time, money. Establishing a matrix of priorities would demand far more science than currently available. By this logic, provitamin A 'Golden Rice' is in a different category from the herbicide-resistant grass for golf courses – recently shown to experience gene flow over very long distances in the US.[42] A developmental calculus between fewer blind children and fewer weeds on golf courses might seem a simple one; but defenders of transgenic grass argue that weeds will be suppressed on golf courses whether we like it or not, and typically with much more dangerous chemicals than glyphosate, thereby justifying RR grass.

Not only are benefits socially disputed, but it is even in principle difficult to calculate risks (Thies and Devare, 2007). *Bt* gene flow from cotton is unlikely to pose great risk. Test trials on *Bt* cotton found that there was no incremental risk: *Bt* plants showed the same characteristics as existing cottons in food and feed, except for the *Cry1Ac* protein, which disappears in pressed oils and seems to be harmful only to *Lepidoptera*. There was no evidence of damage to field ecology. Are these tests adequate? Could data from comparable regions of China or the US have been used to ascertain risk, by-passing field trials in India? Or, are even more fine-grained tests necessary, as agro-ecologies vary continuously rather than clumping into blunt categories?

There is also risk of resistance to the endotoxin developing in *Lepidoptera* – a threat to a public good deployed for decades by organic farmers who use *Bt* foliar sprays to control insects. *Refugia* are proposed for slowing the development of resistance of insects to *Bt* toxin. These *refugia* are costly for small farmers and often are evaded. Is this a dangerous outcome? Every agro-ecology is somewhat different. Mahyco presented to regulators the argument that there may actually be no problem with resistance development at all, as the bollworms will always have natural *refugia* in a great variety of non-cotton plants; there are about 157 hosts for bollworms in India. This argument, while biologically plausible, failed because 'uproar about the terminator' made authorities cautious.[43]

Capital in biotechnology complains that there is too much regulation – even though regulations favour corporate deep pockets over small firms, and thus reinforce technological hierarchy.[44] Environmentalists typically argue that there is too little. By what metric would one recognize the Goldilocks standard of 'just right': not too much, not too little?

Consider the cost of vetting *Bt* cotton in India. Though critics complained that bio-safety testing was done in private and supervised by Mahyco–Monsanto, there were 12 agricultural universities involved in the studies, eight Central Research Institutes, the Indian Council for Agricultural Research, the Indian Council for Medical Research, four Central Ministries and departments of agriculture in six states.[45]

From discussions with Monsanto beginning in 1993, an application by Mahyco to import seed was made on 10 March 1994; a permit was received 3 October 1995. Mahyco imported Bollgard seed in 1996. Imported germplasm was backcrossed into Indian varieties. During this period there were greenhouse and laboratory trials; these and later trials were mandated by the Review Committee on Genetic Manipulation [RCGM] of the Department of Biotechnology. There was only one limited field trial at one location. In 1997–98, a gene-flow study was conducted in limited field trials at five locations. In the same period, toxicological studies were conducted at ITRC in Lucknow. In 1998–99 field and experimental trials were allowed by the RCGM at 40 locations to test allergenicity and toxicity. Regulation in India is based on 'incremental risk' – that is, risk added by the transgenic that is not present in the isogenic variety of the same plant.[46] For this trial, careful controls matching varieties are critical. These were the trials targeted by Operation Cremate Monsanto and from which the dissident farmer Mahalingappa conducted his own test of the terminator hypothesis. All safety data were submitted to the RCGM; the uprooting and burning protests induced authorities to request more trials. In 1999 multi-location trials replicated contained-research trials, at 11 locations. Procedures at this point reflected social turmoil; more caution was applied. It was only in 2000 that the RCGM recommended to the Genetic Engineering Approval Committee of the Ministry of Environment large-scale contained field trials, which entailed seed production for testing. At this point Navbharat transgenic seeds had been in fields for at least one year and were spreading to new acreage, but no regulatory authorities knew it.

Turmoil surrounding court cases and popular resistance caused the GEAC to ask for more field trials under auspices of the Indian Council for Agricultural Research;

these trials were planted late and thus tainted in the eyes of NGO opponents, as peak bollworm infestation was avoided. Consequently, in 2001, a second year of field trials was ordered by the GEAC under the supervision of the ICAR. Before these results could be vetted, the Robin Hood seeds appeared in Gujarat, noticeable only because a bollworm rampage had wiped out other varieties. In 2002, field-trial results were submitted to the GEAC, which subsequently, in March 2002, approved three *Bt* hybrids for commercialisation. This decision was widely and correctly interpreted as a *fait accompli.*

Is this level of social expenditure justified by results? What was lost in these procedures? For every year of delay, some level of unnecessary pesticide application reduced farmers' incomes and contaminated soils, waters, and skins, with an unknown costs in poisonings and loss of fauna. The opportunity costs for Mahyco may have been high: presumably this time and effort could have gone into finding better hybrids than the original three, which were not the best available. Mahyco holds its cash costs to be approximately $8 million for development and testing of three transgenic varieties. The same transgene had been present in cotton in other countries for years, with no known ill-effects. Navbharat bypassed biosafety procedures and created a *Bt* hybrid with early flowering characteristics well suited to Gujarati conditions; if Bollgard varieties are safe, so too, presumably, is Navbharat 151 and its numerous variants/descendents – but the former are legal, the latter are not.

One future consequence of the stealth-seed phenomenon is predictable: real terminator technology. Social institutions will not deal with stealth seeds very well, as they do not deal with any high-value product that is movable across permeable state space: software, pornography, information, drugs, arms. Ajit Singh, Minister of Agriculture at the time of the *Bt* discovery, said: 'see that we cannot even regulate adulterated food; how will we regulate seeds?'[47] Gene-use restriction technologies (GURT) will become the only feasible assurance of restricting gene flow where there is risk and the only way to enforce property claims under many conditions. The biological remedy is arguably more effective than the institutional remedy[48] – if it can be made to work in the field. Yet this conclusion illustrates again the power of discourse to shape and filter interest. Terminator technology was sidelined because of biopolitics. Is terminator technology acceptable as social policy? The clear answer has been no.

Disputes about material interests are demonstrably subject to conciliation, bargaining and compromise – familiar terrain for the political economy of interests. Deals can be struck. Disputes about the nature of the natural, and consequent risks of the unnatural, take on a different politics, dependent on expertise that is asymmetrically distributed, both locally – on the ground within movements – and globally. Biopolitics is less susceptible of ordinary bargaining solutions, but it is an inescapable politics spawned by the genomics revolution.

Acknowledgements

This paper owes much to discussions with numerous political activists, seed producers and farmers in India, as well as helpful interlocutors, of whom only a few can be mentioned here: Sivramiah Shantharam, Suman Sahai, Kameswara Rao,

D. B. Desai, Vikas Chandak, Anil Gupta, Michael Lipton, Maria Jose Sampaio, Andre Goncalves and numerous Cornell colleagues.

Notes

1. On historical uniqueness of agrarian struggles in Kerala, Herring (2003). The Monsanto-terminator story circulated even in India's premier journal of the social sciences, the *Economic and Political Weekly*: Shiva et al. (1999), Sahai (1999).
2. On conjunction between globalisation and terminator technology, Suman Sahai, leader of Gene Campaign, wrote (1999: 84), 'Trials of a genetically altered cotton variety (*Bt* cotton) conducted by the American company Monsanto have provided the trigger because Monsanto also happens to own the terminator technology'. See also Bharathan (2000), Parmar and Vishvanathan (2003). RAFI International 2000 seems to be the origin of the phraseaology in their 'Suicide seeds on the fast track'. On terminator discourse, Gold (2003).
3. Stone (2002), Shiva et al. (2000: 64–110), Srinand Jha (2001). On debt as a cause of suicides, Centre for Environmental Studies Warangal (1998), Department of Agriculture and Cooperation (1998). Government of Karnataka (2002) concluded that debt and agrarian crisis were implicated, but many suicides were tragically personal in nature.
4. How much leaders knew about the truth of the discourse is uncertain. Vandana Shiva et al. wrote in 1999 of the seeds: 'they are in an ecological sense terminator, which terminates biodiversity...' Shiva et al. (2000: 98) wrote: 'Freedom from the second cotton colonisation needs to be based on liberation through the seed.... The freedom of the seeds [is] simultaneously a resistance against monopolies... like Monsanto... The seeds of suicide need to be replaced by the seeds of prosperity'.
5. Joshi (2001), Greenpeace India (2001), Jayaraman (2001), Mathew (2001), Sahai (2002), Shaik (2001), Mehta (2005: 60–79, 130–6), Visvanathan and Parmar (2002), Scoones (2003), Omvedt (2005). These sources are complemented by discussion with Mr Raju Barwale, Managing Director of Mahyco, 28 May 2004 and Dr D. B. Desai of Navbharat and *Bt*-cotton farmers, Gujarat June 2005.
6. 'Cultivation of *Bt* cotton using Navbharat seeds', Government of India, *Rajya Sabha* Unstarred Question No 205 to be answered on 01.03.2002 by Minister for Environment and Forests, Shri T. R. Baalu; author's interviews, New Delhi, June 2002; Ahmedabad, 2005.
7. Interviews in Delhi and Ahmedabad, 2002, 2005; Gene Campaign (2005, Part II: 72 *et passim*).
8. Patil told the Council that this outcome was assured by the Agriculture Secretary in Delhi, a result of successful lobbying by the state government. The question was raised by N P Hirani, a legislator who is also Administrator of the State Cotton-Growers Association. The *Indian Express* (11 December 2001).
9. Herring (2005); also *Indian Express* (Mumbai/Nagpur edition, 13 November 2001; Delhi edition, 25 March 2002), *Financial Express* (3 April 2002), Gene Campaign (2005, Part II: 101, 109 *et passim*).
10. Yamaguchi (2004, Chapter 4), found that farmers with relatives in Andhra Pradesh were more successful in obtaining NB 151 seeds after they were discovered and banned. The text's account was confirmed by conversations with seed producers in Gujarat (June 2005).
11. The English is sometimes used in rural Gujarat to designate what farmers also call *lodhavela biyaran* ('ginned seeds') to designate non-branded F2 *Bt* cotton seeds.
12. I have heard farmers insist that F2s outperform official seeds. Though contrary to conventional wisdom, it is possible that variable agronomic conditions favor variance in phenotypic characteristics other than those for which the hybrid was chosen. See Roy et al. (forthcoming), on farmer logics of seed choice.
13. Personal communication, 16 October 2004; also Seed Quest (2003a), *Indian Express* (28 May 2003), Jayaraman (2004), *The Hindu* (4 November 2004).
14. Minister for Environment and Forests, Shri T. R. Baalu, 'Restriction of production and sale of *Bt* Cotton', Government of India, *Rajya Sabha* Unstarred Question No 206 to be answered on 01.03.2002; Gene Campaign (Part II: 102–3); interview with Ajit Singh, former Minister of Agriculture, New Delhi, 22 June 2005.
15. For positive agronomic results in Gujarat, see Roy et al. (forthcoming), Morse et al. (2005). Mahyco–Monsanto's own data for the first year of legal production of their MECH varieties on 55,000 farms

show an all-India average reduction in pesticide use per acre of 65–70 per cent, worth Rs 920 to the average farmer, an increase in yields of 3.22 quintals per acre and an increase in income of Rs 7520 per acre (unpublished data). The smallest benefits were reported in Andhra Pradesh (Rs 5930), where much of the 'failure of *Bt* cotton' literature originates. Bambawale et al. (2004) in an independent assessment of MECH-162 find positive results. Qaim and Zilberman (2003) created a stir for projecting atypically positive *Bt* cotton results from India.

16. *The Hindu* (14 December 1998). Compare Rediff.com (2001), Srinand Jha (2001). Even the serious journal *Frontline* carries similar stories: Krishnakumar (2004) confidently analyzed 'the disaster wrought by *Bt* cotton'. For useful commentary, Rao (2004). Claims of failure began with field trials: Shiva et al. (1999). Reports of *Bt* failures on a few farms even merited Parliamentary discussion. Responding to insistence of an MP from Gujarat, a central team of scientists was sent to investigate MECH-162; the team found normal agronomic variance. Madhusudan Mistry, MP, Lok Sabha, letter to Dr Murli Manohar Joshi, Ministry of Science and Technology, 18 October 2003; answered in Parliament 8 December 2003 by Mom Raj Singh, Under Secretary, NO.RS.9/6/2003-S&T.
17. The cultivar MECH-184, for example, needs early watering to develop; otherwise, it wilts. Bambawale et al. (2004) reported unpublished data showing superior performance of even this demanding variety, a finding consistent with Roy et al. (forthcoming). Though Stone (2004) argues that biotechnology is especially 'de-skilling', conventional hybrids often require new knowledge and techniques.
18. Interview with Mahalingappa, 7 June 2004. Madsen (2001: 3733–42), Birasal (1998), Visvanathan and Parmar (2002: 2714–24).
19. *The Hindu* (14 December 1998). Similar accounts appear in the *Deccan Herald* (1998, Bangalore edition).
20. Interviews, MMBL, GEAC, cotton-seed producers.
21. Meera Nanda 2003 argues that anti-Enlightenment epistemological relativism, manifest in 'alternative' sciences, is instrumentally deployed by the *Hindutva* right as well as by public intellectuals claiming the mantle of the left.
22. Press release of 20 June 2003, New Delhi. Signatories of the petition range ideologically from the *All-India Bhartiya Kisan Sangh*, a BJP affiliate, to the *All India Kisan Sabha*, a Communist Party of India affiliate.
23. Sanat Mehta (2005: 60–76); interviews with Dr D. B. Desai and the breeder of Navbharat 151, Dr N. P. Mehta in Ahmedabad, June 2005.
24. Conversations of author with Percy Schmeiser, 27–28 January 2004. For a brief history, Federal Court of Appeal Ruling 4 September 2002, http://decisions.fct-cf.gc.ca/fct/2002/2002fca309.html accessed 5 May 2004.
25. Paarlberg (2001: 67–92), Poddar (2004: 24–52), the text also relies on various issues of *Actualización: Un boletín de noticias & análisis mensuales de la Campaña Por Un Brasil Libre de transgênicos* (hereafter *Actualización*).
26. Niiler (1999), Buckly (2000), Paarlberg (2001: 73–4), Hathaway (2002), Fontes (2003).
27. Ironically, the state of Rio Grande do Sul banned cultivation of all transgenic crops effective January 1 2000 (Niiler 1999: 848). On early decisions, Hathaway (2002), Fontes (2003).
28. Sources for this section include *Actualización*; Niiler (1999), Maria Jose Sampaio, personal communications, 2003, 2004.
29. Osava (2004), Associação Nacional de Biossegurança (2001).
30. Estimates of planted area, here and later, are based on extrapolation from area harvested backward to seeds sold; non-transgenic seeds sold were sufficient to plant only about 40 per cent of the state's area (Bensen, 2001). Saved seeds, transgenic and otherwise, are not counted by this technique. See also Nuffield Council (2004: 3.57–3.61).
31. 'En resumo, no había ningún otro objetivo para esta decisión de ahora sino reforzar la estrategia de situación de facto de los pro-OGMs'. *Actualización*, 25 September 2003, Issue 3.
32. INASE opposed 'brown bag' seeds that carry 'no guarantee of quality and identity... and royalties are not paid...' The state's interests are engaged because no taxes are collected on underground sales. 'Illegal GM-crop trading in Brazil and Argentina': http://www.gene.ch/genet/1999/Oct/msg00008.html
33. Ironically, the seeds were produced by local farmers, not Monsanto (Poddar 2004: 32); http://www.organicconsumers.org/monsanto/brazilprotest.cfm

34. Numerical estimates of farmers who have registered to plant transgenic soy vary so much that reporting is premature; there is both compliance and evasion (Ewing, 2003).
35. The mayor of Chapada, Rio Grande do Sul, Carlos Alzenir Catto, was quoted as saying that '98 per cent' of the farmers in his area, including small farmers, were registered to plant transgenic soy: 'Here, people don't have any fear of signing up because they know they are planting a product that is not bad for health or the environment, because they can spray less agrochemicals on the crops...' (Ewing, 2003).
36. Lewis (2004), Reuters (2004).
37. Paarlberg (2001: 70). Leila Macedo Oda, president of the NGO ANBio (National Biosafety Association) draws a parallel the 'market reserve' in computing in the 1980s. By forbidding import of computers, the Brazilian government hoped to favor the local computer industry, but lack of competition only made Brazil lag behind other countries. Computers were then smuggled much as farmers now smuggle seeds (Neto, 2003: 1257–8).
38. The Agriculture Department of Paraná cordoned fields near the town of Pato Branco when crops tested positive for transgenic soy. The crop was ruled illegal because the owner had not signed an agreement to observe new rules for transgenic cultivation. Only 464 farmers had signed the agreement – itself contradicted by a state law, later overruled, banning transgenics. Many farmers refused to sign because the agreement failed to reflect modifications made by Congress to the biosafety bill. The agreement requires that farmers pay royalties on illicitly produced seeds; the new bill rejects this liability (Ewing, 2003; American Soybean Association Weekly Update, Archives, 19 January 2004; Clendenning, 2004).
39. Salazar et al. (2000), Pray and Schimellpfenning (2001), Pray and Naseem (forthcoming), Jayaraman (2004). Luisa Massarani reported on 2 December 2005, on SciDev.Net, an accusation by the state deputy Frei Sérgio Antônio Görgen of transgenic corn being sold in Rio Grande do Sul. Greenpeace announced in mid April 2005 findings of a German laboratory of *Bt* genes in rice seed in markets near Wuhan (*New Agriculturalist Online*, 5 May 2005, http://www.new-agri.co.uk/05-3/newsbr.html#nb7).
40. The parallel to James Scott's *Seeing Like a State* (1998) is clear; Scott's high modernist state needs legibility of its terrain, but finds it hard to attain. The Panopticon is posited, but illusive.
41. There are now much simpler *Bt* detection kits developed in India, costing about Rs 50 (1$US) per kit.
42. Andrew Pollack (2004) reports on a study (forthcoming in Proceedings of the National Academy of Sciences) of transgenic bentgrass created by Monsanto and Scott for golf courses for weed management. Gene flow extended as far as 12 miles, which was unexpected. The US Forest Service is quoted as saying that the grass 'has the potential to adversely impact all 175 national forests and grasslands'.
43. Interview with Mr Raju Barwale, 28 May 2004. MMBL was required to provide seed for the refuge – in a packet of 450 grams of *Bt* seed, 120 grams of non-*Bt* isogenic seed is supplied by the firm for *refugia* planting, along with instructions. The firm is expected to do resistance studies. Cited sources for this section are supplemented by interviews in Delhi, July 2001, June 2002, June 2005. Also Department of Biotechnology (1996).
44. For an argument that approval is excessively slow for biotech in India, see CII (2001).
45. Sources for this section include conversations with Mahyco officials, officers of the Department of Biotechnology, NGO activists and Department of Biotechnology (2002), Scoones (2003), Bagla (2004).
46. Transgenics are governed by the Environmental Protection Act of 1986, Rules of 1989, and specifically Notification No GSR 1037 E dated 5 December 1989.
47. Interview, New Delhi, 21 June 2005.
48. By analogy, a forestry official in Bangladesh once told me: 'Tigers are the best forest guards – utterly incorruptible'.

References

Actualización: Un boletn de noticias & anlisis mensuales de la Campanha Por Un Brasil Libre de transgênicos, various issues, 2003–05.

American Soybean Association (2004) Weekly Update, Archives 19 January.

Asian Social Forum [Hyderabad] Seminar (2003) Beyond Bhopal and Bt: taking on the biotech giants, Research Foundation for Science, Technology and Ecology, Delhi, 4 January.

Associação Nacional de Biossegurança (2001) Report on the commercial approval of transgenic soy in Brazil, accessed at: http://www.anbio.org.br

Bagla, P. (2004) Report says India needs stronger, independent regulatory body, *Science*, 304(5677), p. 1579.

Bambawale, O. M., Singh, A., Sharma, O. P., Bhosle, B. B., Lavekar, R. C., Dhandhapani, A., Kanwar, V., Tanwar, R. K., Rathod, K. S., Patange, N. R. and Pawar, V. M. (2004) Performance of *Bt* cotton (MECH-162) under integrated pest management in farmers' participatory field trial in Nanded district, Central India, *Current Science*, 86(12), pp. 1628–33.

Benson, T. (2001) GM soy planting in Brazil said to be spreading North, *Dow Jones*, 18 December.

Bharathan, G. (2000) Bt-cotton in India: anatomy of a controversy, *Current Science*, 79(8), pp. 1067–75.

Birasal, N. R. (1998) Haveri farmers will resist KRRS destruction trial, *Times of India*, 5 December.

Buckly, S. (2000) Brazil battles over genetically modified soybeans, *Washington Post*, 27 June.

Centre for Environmental Studies Warangal (1998) *Citizens' Report: Gathering Agrarian Crisis—Farmers' Suicides in Warangal District (AP) India* (Kishanpura, Andhra Pradesh, India).

Clendenning, A. (2004) Illegal soy seeds enrich Brazil farms, *The Washington Times*, 6 January.

Conroy, J. (2003) A time to reap, *Spiked Online*, accessed at: http://www.genet-info.org

Confederation of Indian Industry (CII) (2001) Biotechnology on the fast track, unpublished conference proceedings, New Delhi.

Department of Agriculture and Cooperation, Ministry of Agriculture, Government of India (1998) *Report of the Study Group on Distress Caused by Indebtedness of Farmers in Andhra Pradesh* (New Delhi, Ministry of Agriculture).

Department of Biotechnology, Ministry of Science and Technology, Government of India (1996) *Revised Guidelines for Research in Transgenic Plants and Guidelines for Toxicity and Allergenicity Evaluation of Tansgenic Seeds, Plants and Plant Parts* (New Delhi: Ministry of Agriculture).

Department of Biotechnology, Ministry of Science and Technology, Government of India (2001a) *Biotech News*, 2(2), November.

Department of Biotechnology, Ministry of Science and Technology, Government of India (2001b) *Biotechnology: A Vision* (New Delhi: Ministry of Science and Technology).

Department of Biotechnology, Ministry of Science and Technology, Government of India (2002) *Annual Report 2001–2002* (New Delhi: Ministry of Science and Technology).

Deccan Herald (1998) Operation Cremate Monsanto: Raitha Sangha to burn Bollgard cotton in Bellary, *Deccan Herald*, 2 December.

Deshiya Karshaka Samrakshana Samithi (National Agriculturalist Protection Committee) National Farmers' Protection Committee, Plan and Budget (2003) Palakkad, India.

Dow Jones (1998) Agnet, 20 November.

Edelman, M. J. (1964) *The Symbolic Uses of Politics* (Urbana: University of Illinois Press).

Ewing, R. (2003) Brazil farmers declare plans for genetically modified soybeans, *Reuters*, 19 November.

Financial Express (2002) 3 April.

Fontes, E. M. G. (2003) Legal and regulatory concerns of transgenic plants in Brazil, *Journal of Invertebrate Pathology*, 83(2), pp. 100–3.

Gene Campaign (2005) *The story of Bt cotton in India*. CD-ROM (New Delhi: Gene Campaign).

Ghosh, P. K. (2001) Genetically engineered crops in India with special reference to *Bt* cotton, *IPM Mitr*, 1(December), pp. 8–28.

Government of Karnataka (2002) *Farmers' suicides in Karnataka: A Scientific Analysis Report of the Expert Committee for Study on Farmers' Suicides* (Bangalore: Government of Karnataka).

Greenpeace India (2001) *Bt* cotton saga takes a ludicrous turn, *Greenpeace Newsletter*, 11(December).

Gold, A. G. (2003) Vanishing: seeds' cyclicality, *Journal of Material Culture*, 8(3), pp. 255–72.

Gupta, A. K. and Chandak, V. (2005) Agricultural biotechnology in India: ethics, business and politics, *International Journal of Biotechnology*, 7(1/3), pp. 212–27.

Hathaway, D. (2002) Legal controls over GMOs in Brazil, *Actualizacin: Un boletn de noticias & anlisis mensuales de la Campanha Por Un Brasil Libre de transgênicos* [Updates: A Bulletin of News and Analysis of the Campaign for a Brazil free of Transgenics], 25 February.

Herring, R. J. (2000) Authority and scale in political ecology: some cautions on localism, in L. Buck, C. C. Geisler, J. Schelhas and E. Wollenberg (eds), *Biological Diversity: Balancing Interests Through Adaptive Collaborative Management*, (Boca Raton, FL: CRC Press), pp. 187–205.

Herring, R. J. (2002) State property in nature, in J. F. Richards (ed.) *Land, Property and the Environment* (Oakland, CA: Institute for Contemporary Studies).

Herring, R. J. (2003a) The political impossibility theorem of agrarian reform: path dependence and terms of inclusion, in M. Moore and P. P. Houtzager (eds), *Changing Paths: The New Politics of Inclusion* (Ann Arbor, MI: University of Michigan Press).

Herring, R. J. (2005) Miracle seeds, suicide seeds and the poor: GMOs, NGOs, farmers and the state, in R. Ray and M. F. Katzenstein (eds), *Social Movements in India: Poverty, Power, and Politics*, pp. 203–32 (Lanham, MD: Rowman and Littlefield).

Herring, R. J. (2007) The genomics revolution and development studies: science, politics and poverty, *Journal of Development Studies*, 43(1), pp. 1–30.

Hochstetler, K. (2004) Civil society in Lula's Brazil, Working Paper Number CBS-57-04, Centre for Brazilian Studies, University of Oxford, UK.

The Hindu (1998) 14 December.

Indian Express (2001–03) Various issues.

Jain, S. (2002) *Indian Express*, 20 April.

James, C. (2002) Global review of transgenic crops: 2001 feature: *Bt* cotton, ISAAA Briefs No. 26, ISAAA, Ithaca, NY.

Jayaraman, K. S. (2004) India produces homegrown GM cotton, *Nature Biotechnology*, 22(3), pp. 255–6.

Jayaraman, K. S. (2001) Illegal *Bt* cotton in India haunts regulators, *Nature Biotechnology*, 19(12), p. 1090.

Jha, S. (2001) Seeds of death, GMO cotton, *India*, 30 May, accessed at: http://www.tompaine.com

Joshi, S. (2001) Unquiet on the western front, *The Hindu Business Line*, 19 December.

Krishnakumar, A. (2004) Biotechnology. *Bt* cotton, again, *Frontline*. 21(10), pp. 8–21, accessed at: http://www.frontlineonnet.com/fl2110/stories/20040521001408100.htm

Lewis, S. (2004) Brazil's biosafety law stalls in the Senate, *Food Chemical News*, 14 June.

Liptak, A. (2003) Saving seeds subjects farmers to suits over patent, *New York Times*, 2 November, pp. 1–18.

Madsen, S. T. (2001) The view from Vevey, *Economic and Political Weekly*, 29 September, pp. 3733–42.

Mathew, V. (2001) India's GM cotton story gets bigger – 'Uproot and Destroy' begins on Gujarat farms, *The Hindu Business Line*, 20 October.

Mehta, N. P. (2005) Genesis of cotton hybrid Navbharat-151 and *Bt* cottons and their prospects, in S. Mehta (ed.), *Bt Cotton: A Painful Episode* (Vadodara, Gujarat: Gujarat Kapas Utpadak Hitrakshak Sangh).

Mehta, S. (ed.) (2005) *Bt Cotton: A Painful Episode* (Vadodara, Gujarat: Gujarat Kapas Utpadak Hitrakshak Sangh).

Mistry, S. (1998) Terminator gene a figment of imagination: Monsanto chief, *Indian Express*, 4 December.

Morse, S., Bennett, R. and Ismael, Y. (2005) Comparing the performance of official and unofficial genetically modified cotton in India, *AgBioForum: The Journal of Agrobiotechnology Management and Economics*, 8(1), pp. 1–6.

Naik, G., Qaim, M., Subramanian, A. and Zilberman, D. (2005) *Bt* cotton controversy: some paradoxes explained, *Economic and Political Weekly*, 40(15), pp. 1514–7.

Nanda, M. (2003) *Prophets facing backward: postmodern critiques of science and Hindu nationalism in India* (New Brunswick, NJ: Rutgers University Press).

Nanjundaswamy, M. D. (1999) Statement of Professor Nanjundaswamy at the Demonstration Against the World Economic Summit, Cologne, 19 June, GreenFiles, WWF-I, New Delhi.

Niiler, E. (1999) Monsanto remains a magnet for GM opposition, *Nature Biotechnology*, 17, p. 848.

Neto, R. (2003) GM confusion in Brazil, *Nature Biotechnology*, 21(11), pp. 1257–8.

Nuffield Council on Bioethics (2004) *The Use of Genetically Modified Crops in Developing Countries*, (London: Nuffield Council on Bioethics).

Osava, M. (2002) Transgenic soy rampant, despite ban, *Inter Press Services*, accessed at: www.ips.org

Osava, M. (2004) Transgenic soy found guilty by people's court, *Inter Press Services*, accessed at: www.ips.org

Omvedt, G. (2005) Farmers' movements and the debate on poverty and economic reforms in India, in R. Ray and M. F. Katzenstein (eds), *Social Movements in India: Poverty, Power, and Politics* (Lanham, MD: Rowan and Littlefield).

Paarlberg, R. L. and Institute, IFPR (2001) *The Politics of Precaution: Genetically Modified Crops in Developing Countries* (Baltimore, MD: Johns Hopkins University Press).

Parmar, C. and Visvanathan, S. (2003) Hybrid, hyphen, history, hysteria: the making of the *Bt* cotton controversy, IDS Seminar on Agriculture Biotechnology and the Developing World, Sussex, UK, October 1–3.

Poddar, U. (2004, May) Controversy over genetically modified organisms: a comparative study of Brazil and India, MPS thesis, Cornell University, Ithaca, NY.

Pollack, A. (2004) Genes from engineered grass spread for miles, study finds, *New York Times*, 21 September.

Pray, C. E., Huang, J. Hu, R. and Rozelle, S. (2002) Five years of *Bt* cotton in China – the benefits continue, *The Plant Journal*, 31(4), pp. 423–30.

Pray, C. and Schimmelpfennig, D. (2001) The impact of biotechnology on South African agriculture: first impressions, College of Agriculture Seminar, University of Pretoria, South Africa.

Pray, C. and Naseem, A. (2007) Supplying crop biotechnology to the poor: opportunities and constraints, *Journal of Development Studies*, 43(1), pp. 192–217.

Qaim, M. and Zilberman, D. (2003) Yield effects of genetically modified crops in developing countries, *Science*, 299(5608), pp. 900–2.

RAFI International (2000) Terminator 2 years later: Suicide seeds on the fast track, RAFI Communique of February–March and news release of February 20 Winnipeg, Manitoba, Canada, RAFI International Office.

Rao, C. K. (2004a) Frontline mischief, Foundation for Biotechnology Awareness and Education, Bangalore, India, 18 May.

Rao, C. K. (2004b) One swallow does not make the summer – *Bt* cotton in India, Foundation for Biotechnology Awareness and Education, Bangalore, India.

Rediff.com (2001) *Bt* cotton will kill farmers, financially and literally, an interview with Devinder Sharma. 12 December, accessed at: http://www.rediff.com/money/2001/dec/12inter.htm

Reuters (2001) Brazilian farmers seize Monsanto facilities in anti-GE protest, *Não Me Toque*, 26 January, accessed at: http://www.organicconsumers.org/monsanto/brazilprotest.cfm

Reuters (2003) Brazil suspends anti-GM law, *Reuters*, 11 December.

Reuters (2004a) Brazil farmers fight for GMOs, *Reuters*, 12 February.

Reuters (2004b) Brazilian scientists develop new GMO soybean, *Reuters*, 10 March.

Roy, D., Herring, R. J. and Geisler, C. C. (2007) Naturalizing transgenics: official seeds, loose seeds and risk in the decision matrix of Gujarati cotton farmers, *Journal of Development Studies*, 43(1), pp. 158–76.

Salazar, S., Falconi, C., Komen, J. and Cohen, J. I. (2000) The use of proprietary biotechnology research inputs at selected Latin American NAROs, Briefing Paper No. 44, ISNAR, The Netherlands.

Sahai, S. (2002) *Bt* cotton: confusion prevails, *Economic and Political Weekly*, 25 May.

Sahai, S. (1999) What is *Bt* and what is terminator?, *Economic and Political Weekly*, 16–23 January, pp. 84–5.

Scoones, I. (2003, August) Regulatory manoevres: the *Bt* cotton controversy in India, IDS Working Paper No. 197, Brighton, UK.

Scott, J. C. (1998) *Seeing Like a State* (New Haven: Yale University Press).

SeedQuest (2003a) Brazil introduces bill to regulate GM crops, Rio de Janeiro, Brazil, 30 October.

SeedQuest (2003b) GM is cottage industry; hybrid seeds flood Gujarat fields, *Rajkot/Junagadh, NewsRelease*, 28 May.

Shaik, S. (2001) Farmers decide to defend their *Bt* gene cotton crops, *The Times of India*, 31 October.

Sharma, D. (2001) The introduction of transgenic cotton in India, *Biotechnology and Development Monitor No. 44*, March, pp. 10–3.

Shiva, V. (1997) *Biopiracy: The Plunder of Nature and Knowledge* (Boston, MA: South End Press).

Shiva, V. (1999) Ecological balance in an era of globalization, in N. Low (ed.), *Global Ethics and Environment*, pp. 47–69 (London: Routledge).

Shiva, V., Emani, A. and Jafri, A. H. (1999) Globalization and threat to seed security: case of transgenic cotton trials in India, *Economic and Political Weekly*, March, 6–12, 13–9.

Shiva, V., Jafri, A. H., Emani, A. and Pande, M. (2000) *Seeds of Suicide: The Ecological and Human Costs of Globalisation of Agriculture* (New Delhi: Research Foundation for Science Technology and Ecology).

Stone, G. D. (2002) Biotechnology and suicide in India, *Anthropology News*, 43(5), p. 5.

Stone, G. D. (2004) Biotechnology and the political ecology of information in India, *Human Organization*, 63(2), pp. 127–40.

Thies, J. E. and Devare, M. (2007) An ecological assessment of transgenic crops, *Journal of Development Studies*, 43(1), pp. 97–129.

Visvanathan, S. and Parmar, C. (2002) A biotechnology story: notes from India, *Economic and Political Weekly*, 6 July, pp. 2714–24.

Yamaguchi, T. (2004) Discourse perspective on agrifood biotechnology controversies: *Bt* cotton in India, doctoral dissertation, Michigan State University, East Lansing, Michigan.

Naturalising Transgenics: Official Seeds, Loose Seeds and Risk in the Decision Matrix of Gujarati Cotton Farmers

DEVPARNA ROY, RONALD J. HERRING, &
CHARLES C. GEISLER

I. Indian Cotton, *Bt* and Risk

India has more area under cotton than any other country in the world, but cotton lint yield has been among the lowest. India's 22 million acres represents about one quarter of the world's total cotton area and occupies 5 per cent of India's cultivated area (for comparative data, James, 2002b). Farmer suicides in India's cotton belt, especially in 1998 and especially in Warangal district, Andhra Pradesh, underlined the risk of growing cotton. Debts sometimes overwhelmed farmers caught in a cost-price squeeze.[1] Farmers most at risk are those with fewest resources. To the extent risk-reducing innovations are scale-neutral, there is considerable potential to improve the condition of small producers (DuPuis and Geisler, 1988). This is the pro-poor potential claimed for biotechnology. But no insurance is risk-free (Eyzaguirre et al., 2004; Smil, 2004). Agricultural history is replete with cases in which early adoption of technology disproportionately advantaged more prosperous farmers (Dovring, 1973). Risks tend to intensify during experimental phases of new technologies. Thus, as Ortiz (1990) observes, peasant farmers cope by continually modifying their 'insurance' strategies and often pay a premium for seeds and other inputs to avoid worst-case income or property losses. Michael Lipton (1968) emphasised in a classic treatment that it is a rare farmer who does otherwise.

Among the most serious risks to cotton farmers in India are unmanageable pests, particularly the 'American' (green) bollworm (*Helicoverpa armigera).* Pesticides are costly and not always effective; resistance has developed in major pests. Biotechnology firms argue that their products can reduce this risk; *Bt*[2] cotton plants produce an endotoxin effective against bollworms. They are appealing to farmers if enhanced seed cost is compensated by savings of reduced pesticide applications and higher yield, improving net gains and insuring against risk of crop loss. Indian farmers are subjected to a tempest of claims and counter-claims, from extremely positive results to press reports of 'the failure of *Bt* cotton' (Parmar and Vishvanathan, 2003; Herring, 2005).

Gujarat is a major cotton-growing state in India, agro-ecologically falling into the central zone with the states of Maharashtra and Madhya Pradesh. Area under cotton was about 1.61 million hectares in 2001. The Asiatic types of cotton have been cultivated in Gujarat for thousands of years, and Gujarat has historically been an important centre of cotton production and trade. Attempts to introduce the New World species *G. hirsutum* were made in the eighteenth century. The world's first hybrid cotton *Shankar*-IV (or H-4) was released in 1971 by Gujarat's Cotton Research Station in Surat. Thus, Gujarat's farmers were among the first to experiment with hybrid cottons. Cottons are grown with intensive agro-chemical inputs; many farmers in Gujarat believe that heavy applications of pesticides and inorganic fertilisers threaten soil exhaustion. Farmers who exhaust their soils must change management practices or relocate to new lands, an option infeasible in a country where demand for land is endlessly backlogged. India's land per agricultural capita over 30 years has fallen by half, from 0.44 ha in 1970 to 0.22 ha, in 2000. Farmers are pressured to intensify production on what land they have through new management strategies and technologies. There are good reasons to consider new options, and *Bt* cotton has some attractive features.[3]

Seed choice is fundamental to crop risk; unreliable seed supply and adulteration plague Indian cotton farmers. There is frequently a large gap between demand for certified seed and supply (Siddhu, 1999). Shortage of certified seeds necessitates seed trading with fellow farmers and unauthorised seed sellers and traders (Lalitha, 2003). Cotton farmers typically buy hybrid seeds from recognised companies for reliability and to avoid loss of 'hybrid vigour', despite the lower cost of saved seeds. Compared to traditional indigenous (*desi*) seeds, hybrid seeds also require more industrial approaches to farming – more chemical fertilisers and pesticides, with attendant risks – and increased irrigation – a continuing crisis in Gujarat.

The diffusion-adoption literature on agricultural technologies overflows with examples of 'treadmill' behaviours. That is, structural pressures lead to adoption of state-of-the-art technologies to avoid ceding ground (often literally) to even earlier adopters (Cochrane, 1965). Today, this competition for 'adopter rents' is not limited to the farmer next door. In a world of receding protective tariffs and continued subsidies for farmers from richer countries, Indian cotton farmers find themselves competing with cotton farmers from around the globe. There is thus pressure from within – the stressful agricultural environment of Gujarat – and from external movements in prices and yields to experiment with new solutions. Gujarati cotton farmers cannot afford to be conservative.

Severe risk introduced by inadequacy of existing methods of pest control was manifest in September 2001, when a massive bollworm infestation struck Gujarat, devastating hybrid cotton varieties. There was one notable exception – *Navbharat* 151 (NB151), an unapproved *Bt* variety produced locally. Subsequently, Mahyco–Monsanto Biotech Ltd (MMB) found that NB151 contained the *Cry1Ac* gene for pest control present in the experimental varieties that firm simultaneously had undergoing politically charged biosafety tests. MMB claimed that Navbharat Seeds Limited – a small company based in Ahmedabad – had been selling NB151 seeds for the previous three years and demanded punitive action against the firm.[4] Once the transgenic pedigree of NB151 was discovered, the Genetic Engineering Approval Committee in Delhi ordered the Gujarat Biotechnology Coordination Committee to burn all illegal plantations, sequester the crop and sterilise the fields. These orders were not carried out. Resistance from mobilised farmers and the political importance of Gujarati farmers to the state government over-ruled Delhi.[5]

The discovery of illegal *Bt* cotton eased the way for approval of official *Bt* cotton: the crop was a *fait accompli*. In June 2001, Mahyco–Monsanto had sought approval from the government's Genetic Engineering Approval Committee (GEAC) for commercial cultivation of *Bt* cotton, but GEAC ruled that existing field-trial data were unreliable. Specifically, the crops had been grown when pest incidence was low (Iyengar and Lalitha, 2002). GEAC required MMB to repeat the field trials on a larger scale under the supervision of the Indian Council of Agricultural Research (ICAR). Ten months later, in March 2002, the government granted permission to MMB to sell three *Bt* varieties, provisionally for a three-year period. Deciding otherwise would have been pointless: the state government of Maharashtra had already legalised the seeds from 1 January 2002 and the state government of Gujarat had essentially told Delhi that its farmers would grow this crop if they wished. MMB released three transgenic cotton varieties under the brand name Bollgard (MECH-12, MECH-162, and MECH-184). In 2002–03, against a background of wide-spread crop failure produced by 'bollworm rampage' (Joshi, 2001) and now-legal transgenic cotton cultivars, Gujarati farmers had new planting options before them: Bollgard varieties; illegal non-Bollgard *Bt Navbharat* 151; offspring of NB 151 – the so-called 'loose' seeds of uncertain parentage;[6] and emergent branded (e.g., *Vaman*) *Bt* varieties locally but illegally produced (Jayaraman, 2004). How should new varieties fit into cropping strategies, if at all? Does official approval of some *Bt* seeds and not others matter? How did farmers think about managing new risks that might accompany whatever insurance strategy they pursued – for example, possible crop failure, heightened supplier dependency, adulterated seeds, higher production costs?

II. Mahyco–Monsanto's Survey of its Official Seeds: Comparisons Across States

At the time Gujarati farmers were making cropping decisions in 2002, the national picture was still dominated discursively by the 'Monsanto-terminator-suicide-seed' hoax (Herring, 2005). Monsanto was portrayed widely as a threat to India and to Indian farmers: why would farmers buy seeds from such a firm? The answer found in a large all-India marketing survey, and consistent with the in-depth Gujarat survey discussed below, is quite simple: higher profits.

Having marketed its three varieties of Bollgard seeds widely in 2002 and 2003, Mahyco–Monsanto sought to discover what farmers thought of their product. ACNielsen was retained by MMB for a nationwide survey.[7] That survey found improved yields, reduction in pesticide sprays against bollworms and higher profits in five cotton-growing states: Maharashtra, Madhya Pradesh, Andhra Pradesh, Karnataka and Gujarat for the growing season 2003–04 (Table 1). The study concluded that:

> ... there has been an approximately 30 per cent or 1.7 *quintals* per acre yield increase in Bollgard fields, when compared with conventional cotton fields. The net profit to farmers from Bollgard cultivation has increased significantly by nearly 80 per cent or Rs. 3126 per acre.

Moreover, 'the reduction in bollworm pesticide sprays . . . translates into an average savings of Rs.1294 per acre (reduction of 2–3 sprays per acre) for Bollgard farmers'. Of the surveyed farmers, 'more than 90 per cent of Bollgard users and 42 per cent of non-users express their intention to purchase Bollgard in 2004' (ACNielsen 2004a: 1). These results are not unlike findings in China and elsewhere (James, 2002b).

Selected performance measures for Gujarati cotton farmers using Bollgard seed are summarised in Table 1. In this data, reduction of pesticide use for bollworm reached 70 per cent (an average savings of Rs. 1392 per acre, the second highest reported among states). Average yields increased by 18 per cent (lowest among the six states), and net profits increased by 164 per cent (highest among the states). These results also indicate a strong inverse relationship between size of holding and net benefits of transgenic technology: marginal cotton farmers reported over twice the

Table 1. Bollgard outcomes across cotton-growing states

	Bollworm pesticide reduction		Yield increase		Increase in net profit	
State	%	Rs.	%	qu/ac	%	Rs/ac
Andhra Pradesh	58	1856	24	1.98	92	5138
Karnataka	51	1184	31	1.36	120	2514
Maharashtra	71	1047	26	1.48	66	2388
Gujarat	70	1392	18	1.20	164	3460
Madhya Pradesh	52	889	40	2.2	68	3876
All India weighted	60	1294	29	1.72	78	3126

Source: ACNielsen (2004a: 1).

net benefits of large farmers using Bollgard seeds.[8] The smallest producers in this data do especially well, contrary to much conventional wisdom concerning biotechnology. This outcome, if sustained in more detailed studies, conforms to a classic relationship widely reported in the agrarian-reform literature: acres in the hands of small farmers are more productive than those in the hands of large producers, primarily for reasons of labour-intensity and superior oversight (Herring 1983: 239–67).

The ACNielsen survey also found that there was a premium for Bollgard cotton lint in the market, averaging 8 per cent over conventional cotton.[9] More important for consideration of poverty, the survey found that agricultural labourers (who are paid according to the weight of cotton picked)[10] earned a significant increase in income because of the increased Bollgard yield. Though Bollgard was applied to just 1 per cent of the country's total cotton acreage in 2003, the survey projected that its use accounted for a 304,690-litre reduction in formulated pesticides used – saving some water and significant amounts of toxins.

The ACNielsen survey is widely distrusted in civil society in India because of corporate sponsorship; interpretive caution is warranted.[11] Nevertheless, it is the only nation-wide survey available. Highly aggregated data of necessity loses variance in favour of broad coverage. Probing beyond the large-scale survey in individual cotton-growing states will be necessary to interpret these results. In the following section, we explore nuances. Do the summary conclusions of the national survey accord with farmer decision-making on the ground in Gujarat?

III. The Gujarat Small-Scale Sample: *Bt* Adoption Logic

Our primary data source for this section is intensive in-depth interviews with 45 farmers conducted by the senior author in 2002–03 and supplemented by fieldwork the following summer.[12] The 45 farmers were from three different regions of Gujarat. Fifteen were from Saurashtra (Junagadh and Bhavnagar districts), 15 from north Gujarat (Sabarkantha district), and 15 from central Gujarat (Vadodara district). A summary of varieties grown by farmers in these regions is provided in Appendix 1. The sample is not representative, nor can one imagine any small sample that could be in so varied an agro-eological setting.

Of the 45 farmers in this sample, 35 chose to grow some *Bt* cotton variety in 2002–03, the first year of official seeds. Their reasons clustered around bollworm risk management and profitability. These farmers are continuously trying new varieties of cotton; most who had not yet tried *Bt* varieties planned to do so the following season, based on their assessment of local *Bt* results. After the devastating bollworm attack of 2001, which only the un-official transgenic NB151 survived well, one might wonder why some farmers rejected *Bt* technology.

Of the 10 farmers who did not grow *Bt* cotton in 2002–03, there were six in north Gujarat, two in central Gujarat, and two in Saurashtra. Of the six in north Gujarat, four were organic farmers who grew non-hybrid *Lalsanthi* seeds in 2002–03. This indigenous variety is considered less susceptible to diseases and requires less water. Two brothers who make their own herbal pesticides for organic cultivation were planning in 2003–04 to grow two-and-a-half acres of *Lalsanthi* and three acres of *Bt* cotton F_2 (loose seeds).[13] They planned to use only organic fertilisers and herbal pesticides to test loose *Bt* seeds on their land. They saw no contradiction between organic farming and transgenic seeds, though other farmers did. One organic farmer said 'My opinion of *Bt* is that the farmer likes it now. But my mind says it is not sustainable technology'. He cited research in China showing that 'American' (green) bollworm develops full resistance to *Bt* cotton in the twentieth generation. He said that the roughly 5 per cent of the farms in his area which were growing *Bt* cotton – Bollgard and *Navbharat* 151 – were susceptible to green bollworm attack. Moreover, though cotton is generally susceptible to wilting, he felt that 'in *Bt* cotton the wilting problem comes so early that fields turn dry', though this problem was more pronounced in Bollgard than in *Navbharat* 151. He worried that there is no scientific analysis of what kind of losses will be incurred and what kind of environmental pollution will occur.

Non-organic farmers had various reasons for rejecting *Bt*. One farmer growing *Shankar* VI regretted his low yields but declined *Bt* seeds because of price. He added that many farmers in his village had grown *Bt* cotton, with good results. However, the *Bt* cotton crop in his village was susceptible to 'American' (green) bollworms. He added that even though F2 seeds of *Bt* cost less, they give less yield and are less pest resistant. Contrary to the general acceptance of gray-market seeds and casual approach to regulatory systems in Gujarat, he felt that:

> ... we should not oppose any technique, but we should not let any technique come in without proper testing. We don't know the effects of *Bt* on other cotton or other crops. Nobody knows that. We should have given permission to *Bt* only after proper testing. The way it came into Gujarat was wrong. The government also came to know of it later. It shouldn't take place in a democracy.... There is difference in the environment of other countries and ours. We have many cotton varieties. Why doesn't the government produce *Bt* itself as has been done in China?

Other reasons for not growing *Bt* cottons in 2002–03 were as varied as are farmers' experiences, but most who did not try *Bt* in the first year of legal growing did plan to try it in the following season. One farmer in Saurashtra was fairly typical: he was planning to experiment with one packet of Bollgard seeds in 2003–04 after asking neighbouring farmers, the government extension agent, and dealers about which Bollgard variety to buy. He said, 'Bollgard variety requires less pesticide, we have seen that with our own eyes', but there is always the question of how well any given variety will do in one's own soil – hence experimentation. A colleague in Saurashtra said, 'I do three to four acres of cotton every year. I put one variety for experience – to see if it does well in my soil. So I use different varieties of cotton for experience. This year I did not do any research varieties because I had found Vikram 5 to be good for my field'. He planned to grow *Bt* cotton on seven acres in 2003–04, and specifically MECH-12 because of its short duration and bollworm resistance. The Bollgard transgenic will allow resuscitating his earlier strategy of crop rotation. He used to grow Mahyco 4, which would be followed by groundnut and wheat, but this variety is no longer used in his area. He said that perhaps he could grow groundnut and wheat after taking a crop of short-duration MECH-12.

In this sample, *Bt*-non-*Bt* adoption derives from agronomic differences across farms, ideological considerations, and differences in knowledge. The 10 farmers who did not grow *Bt* cotton in 2002–03 were somewhat less likely to have irrigation (80 per cent, compared to 91 per cent of the *Bt* cotton farmers). Fewer used both synthetic and non-synthetic fertilisers (60 per cent, compared to 91 per cent of the *Bt* cotton farmers), and fewer combined both chemical and non-chemical pest control measures (10 per cent, compared to 80 per cent of the *Bt* cotton farmers). Almost 63 per cent of the *Bt* growers used market seeds, because they believe that purchasing seeds each year guarantees 'hybrid vigor' in their plants – and better yield. Half of the conventional cotton growers use market seeds, for the same reasons. Conventional cotton growers were all members of cooperative societies, compared to about two-thirds of the *Bt* cotton growers. This outcome may reflect the weakening of the cotton cooperative movement in Gujarat: farmers at the cutting edge of new technology are less likely to be members.

Though seed-saving is often presented as the antithesis of biotechnology, *Bt* cotton farmers *of all sizes* prefer to use a mix of saved and market seeds. *Bt* cotton farmers were almost as likely to use saved seeds as conventional farmers: 34 per cent of the *Bt* farmers used saved seeds, compared to 50 per cent of the conventional cotton farmers. Several farmers used saved 'loose' *Bt* seeds – mostly F_2 seeds derived from NB151, but unmarked and of uncertain lineage – because the illegal but popular *Navbharat* 151 seeds were no longer available in the market. Farmers were using new illegally marketed *Bt* seed varieties as well (with brand names such as *Vaman* and *Rakshak*), along with both Bollgard and non-transgenic cotton varieties from the market. Table 2 presents the cotton choices in this sample

In this sample, loose *Bt* was preferred by more farmers than official *Bt* [Bollgard]. Far from constituting monopoly power over farmers, Monsanto's *Bt* cotton is not winning the competition on the ground in these data. The choice matrix is in part

Table 2. Variety choices of 35 *Bt* and 10 non-*Bt* cotton farmers, 2002–03 growing season

Varieties	Number of farmers
Farmers growing *Bt* cotton	
Official *Bt* (i.e., Bollgard varieties) only	7
Loose *Bt* only	11
Official *Bt** + Loose *Bt*	6
Branded but illegal *Bt* (*Vaman Bt*) + Official *Bt*	1
Loose *Bt* + non-*Bt* varieties	6
Loose *Bt* + one *Bt* official + one non-*Bt* variety	1
Official *Bt* + non-*Bt* varieties	3
Farmers not growing *Bt* cotton	
Lalsanthi (using organic techniques)	4
PAC 135	1
Shankar-VI	1
JK Durga + Banni	1
JK Durga + Banni + Shankar-VIII + 555	1
Shankar-VI + Sanju	1
Vikram 5	1

Note: 'Official *Bt*' is defined as growing at least one Bollgard variety.

clear: the cheapest seeds are 'loose *Bt*': the F_2 generation of *Navbharat* 151. Though the government has banned the sale of NB 151, the F_2 seeds were available at the rate of Rs. 50 to Rs. 100 per kilogram. 'Loose' varieties were sold or traded informally among farmers. Though cheaper, they may give weaker expression of the *Bt* toxin lethal to bollworms. Moreover, there is no authentication of the seeds: they may be of low quality – or even fraudulent in their *Bt* claim. What stands out is the experimentation among these farmers, and the mix of *Bt* and non-*Bt* in cropping strategies.

IV. Elements of the Decision Matrix

Coping with risk begins with information. How did farmers find out about seed options? All 45 farmers in the sample were informed about *Bt* cotton and its claim to mitigate bollworm attacks. Further, they knew that both legal and illegal varieties of *Bt* cotton were available in Gujarat. Most farmers said that they depended on their own experience when deciding which varieties of seeds to purchase. A second important factor was the experience of other farmers (often those belonging to the same cooperative society). Farmers also paid attention to the news from articles in newspapers such as *Gujarat Samachar* and articles in farming magazines such as *Krushi Jeevan* and *Krushi Gauvidya*. A fourth important source was radio and television programs (for example, *Annadata*). Some farmers also mentioned that government extension agents were important sources of information about the qualities of different seed varieties, while others said they depended on private company extension agents, demonstrations and literature to inform their choices; others cited agro-dealers as important sources of information. Trust in the brand name and company were important. Some said they went to agricultural fairs to get more information about seed varieties. Others learned about varieties from their cooperative societies or from NGOs. NGOs were especially important for organic farmers. A few farmers went to the agricultural universities in Gujarat to get more information about seed varieties.

There is nothing passive about this behaviour. These farmers were actively experimental, continually testing different varieties to see how well they would grow in their particular soil and water conditions.

After collecting information about different seed varieties from various sources, farmers faced two decisions: (1) whether to grow *Bt* cotton or not; (2) if yes, whether to grow legal varieties of *Bt* cotton or not. Those who decided to grow *Bt* cotton (whether legal varieties or illegal varieties) considered water and electricity to be necessary conditions. Most farmers believed that all *Bt* cotton varieties (whether legal or illegal) needed more water compared to non-*Bt* varieties: both opinions and experience varied on this dimension.

If farmers chose to grow *Bt* cotton, they then had to decide whether to grow legal or illegal varieties. Again water came first. Many farmers believed that the water requirement of *Navbharat* 151 was less than that of Bollgard varieties. Several believed that pesticide costs of *Navbharat* 151 for bollworms and sucking pests were less than those required for Bollgard varieties; they preferred F_2 offspring of NB 151 to Bollgard. Pest-resistance management also figured into the decision of some farmers: they believed that there were no refuges required when growing

Navbharat 151, while those planting Bollgard had to grow refuges. These farmers did not wish to grow refuges, considering it a waste of land. The result of these multiple considerations was the predominance of loose seeds in Table 2. Farmers who rejected loose seeds despite their low cost cited loss of hybrid vigour in F_2 or F_3 seeds. Reinforcing this logic was a feeling of some farmers that seeds available in the open market were preferable to black-market seeds; they chose Bollgard varieties precisely because of their official, government-approved status. One farmer chose a Bollgard variety because of its comparatively long staple length. These preferences by their nature are unstable; all are subject to testing in the field. For example, several farmers who grew MECH-184 in 2002–03 found it requires early watering to avoid wilting and thus is inappropriate for their farms.

Supply influences choice as well. Some farmers said that in 2002–03, their agro-dealers had few packets of particular *Bt* varieties; they were not able to sow *Bt* seeds according to their choice. These farmers said that the government should give permission to many companies, including Navbharat, to sell *Bt* cotton seed legally, to ensure availability of multiple varieties of *Bt* seeds.

V. Selected Farmers' Experiences with Yields and Income

In this sample, there were only two farmers (both from Junagadh district) who had planted Bollgard MECH-12 (hereafter Bollgard will be assumed; MECH is Mahyco's name for their hybrid varieties). S. B. Patel of Junagadh district cultivated MECH-12 on five acres in 2002–03 and obtained a yield of 14 quintals per acre. He approves of MECH-12 because of its yield, and because the boll size was good and the staple length was 32 mm. The other farmer, J. V. Barot, reported a significantly lower yield of seven-and-a-half quintals per acre. Mr. Barot had also grown a branded unofficial *Bt* variety called *Vaman Bt*, which gave the same yield. He reported that MECH-12 had some problems from sucking pests and American bollworm, while there were no such problems with *Vaman Bt*. For this farmer, MECH-12 had certain disadvantages compared to the gray-market transgenic competitor.

Another farmer in the Junagadh district, J. Chauhan, reported that he had grown Mahyco 1 (which is non-transgenic) on six acres, MECH-184 on one acre, and loose *Bt* on one acre. His yield was five quintals per acre from Mahyco-1, 5 quintals per acre from MECH-184, and 10 quintals per acre from loose *Bt*, though he gave the same irrigation to all three varieties. Sucking pest attacked all three varieties. MECH-184 suffered the least sucking-pest attack. Mahyco 1 had the most serious bollworm attack; MECH-184 had few bollworms, whereas loose *Bt* had none. On all counts, for this farmer, the loose-seed transgenic proved superior, even leaving aside the seed cost differential, which is huge.

Mahyco–Monstanto's MECH-184 also had problems with wilting, as mentioned by a number of farmers from different parts of Gujarat, including S. B. Patel of Junagadh. In addition to being a farmer, S. B. Patel is a seed trader. He said he did not sell much MECH-184 because it is an early-maturing variety, and thus prone to wilting. He concluded that this variety must get water and fertiliser in its early stages or the crop can fail. Patel compared MECH-184 to a small baby that should be carefully fed milk during the early stages of life. He believed that farmers had problems with this variety because they did not know this critical agronomic fact. In a similar vein, S. J. Jamod of Vadodara said that MECH-162 was successful in his area, but not MECH-184. He obtained 14 quintals per acre from MECH-162, but 11 quintals per acre from MECH-184. He had chosen not to grow NB151 in 2002–03 because only F_2 and F_3 generations were available to him, and he believed these seeds do not yield well. But he maintained that NB151 was better than official varieties because it grows faster and needs less water than MECH-184. He was advised by the representative of Monsanto to water MECH-184 every 10 to 15 days. As he had no water source of his own and had to buy water from others, 184 was an inferior choice to 162. Not trusting F_2 and F_3 generation seeds, Mr. Jamod was willing to invest in costly Bollgard seeds, but found only one variety appropriate for his circumstances.

L. Savjibhai of Junagadh experimented with all available transgenics: he had grown MECH-184 on two acres, MECH-162 on three acres and *Navbharat* 151 (F_2) on one acre. He owns 25 acres of land and a tractor and irrigates from his own well. Savjibhai reported a yield of seven-and-a-half quintals per acre with MECH-184, six quintals per acre with MECH-162, and only three quintals per acre with F_2 seeds. Water figured in his understanding of differences: Bollgard plants needed more water than NB151, and MECH-184 needed more water than MECH-162. He felt that NB151 had less sucking pest attacks and fewer bollworms than MECH-184 and MECH-162. Despite some pest attacks, Savjibhai thought that there was a 'difference of earth and sky' (*zamin aasman no farak chhe*) when comparing *Bt* cotton with non-*Bt* cotton; *Bt* cotton was profitable because it lowered the incidence of pest attack. Because Gujarati farmers generally face a shortage of water and NB151 required less water than MECH varieties, and handled pest attacks with less expenditure, Savjibhai preferred NB151. The conclusion of this farmer and seed merchant has negative aggregate consequences: Savjibhai prefers a variety that gives less than half the yield of the best MECH variety, but the upfront costs of seeds are a fraction of those of MECH seeds and water requirements are less. The profit margin may be higher, but cotton yield suffers and labourers who are paid by the kilogram picked have less income than would be the case with higher yields. Moreover, his preferred variety is illegal.

N. Patel of Vadodara had grown MECH-162 on about three acres and loose *Bt* on 45 acres in 2002–03. He obtained similar yields (about 10 quintals per acre) for both varieties. Attacks of bollworms and other pests had been so severe in his area that yields had not exceeded 7.5 quintals per acre during the previous five years. He attributed his highest yield in five years specifically to transgenic cotton. Nevertheless, N. Patel said that he would not be growing MECH in 2003–04 because the seeds were costly and the yields of loose *Bt* were comparable. In contrast, K. Bhatt of Vadodara had been growing *Navbharat* 151 for the previous three years. But it was only in September 2001 that he realised that he had been growing a transgenic crop. In 2002–03, he planted NB151 on 11 acres and MECH-184 on one acre. The weather was good in 2002–03 and there was less incidence of pests; attacks of sucking pests were less on NB151 compared to MECH. He obtained a high yield of 13.6 quintals per acre from MECH-184 and a yield of 10 quintals per acre from NB151. B.R. Patel of Vadodara grew NB151 in 2001–02 and obtained yields of eight quintals per acre. In 2002–03, he obtained yields of 13 quintals per acre from NB151. Most cotton farmers reported that 2002–03 was a lower pest-incidence year. Still, he noted an attack of sucking pests on NB151, but no bollworm attack. A similar conclusion on pests was reached by A.R. Bhanderi of Vadodara district. He grew *Gujarat 23* (an indigenous variety) and a loose *Bt* variety. He obtained a yield of 2.3 quintals per acre for Gujarat 23 and 4.5 quintals per acre for loose *Bt*. Loose *Bt* seeds did better than *Gujarat 23* in terms of yield, but required two sprayings for sucking insects.

VI. Small Farmers' *Bt*

Are small farmers experiencing the introduction of transgenics in any systematically different way when compared to their larger neighbours? In general, the answer is no. The 14 smallest farmers in the sample split between *Bt* and non-*Bt* in ways similar to the larger farmers. Six of these farmers chose not to plant *Bt* in 2002–03, but of these, four said they would plant *Bt* in 2003–04. The other two farmers preferred organic farming and ruled out *Bt*.[14] Of the four planning to plant *Bt* in 2003–04, three said they would plant MECH while one said he would plant loose *Bt*. Of the smallest 14 farmers, eight planted *Bt* cotton in 2002–03: a combination of loose *Bt* F_1, loose *Bt* F_2, MECH-184, MECH-162 and MECH-12. All eight said they would plant *Bt* cotton (either loose or official) in 2003–04.

An example may help illustrate their logic. M. D. Patel, from Saurashtra, owned five acres of land; he two acres to *Navbharat* 151 F_2 seeds and obtained a yield of 300 kg per acre. He sowed MECH-184 seeds on 1.3 acres but he did not get good results. He said, 'Sucking pest attack and bollworm attack was more on MECH-184 compared to NB151. I sprayed Confidor (for sucking pests) one time on NB151 and I had to spray Confidor six times on MECH-184'. He said he would sow NB151 in 2003–04. Another small farmer from Central Gujarat who owns 5.5 acres of land said he grew MECH-162 and MECH-184 in 2002–03.

> I decided that since no other company has permission to sell *Bt*, I will buy Monsanto's seeds. The seeds were costly – Rs 1600 for 450 grams, but I got enough yield, so MECH-162 suited me. MECH-162 is a success for our soil but MECH-184 has not suited the farmers at all.

He obtained a yield of 1439 kg per acre for MECH-162 in 2002–03. His yield of 1136 kg per acre for MECH-184 was termed a 'failure'. He planed to sow MECH-162 in 2003–04. This farmer did not use NB151 F_2 because he claims that yields decrease by 50 per cent with the second generation of hybrids. He used to spray pesticides 15 to 17 times on *Shankar-VIII* and *Shankar-X*; with MECH he sprays pesticides only four to five times.

The smallest farmer in this sample – Mr Barot – owns four acres of land and grew cotton on three acres in 2003–04. He has been practicing organic farming for seven years. He grew an indigenous variety of cotton (*Lalsanthi*) in 2002–03, as he had for the previous three years; its yield is only 150 kg per acre. Despite this performance, the benefits of *Lalsanthi* are many in this farmer's mind: it is immune to pests, it gives 'good yields', it is early maturing – ready in 160–80 days like the hybrid varieties – and picking is easy because the bolls open well, though they are small.

VII. Conclusion: Risk, Transgenics and Gujarati Cotton

Gujarati farmers use an array of information points and risk assumptions in deciding which cotton seeds to combine in their fields. The pattern was similar for large farmers and small farmers in field interviews. Most are comfortable with transgenic seed materials as part of their familiar risk-aversion strategies. *Bt* cotton varieties, if accessible and affordable, contribute to risk management. Cotton farmers in Gujarat are wary of dependence on a single variety: even if *Bt* cotton monocultures have a reputation for high yields, the dominant ground strategy is continual experimentation and mixing of varieties. These farmers say that even though *Bt* varieties reduce pesticide costs substantially and improve protection from bollworms, they are not without risk. Seed traders can be unscrupulous, passing along adulterated seeds of uncertain parentage. Risky seeds include the inexpensive 'loose' seeds (unpackaged, unlabeled transgenic varieties) with which farmers continue to experiment.

But if loose seeds entail risk, farmers also feel that it is risky to incur the very high cost of transgenic seeds officially approved by the government. *Navbharat* 151 had given good results in the preceding year (September–October 2001), when the bollworm infestation was particularly heavy, motivating farmers to try filial generations of NB151 in 2002–03.

Seed cost alone cannot be the determining factor in choice, however. There is considerable risk in tying up land and resources in any seeds, inexpensive or dear. Most important, farmers' experiences with *Bt* cotton cultivars were quite variable – contrary to the claims of public intellectuals claiming *Bt* cotton to have been a disaster.[15] Experientially, *Bt* cotton has done well for most farmers in both data sources used in this study, but there are many choices to be made within the *Bt* frame and across other cotton varieties:

(a) *Seed price*: For critics of transgenic technologies, the potential of monopoly and price-gouging from dependency makes the new seeds unattractive for the poor. In Gujarat, something like the obverse of monopoly is evident in the fields – a rare competitive market. Some farmers complained about the high cost of Bollgard varieties, but demand for official seeds continues to grow. The price of *Bt* seeds can be considered an insurance cost; buying bollworm insurance in 2002–03 would have been a far less wise investment than in 2001, but farmers could not know this in advance. For many farmers in 2001, the choice to stick with tried and true traditional varieties resulted in disaster. In the ACNielsen survey, smaller farmers benefited more from *Bt* than large in revenue enhancement per acre. Yet limited-resource farmers, for whom the higher cost of official seeds must be subjectively set against their insurance value, have not coalesced on any particular strategy.

(b) *Legal risk*: Both official and loose *Bt* seeds performed reasonably well in 2002–03. Farmers who preferred the cheaper grey-market seeds perceived no legal risk in buying them.

(c) *Yields*: There was no general pattern; agronomic variation is telling. Some farmers obtained higher yields with loose-seed transgenics, others with Mahyco–Monsanto MECH varieties. If there is any conclusion, it is that *Navbharat* 151 – the 'Robin Hood' seed[16] – may have been the best variety, though it has been officially banned. MECH-184 seems to be the most finicky about water timing. All *Bt* seeds proved capable of producing profits under typical conditions.

(d) *Safe seeds*: Though yields achieved with unofficial seeds were high, there is risk in buying loose seeds. Farmers feel their choices would improve if the government legalised NB151; they would then have better assurances of unadulterated seeds from licensed firms. Most farmers felt that the government should allow Navbharat and other companies to sell *Bt* cotton, providing wider varietal selection and competition to reduce prices.

(e) *Sucking pests and bollworms*: *Bt* cottons provided a fairly high degree of protection from the American bollworm, the spotted bollworm and the pink bollworm in Gujarat; sucking pests such as aphids, jassids and whiteflies remain a threat. Farmers would like to see new technology for protection against sucking pests and complete protection against bollworms to reduce pesticide spraying further.

(f) *Pest insurance*: On some farms, pest incidence was much lower in 2002–03 compared to 2001–02. Therefore the performance difference between *Bt* cotton and non-*Bt* cotton declined somewhat in 2002–03. Most *Bt* farmers saved money by reducing pesticide application. If *Bt* seeds command a premium price, the value of implicit insurance purchased thereby is difficult to calculate, as farmers cannot predict pest incidence. This uncertainty helps explain the widespread preference for loose *Bt* seeds: the insurance is cheap and works reasonably well. It is more a satisficing than an optimising strategy.

(g) *Wilting and agronomic knowledge*: Farmers who cultivated MECH-184 reported problems with wilting. Some lacked knowledge of agronomic characteristics of the variety: they were not aware that more water was required early in the growing cycle. Others simply could not supply enough water. Nevertheless, some farmers had excellent yields from this most problematic of the MECH varieties.

(h) *Resistance management*: The *Cry1Ac* toxin is a public good. If *Bt* crops speed development of resistance to this natural pesticide, there is enhanced collective risk. *Refugia* are the proposed remedy. All farmers growing MECH varieties in the small sample claimed to have planted refuges, but those cultivating unofficial *Bt* varieties did not. It is unknown whether bollworms will develop resistance more quickly in Gujarat because of lack of refuges or if the combination of small farms and polycultural practices, as well as alternative hosts for bollworms provided by India's specific plant diversity, are sufficient to offset poor refuge management.[17] Most farmers were unconcerned with this long-term public good in their seed choices.

The conclusion of the national study by ACNielsen – that Indian farmers will prosper if they adopt Bollgard seeds due to remarkably high net returns – is not contradicted by this analysis, but caveats are introduced. Likewise, the structure of decisions is consistent with the model and outcomes suggested by Zilberman et al. (this issue), though with more complexity of choice. Specific MECH varieties may not prosper in specific fields. It was highly unlikely that three officially approved varieties of *Bt* cotton could serve the quite variable agronomic conditions Indian farmers face; complaints a bout Mahyco–Monsanto followed. Moreover, the official seeds carry a high cost that must be borne upfront, with the possibility of indebtedness if the crop fails. This cost could deter poorer farmers if credit institutions are not effective. Despite national-level rhetoric, Gujarati farmers see transgenic cotton varieties not as 'miracle seeds or suicide seeds', but as useful additions to strategies of adapting to adverse biotic and abiotic risks. Unlike lumpy capital investments that systematically favour wealthy farmers, transgenic seeds – both in theory and on the ground – exhibit what Rogers (1995) calls 'trialability' – defined as the degree to which experimentation with an innovation may be done on a limited basis. Trial and error experimentation is consistent with the way Gujarati farmers approach cotton.

Despite globalisation and industrialising pressures in agriculture, Gujarati farmers are empirical and experimental in their struggle with the cotton economy and its risks. Their attitude is essentially empirical: what works? Larger ideological constructions of transgenics are missing in the fields (Herring, forthcoming b). Large and small farmers alike used saved and 'loose' seeds, officially approved and illegal transgenic seeds in 2002–03. The high price of official seeds distressed many; the uncertain quality of loose seeds worried others. Accordingly, they mixed and matched seeds according to their needs, not all of which were economic.[18] Ongoing experimentation is an elemental part of their risk-management strategies; transgenics have simply added new possibilities to the mix. It is in this sense that most Gujarati cotton farmers have naturalized transgenics, fitting them into traditional strategies of conceptualizing and managing risk and assuring a livelihood.

Most discussion of agricultural risk centres around capital: who has put up money and may lose it? But there are many at risk in an agrarian economy who have no voice in technology choice. As in much of the world, the most precarious poor in Gujarat are the rural landless labourers. The ACNielsen study reports very large aggregate benefits for agricultural labour.[19] This finding is consistent with the agronomic findings in those agrarian settings in which labourers are paid by the weight of the harvested crop. As yields increase, so too does the harvest wage bill, assuming farmers cannot exert sufficient power to capture all the benefits of technical change. There is some support for the ACNielsen findings in Gujarat: some labourers interviewed in summer 2004 found *Bt* cotton beneficial, as they could pick more in a given day.[20] Though they lacked knowledge of the different varieties, they did not fear displacement by the reduction of pesticide applications, as this happened at a time of year in which other employment was available. There can be no firm conclusion here, but it is important to note that not all risk is borne by farmers: in the event of crop failure, such as the bollworm rampage of 2001, there is no cotton to harvest, and hence no income for cotton workers. In bumper crop years, labourers have a somewhat easier time making subsistence. Labour has a stake in the risk strategies of farmers.

It is widely recognised in the development community, at least rhetorically, that for both pragmatic and ethical reasons, farmers should have a say in agricultural developments that affect their welfare (Toennissen, 2004). We would add other elements of rural society, including farm workers. The views and voices of real agriculturalists have frequently been drowned out in the largely metropolitan conflict over biotechnology – more often a discourse of cities than of villages. *Bt* cotton is already altering the landscape and risk calculus of rural India. What we learn from farmers of Gujarat should help understand both policy challenges of a technology rapidly spreading across India and the complexity of decision-making among farmers facing new options.

Notes

1. Vasavi (1999), Stone (2002), Parmar and Visvanathan (2003). Reports of a suicide epidemic have not involved Gujarat.
2. In 1998, the Indian firm Maharashtra Hybrid Seeds Company Ltd (Mahyco) partnered with Monsanto to produce a joint venture – Mahyco–Monsanto Biotech Ltd (MMB). Breeders back-crossed Monsanto germplasm with local cultivars to obtain varieties that would suit Indian agronomic conditions. On *Bt* biology and agronomics, see Thies/Devare, forthcoming.
3. Clive James (2002a: 19) reports that 'In China the economic gain for resource-poor *Bt* cotton farmers [in 2001] was $500 per hectare. ... In South Africa, where 50 per cent of cotton farmers are women, cultivation of *Bt* cotton allows them more time to care for children, the sick, and/or generate additional income from other activities'.
4. Navbharat did not market NB 151 as a transgenic, only as a hybrid especially resistant to bollworms. It resulted from a cross of plants containing the *Cry1Ac* gene (the origin of which is disputed) with a local variety (Gujarat Cotton 10). Navbharat still faces charges for biosafety violations in selling un-approved transgenic seeds, but there is no patent protection for the Monsanto gene in India. (See Herring, forthcoming a).
5. Joshi (2001), Parmar and Vishvanathan (2003), Herring (2005: 210–7); Herring (forthcoming a).
6. The English word is in use locally; farmers also call these seeds *lodhavela biyaran* ('ginned seeds' in Gujarati) to designate non-Bollgard *Bt* seeds. These are sold loose – unpackaged and unbranded – by farmers to each other or by traders to farmers. Gupta and Chandak (2005) report widespread crossing of hybrids with *Bt* varieties by farmers for informal sale among neighbours.
7. The authors thank researchers at ACNielsen ORG-MARG for sharing unpublished information on their research findings. Their study interviewed 1672 Bollgard cotton farmers and 1391 conventional cotton farmers in five states. For research design, ACNielsen (2004).
8. Using size categories standard for India, the following net revenue *changes* [in Rs/acre] for *Bt* adoption are reported (ACNielsen 2004b): marginal farmers: 11,000 to 12,000; small farmers: 7000 to 8000; medium farmers: 6000 to 7000; large farmers: approximately 5000.
9. The reason for a premium is that bolls are less damaged. Raju Barwale, MMBL, 2004, personal communication to Herring described the same result, of somewhat lower magnitude; the small Gujarat sample discussed below finds no premium, nor did Herring's interviews with cotton farmers in Gujarat in 2005.
10. An estimate of Rs. 75,00,000 ($1.67 million). This payment practice is not universal. Given differential class power, it is uncertain that labourers can count on sharing gains in labour productivity from technological change.
11. Even if not tainted by MMB funding the ACNielsen data face potential validity and reliability problems in their size categories, imputed labour values and other issues. Farmer responses to questions about acres farmed often confuse operational and owned holdings and variable interpretations of joint-family versus nuclear-family holdings. Rented land takes many different form. Not all rented lands involve cash leases and may mix elements of lease and mortgage. In the field study discussed below, 8 of 45 farmers (17 per cent of the sample), had leased land. Finally, there is a problem of comparing isogenic varieties for scientific assessment of the *Bt* added value.
12. The 45 farmers are part of a larger snowball sample of 120 farmers for Devparna Roy's dissertation (Roy, 2006). Non-random snowball sampling was used to identify respondents, based initially on information of local extension agents. All farmers' names used in this paper have been changed to protect their privacy.
13. When two different parent lines (whether pure lines or not) are crossed, the first generation of seeds is called F_1. When resultant seeds are sown again, that generation is called F_2; the third generation, F_3, and so on.
14. Organic cultivation of cotton has emerged as an alternative, but biopesticides account for a small fraction of the market – about 3000 tonnes of the 0.1 million tonnes of annual pesticide trade in India (Singh, 2002).

15. For example, Krishnakumar (2004) reports that a six-member panel set up by the Gujarat government to evaluate the performance of *Bt* cotton concluded that Bollgard is unfit for cultivation and should be banned in Gujarat. Gupta and Chandak's (2005) survey of Gujarat's farmers found that Mahyco–Monsanto Biotech (MMB) cotton gave higher yields than *Navbharat* 151, but their preference remains the latter because of its ability to yield for longer duration, earliness in flowering and lower seed costs.
16. On farmer organisations' role in resisting official orders to destroy the crop (Herring, forthcoming a).
17. See Thies and Devare (forthcoming) for explanation of why the effects of *refugia* on resistance development in *Lepidoptera* are unknown.
18. Though not part of the sample under discussion, Roy has discussed seed choice with farmers who prefer traditional values of agriculture, including organic methods. There is in India a distinctive Gandhian undertone to preference for established over novel technologies.
19. But their methodology prevents a necessary disaggregation: 'labor' is imputed a value in their findings whether performed by the farmer or by hired wage labourers.
20. Laborers said that they could pick more cotton because of the shape of the boll. The *Bt* gene should do nothing to the morphology of the boll; this must be a varietal characteristic, perhaps more common in the varieties to which the *Bt* technology has been applied, but not a *Bt* trait. Gupta and Chandak (2005) report that several farmers have crossed (illegal) Navbharat 151 with local hybrids and produced economically beneficial results, extending the life of cotton in the field from six to nine months to take advantage of continuous flowering and thus higher yield.

References

ACNielsen (2004a) Nationwide survey by ACNielsen ORG–MARG underscores benefits of Bollgard™ cotton, Press release, ACNielsen ORG–MARG, Mumbai.

ACNielsen (2004b) Performance of Bollgard cotton in 2003, PowerPoint presentation produced by ACNielsen ORG Centre for Social Research, Mumbai, 26 March.

ACNielsen (2004c) Research design, AC Nielsen ORG–MARG, Mumbai, accessed at: http://acnielsen.com

Bambawale, O. M., Singh, A., Sharma, O. P., Bhosle, B. B., Lavekar, R. C., Dhandhapani, A., Kanwar, V., Tanwar, R. K., Rathod, K. S., Patange, N. R. and Pawar, V. M. (2004) Performance of *Bt* cotton (MECH-162) under integrated pest management in farmers' participatory field trial in Nanded district, Central India, *Current Science*, 86(12), pp. 1628–33.

Cochrane, W. (1965) *The City Man's Guide to Farm Problems* (Minneapolis, MN: University of Minnesota Press).

Directorate of Agriculture, Gujarat State (2002) Districtwise area, production and yield per hectare of important food and non-food crops in Gujarat State, Krishi Bhavan, Gandhinagar, Gujarat.

Dovring, F. (1973) Land reform: ends and means: a background study to the IBRD policy paper on land reform, International Bank for Reconstruction and Development/International Development Association, University of Illinois – Urbana.

DuPuis, E. M. and Geisler, C. (1988) Biotechnology and the small farm: what can green-revolution studies tell us about the social impacts of bovine growth hormone?, *BioScience*, 38, pp. 406–11.

Eyzaguirre, P., McCarthy, N., Di Gregorio, M. and Dennis, E. (2004) Property rights, collective action, and plant genetic resources, FOCUS Brief 10, in R. S. Meinzen-Dick and M. Di Gregorio (eds), *Collective Action and Property Rights for Sustainable Development.*

Gupta, A. K. and Chandak, V. (2005) Agricultural biotechnology in India: ethics, business and politics, *International Journal of Biotechnology*, 7(1–3), pp. 212–27.

Herring, R. J. (2001) *Promethean science, Pandora's jug: conflicts around genetically engineered organisms in India,* Mary Keating Das lecture, Columbia University, New York, 3 December. Transcribed and web-published, http://www.einaudi.cornell.edu/southasia/workshop/pdf/columbia2001b.pdf

Herring, R. J. (2005) Miracle seeds, suicide seeds and the poor: GMOs, NGOs, farmers and the state, in R. Ray and M. F. Katzenstein (eds), *Social Movements in India: Poverty, Power, and Politics,* pp. 203–32 (Lanham, MD: Rowan and Littlefield), pp. 203–32.

Herring, R. J. (2007a) Stealth seeds: bioproperty, biosafety, biopolitics, *Journal of Development Studies*, 43(1), pp. 130–57.

Herring, R. J. (2007b) The genomics revolution and development studies: science, poverty and politics, *Journal of Development Studies*, 43(1), pp. 1–30.

Herring, R. J. (1983) *Land to the Tiller: The Political Economy of Agrarian Reform in South Asia* (New Haven: Yale University Press).

Iyengar, S. and Lalitha, N. (2002) *Bt* cotton in India: controversy visited, *Indian Journal of Agricultural Economics*, 57(3), pp. 459–66.

James, C. (2002a) Preview: global status of commercialized transgenic crops: 2002, ISAAA Briefs, No. 27, Ithaca, NY.

James, C. (2002b) Global review of commercialized transgenic crops: 2001 feature: *Bt* cotton, ISAAA Briefs, No. 26, Ithaca, NY.

Jannuzi, F. T. (1994) *India's Persistent Dilemma: The Political Economy of Agrarian Reform* (Boulder, CO: Westview).

Jayaraman, K. S. (2004) India produces homegrown GM cotton, *Nature Biotechnology*, 22(3), pp. 255–6.

Joshi, S. (2001) Unquiet on the western front, *The Hindu Business Line*, 19 December.

Krishnakumar, A. (2004) Biotechnology. *Bt* cotton, again, *Frontline*, 21(10), accessed at: http://www.frontlineonnet.com/fl2110/stories/20040521001408100.htm

Lalitha, N. (2003) *Intellectual property protection to plant varieties: issues in transgenic cotton cultivation*, Monograph (Ahmedabad, India: Gujarat Institute of Development Research).

Lipton, M. (1968) The theory of the optimizing peasant, *Journal of Development Studies*, 4, pp. 327–51.

Main Cotton Research Station, Gujarat Agricultural University (GAU) (1996) *Souvenir of the National Seminar on Century of Cotton in India: 1996* (Surat, India: GAU).

Ortiz, S. (1990) Uncertainty reducing strategies and unsteady states: labor contracts in coffee agriculture, in E. Cashdan (ed.), *Risk and Uncertainty in Tribal and Peasant Economies*, pp. 303–17 (Boulder, CO: Westview).

Parmar, C. and Visvanathan, S. (2003) Hybrid, hyphen, history, hysteria: the making of the *Bt* cotton controversy, IDS Seminar on Agriculture Biotechnology and the Developing World, Sussex, UK, 1–3 October.

Qaim, M. and Zilberman, D. (2003) Yield effects of genetically modified crops in developing countries, *Science*, 299(5608), pp. 900–2.

Rogers, E. M. (1995) *Diffusion of Innovations* (New York: Free Press).

Roy, D. (2006) Farming 'white gold': early experiences with genetically engineered cotton production in Gujarat, India, doctoral dissertation, Cornell University, Ithaca, NY.

Sidhu, M. S. (1999) Impact of intellectual property rights on the Indian seed industry, *Indian Journal of Agricultural Economics*, 54(3), pp. 370–9.

Singh, R. V. (2002) Farming for the future, *Down to Earth*, 10(18), pp. 38–43.

Smil, V. (2004) Food production and the environment: understanding the challenge encountering the limits and reducing the waste, paper prepared for the World Bank, Washington, DC, March.

Stone, G. D. (2002) Biotechnology and suicide in India, *Anthropology News*, 43(5), p. 5.

Toenniessen, G. (2003) Opportunities for and challenges to plant biotechnology adoption in developing countries, in A. Eaglesham, S. Ristow and R. W. F. Hardy (eds), *Biotechnology: Science and Society at a Crossroad*, NABC Report 15, Ithaca, NY, National Agricultural Biotechnology Council, pp. 245–61.

Vasavi, A. R. (1999) Agrarian distress in Bidar: market, state and suicides, *Economic and Political Weekly*, 34(32), pp. 2263–8.

Appendix 1. Details of varietal mixes in cotton farming, 2002–03

No.	Region	Total acres	Acres of cotton	Variety1	Variety2	Variety3	Variety4	Variety5	Variety6
1	1	55.00	5.00	lalsanthi					
2	1	30.00	20.00	**loose Bt**					
3	1	55.00	5.00	lalsanthi					
4	1	22.00	8.00	lalsanthi					
5	1	15.00	1.00	**loose Bt**					
6	1	33.00	9.00	**loose Bt**					
7	1	4.00	3.00	lalsanthi					
8	1	25.00	1.00	**Mech162**					
9	1	13.20	4.00	S-VI					
10	1	20.00	1.00	**Mech184**					
11	1	20.00	2.00	PAC135					
12	1	20.00	2.00	**Mech162**					
13	1	8.00	1.48	**loose Bt**					
14	1	18.00	1.00	**loose Bt**					
15	1	10.00	1.00	**Mech12**					
16	2	30.00	15.00	**loose Bt**					
17	2	25.00	10.00	**loose Bt**	S-VIII				
18	2	16.50	15.80	**loose Bt**	S-X				
19	2	10.50	6.00	**loose Bt**					
20	2	11.90	8.00	JKDurga	Banni	555	S-VIII		
21	2	13.20	6.00	JKDurga	Banni				
22	2	21.00	12.00	**loose Bt**	**Mech184**				
23	2	55.00	12.00	**loose Bt**					
24	2	5.50	5.5	**Mech162**	**Mech184**				
25	2	20.00	20.00	**loose Bt**	S-VIII	VCH9			
26	2	15.00	10.00	**loose Bt**					
27	2	75.00	45.00	**loose Bt**	**Mech162**	**Mech184**	S-VIII	VCH9	Guj.23
28	2	115.00	40.00	**Mech162**	**Mech184**	S-VI	S-VIII	**LooseBt**	
29	2	138.00	48.00	**loose Bt**	**Mech162**				

(*continued*)

Appendix 1. (*Continued*)

No.	Region	Total acres	Acres of cotton	Variety1	Variety2	Variety3	Variety4	Variety5	Variety6
30	2	50.00	8.00	**loose Bt**					
31	3	9.9	2.50	S-VI	Sanju				
32	3	66.00	6.00	**loose Bt**	Sandrocot	Ajeet9	JK99		
33	3	40.00	5.00	**Mech12**					
34	3	59.00	18.80	**Mech12**	**VamanBt**				
35	3	10.00	8.00	**loose Bt**	**Mech184**	Mahyco1			
36	3	25.00	6.00	**loose Bt**	**Mech184**	**Mech162**			
37	3	5.00	3.30	**loose Bt**	**Mech184**				
38	3	13.20	4.00	**Mech162**	Rasi 2				
39	3	10.00	3.00	Vikram5					
40	3	13.20	5.00	**loose Bt**					
41	3	28.50	6.50	Mech162	Mech184				
42	3	29.70	4.00	**Mech184**	Sanju				
43	3	33.00	12.50	**Mech184**	Sanju	S-VI	Dhanno		
44	3	26.40	18.00	**loose Bt**	**Mech184**	Vikram9			
45	3	33.00	10.00	**loose Bt**	**Mech162**	Vikram9	Vikram11		

Note: Transgenic varieties in **bold type.** (a) Region 1: north Gujarat; Region 2: central Gujarat; Region 3: Saurashtra; (b) S-VI = Shankar-VI; S-VIII = Shankar-VIII; S-X stands for Shankar-X; Guj. 23 = Gujarat 23; (c) Mech 12, Mech 162 and Mech 184 are the three Bollgard varieties legally approved for use in Gujarat during 2002–03. Loose *Bt* includes the technically illegal *Navbharat* 151 and other unbranded varieties of *Bt* cotton.

The Adoption of Genetically Modified Papaya in Hawaii and Its Implications for Developing Countries[1]

C. GONSALVES, D. R. LEE, & D. GONSALVES

I. Introduction

The international debate on agricultural biotechnology continues, but the terms are progressively shifting as evidence on adoption increasingly becomes available. This shift is evident, for example, in recent summary reports by the International Council for Science (2003), the Nuffield (UK) Council on Bioethics (2003) and the Food and Agricultural Organization of the United Nation's *State of Food and Agriculture* report (2004) on agricultural biotechnology. All these reports find no verifiable evidence of toxic, health-related or nutritionally harmful effects resulting from consumers' use of genetically modified foods over the past decade or more. The shift is increasingly from 'whether' or 'if' genetically modified crop varieties should be approved for use by farmers in many countries to how best to assure broad-based adoption, equitable sharing by farmers in the social benefits accruing from technology adoption (given the associated market and price effects), and maximising the likelihood that these benefits are pro-poor.

The FAO's report – *Agricultural Biotechnology: Meeting the Needs of the Poor?* – deserves particular note, as this organisation had previously been 'sitting on the fence' in the debate over genetically modified agriculture (Paarlberg, 2005: 104). The report concludes that 'biotechnology is capable of benefiting small, resource-poor farmers. The key question is how this scientific potential can be brought to bear on agricultural problems of developing-country producers'. Even the title of the report is revealing: first, because distributional consequences are front and centre, and second, because the issue is phrased as a question, suggesting that the answer is not readily apparent. In part, the reason for this uncertainty is the very limited empirical evidence on the commercialisation of GM crops. Only a decade after the first commercialisation of genetically modified varieties, just four crops – soybeans, maize, cotton and canola – grown in six countries (the US, Argentina, Canada, Brazil, China, and South Africa) account for 99 per cent of GM crop production globally (FAO, 2004). Empirical studies of farmer adoption and the distribution of benefits accruing from these crops have consequently been highly limited, compared, for example, to the many such studies of Green Revolution technologies.

This essay examines the experience of genetically modified papaya in Hawaii and discusses the relevance of this case study to the adoption of GM crops more generally. We begin by identifying some of the key factors that have been suggested in the literature as prerequisites for the successful adoption and dissemination of GM varieties by low-income farm households in developing countries. We then introduce a unique case study of the recent development and commercialisation of genetically modified papaya in Hawaii. Results from a recent survey of Hawaiian papaya producers conducted by the authors are reported. We discuss the key factors leading to the unprecedented rapid adoption of GM papaya in Hawaii. Finally, we find that the distinctive case of GM papaya adoption in Hawaii holds many implications for developing countries and elaborate on these implications.

II. Constraints Facing GM Crop Adoption in Developing Countries

Although evidence regarding GM crop adoption in developing countries is limited, results to date have consistently identified a number of critical constraints facing farmers. This evidence stems from several sources. First, a number of empirical adoption studies have been conducted over the past several years of several technologies, particularly *Bt* cotton, in selected countries. Second, from abundant evidence regarding Green Revolution technologies, knowledge about the crucial regulatory and institutional environments provides insights into likely adoption constraints surrounding GM crops. Finally, formal ex ante research on GM crops has highlighted a number of critical factors likely to influence farmer adoption. Together, these sources suggest a consensus set of limiting constraints which affect developing country farmers' access to and adoption of genetically modified crops.

Much of the debate concerns the institutional, policy and infrastructural factors that influence farmers' *access* to GM crop varieties. However, as suggested in the 2004 FAO report on agricultural biotechnology, even when the constraints to access are surmounted, farmer adoption rates for GM crops, like those for any technology, will vary widely as a function of heterogeneous production environments (and associated profitabilities), the diversity of management objectives and constraints facing farm households, and varied marketing and institutional environments. For this reason, adoption paths for GM crops can be expected to follow the widely applicable logistic or S-type path of slow initial adoption, followed by more rapid adoption as the benefits of the new technology become apparent (and before associated price impacts are fully felt), and culminating in a lengthier period of slow adoption by late adopters.

Bt cotton is the most widely studied GM crop to date; while cotton pests are similar around the world (Huesing and English, 2004), the particular pests that cause damage to farmers in specific production environments are diverse: bollworm, budworm, armyworm, boll weevil, and so forth. Consequently, pest impacts are variable, and the potential gains from introduction of *Bt* cotton vary from region to region (FAO, 2004); thus farmer adoption rates inevitably vary. For crops such as maize, which are more highly sensitive to local agronomic and environmental constraints, these locally relevant factors are even more important. Factors that will positively influence the likelihood of farmer adoption, in addition to institutional and policy factors, include the availability of crop varieties developed for local conditions, the severity of yield impacts resulting from pests and disease (for example, a high opportunity cost to *not* adopting), and in labour-abundant countries, the labour savings associated with adoption (Bennett et al., 2004). As the FAO report concludes, one condition favouring the adoption of GM crops is the 'simplification of the management process' in ways that may favour small farmers. This has been evident from studies in countries such as China, India, South Africa, and Mexico.

From the individual farmer's perspective, the key factors influencing adoption are whether the crop is relevant to farmers' food and income needs, and the impacts on yields (and stability of yields) and net economic costs, specifically, the benefits of lower pesticide and labour costs balanced against the higher costs of GM seeds. One of the compelling factors accounting for the growth in *Bt* cotton adoption across several countries is the ability to simultaneously achieve higher yields (11 to 80 per cent in the countries reviewed), lower pesticide use and costs (−47 to −77 per cent), and higher overall profit (FAO 2004). There remains concern about variability both year-to-year and field-to-field. *Bt* maize has achieved yield gains of at least 10 per cent – in China and the Philippines, much higher – in those few countries where it is in the initial adoption stages (Huesing and English, 2004). Herbicide-tolerant soybeans are grown widely in Argentina and Brazil, in addition to the US, with benefits to farmers stemming from lower herbicide and tillage costs (the latter stemming from reduced weed growth), including labour and machinery costs. Overall variable costs of production are estimated to have been reduced about 8 per cent per hectare in Argentina, a similar level of cost savings as in the US (FAO, 2004). One benefit of GM seeds in many environments is that they are highly divisible (Lele, 2003), and their adoption appears in many cases to be scale-neutral.

Early evidence suggests successful and spreading farmer adoption of GM varieties; there remain significant institutional and policy obstacles to farmers' access to GM crops. These begin with research and development. To begin, most of the agricultural research conducted by international and national organisations takes place in industrialised nations, which have the requisite resources – both human and financial – and research infrastructure. Very few developing nations – China, India, Brazil,

Mexico and South Africa are notable exceptions – have a critical mass of scientific capacity, research infrastructure, public sector support, and a sufficiently large agricultural sector to succeed (Byerlee and Fischer, 2001; Kent, 2004). Moreover, these resources have historically been directed mostly toward research innovations for favoured environments where modern varieties respond well to water and fertiliser inputs; marginal areas where yields are constrained by climate, soils or infrastructure have traditionally not been the focus of public-supported agricultural research (Huang et al., 2002). Additionally, compared to the modern crop varieties associated with the Green Revolution, the development of GM crops is much more dependent on the private sector, which is typically less interested in the generation of the types of unappropriated spillover benefits that stem from public financing of research that characterised the Green Revolution (Pingali and Traxler, 2002).

Closely related to research-related constraints and increasing privatisation of GM technology research and development is the question of intellectual property rights. As bioscience companies that dominate agricultural biotechnology are primarily those of industrialised countries, most developing nations are unable to take advantage of Article 27 of the TRIPS Agreement (trade-related aspects of intellectual property rights), which since 1994 has governed the patenting of biotechnical processes and products. To cite just one example, Byerlee and Fisher (2001), document the dominant position in patent protection held by the US and Japan in the case of vitamin A rice: as of 2000, key rice-producing developing countries such as Thailand and Bangladesh held no patents at all on vitamin A rice. Industrialised countries argue that without such protection, insufficient incentives will exist for private companies to invest in product development (Beadle, 2000). Yet without access to intellectual property rights, poor producers living in marginal environments where the appropriable benefits are insufficient to justify initial research investments are likely to be disadvantaged (Pingali and Traxler, 2002). Although the public sector and the publicly supported research institutions like those of the CGIAR continue to play a role in new technology development, alternative methods must be developed to access these technologies such as purchasing agreements, material transfer and licensing agreements, alliances and joint ventures (Byerlee and Fischer, 2001). The development of these and other forms of public–private partnerships is widely seen as key to surmounting this constraint.

Perhaps the key institutional and policy constraint to farmer adoption of GM crops in recent years has been whether a country has in place a biosafety regulatory framework governing field testing of GM crops and the subsequent commercial distribution of GM seeds. Because of liability concerns, their commercial reputation, and simply good business practices, large international bioscience companies are unwilling to sell their products in countries that have not enacted biosafety and food safety regulatory frameworks (Kent, 2004). Establishing regulatory frameworks to permit field trials is time-consuming, complex, data-intensive, and expensive. It is also sensitive politically, particularly because in most developing countries, the public debate over GM crops has been relatively recent, the technological developments have occurred elsewhere, and there is a paucity of locally available information to backstop arguments on either side of the public debate (de Kathen, 2000). As a result, relatively few developing countries have biosafety regimes in place. The effect has been to delay field trials and thus subsequent adoption and commercialisation of GM products.

A final constraint facing the adoption of agricultural biotechnology are regulations governing the importation of GM crops in markets such as Europe and Japan. World Trade Organization regulations permit technical barriers to trade, including GM crop-targeted import restrictions, when 'information about health, hedonistic, or ethical attributes of agricultural products is either unknown or asymmetrically distributed between producers and consumers, and the transaction costs of obtaining this information are prohibitively high for consumers' (Roberts et al., 1999). In Japan, the import market of particular concern in this case, consumers are particularly concerned with two aspects of foods – country of origin and the GMO content of the product (Lohr, 2001). Japanese consumers have, in recent years, been willing to pay large price premia for domestically produced soybeans compared to imported soybeans and for imported organic soybeans compared to imported GM soybeans (Lohr, 2001). This is one example of the growing phenomenon of 'identity preservation, production and marketing' (IPPM) for GM crops. By facilitating growth of segmented markets, GM crops can be accommodated in the market, while maintaining guaranteed markets for non-GM crops in countries where this is important to consumers. At the extreme, however, import restrictions lacking any scientific basis can be viewed simply as technical barriers to trade.

III. Papaya Ringspot Virus and Development of the Genetically Modified Papaya

Papaya is primarily produced in the developing world; Brazil is the world's largest producer, followed by Nigeria, Mexico, Indonesia and India among the top five producing countries (FAO, 2001). In most developing countries, papaya is primarily produced in home gardens for household consumption. Though an important commercial crop in Hawaii – papaya is the state's second most important commercial fruit crop, with a market value of about $18 million in 2000 – the US share of the global papaya market is small.

The fruit contains high levels of three carotenoids (beta-crypto-xanthin, beta-carotene and lypocene) which are a major source of vitamin A (Dashwood and Bertram, 1994). An estimated 100–140 million children worldwide are vitamin A-deficient, leading to an estimated 250,000–500,000 children developing blindness annually (WHO, 2001); papaya can be a primary affordable source of naturally occurring vitamin A which can lead to improved health for many children in tropical countries. Further, it is a cheap, locally available food which can be consumed in many different forms (C. Gonsalves, 2001). Among high-income country consumers, papaya is one of a group of tropical fruits which have enjoyed large consumption increases in recent years as consumers have increased their demands for fresh fruits and vegetables.

Most of Hawaii's papaya (95 per cent) is grown in the Puna area on the eastern side of the island of Hawaii (the 'Big Island'). Nearly all (95 per cent) of the product is destined for the fresh market; historically 40 per cent or more of production has been exported to Japan (Sebastiano, 2003). Most production is by small farmers, many of whom are ethnic Filipinos and immigrants. There are few packing companies and marketers, and growers work closely with packers in the production, harvesting, and marketing of their crops. Hawaiian production accounts for virtually all the commercial US papaya market. Figure 1 shows some of the key characteristics in supply and demand in that market between 1970 and 2000. Demand and imports – 80 per cent or more of which is from Mexico – grew rapidly over the 1990s due to changes in consumer preferences, a rapidly expanding immigrant Latino population and trade liberalisation (notably, the North American Free Trade Agreement) which have increased Western Hemisphere trade in horticultural products. US domestic production and exports were erratic over the 1990s, declining in most years due largely to devastation from the spread of papaya ringspot virus in Hawaii.

Papaya ringspot virus (PRSV) is a virulent plant pathogen that has severely affected papaya production globally, in major producing countries such as Malaysia, Thailand, Australia, and Brazil, as well as Hawaii. PRSV is generally carried by aphids from infected to healthy plants, which display various symptoms before eventual mortality: a yellow or mosaic colouring on the leaves, water-soaked oily streaks on the upper trunk, and distortion of young leaves (Swain and Powell, 2001). An infected plant also typically displays stunted growth and deformed and inedible fruit. Prior to the development of transgenic papaya, there has been no broadly effective treatment for an infected papaya plant. Farmers try to prevent spread of the virus, usually unsuccessfully, through practices such as pruning young plants to build plant strength, netting and heavy chemical sprays to prevent aphid infestation, using tolerant plant varieties, and, as last resort, quarantining and burning infected and surrounding plants (Swain and Powell, 2001).

Hawaii's main production region had remained free of papaya ringspot virus for over 30 years until, in mid 1992, PRSV was discovered in the Puna region (C. Gonsalves, 2001). By late 1994, PRSV had spread throughout the Puna area, to the extent that farmers began to abandon traditional management practices to halt the spread of the disease. Within five years (1997), Hawaii's production of papaya

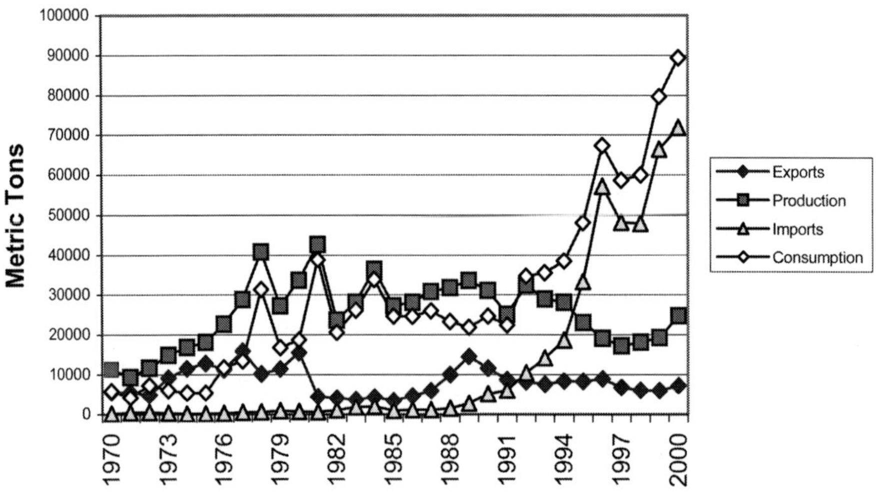

Figure 1. US trends in Papaya production, consumption and trade. *Source*: FAOSTAT database

was cut virtually in half: from a level of 71.3 million pounds in 1992, state-wide production fell to 38.8 million pounds in 1997 (Hawaii Department of Agriculture, 2003). Exports declined and many growers went out of business. Exports to Japan dropped by over half (52.4 per cent) between 1995 and 2000 (USDA Foreign Agricultural Service, 2002).

In anticipation of PRSV eventually reaching Hawaii, scientists at Cornell University and the University of Hawaii began a research program in the late 1980s to develop a virus-resistant papaya with parasite-derived resistance, PDR (D. Gonsalves, 1998). Plants with PDR express the gene of the pathogen and demonstrate resistance against that pathogen. This can be done through conventional plant breeding, but combines many genes from both parents at once and may require several backcrosses to remove undesired genes. Instead, scientists used genetic engineering techniques to develop a transgenic papaya by inserting the coat protein gene of PRSV directly into the papaya genome (D. Gonsalves et al., 2000). The resulting line, 55-1, was developed using the commercial Hawaiian cultivar Sunset, and demonstrated a high degree of resistance to PRSV (D. Gonsalves, 1998). Subsequent research led to the transgenic PRSV-resistant lines of the preferred Hawaiian papaya variety, Kapoho Solo, called 'SunUp' and 'Rainbow' (D. Gonsalves et al., 2000; Swain and Powell, 2001). The Rainbow hybrid was shown to be highly resistant to PRSV and possesses several desirable production characteristics: very high yields of marketable fruit, high individual fruit weight and sugar content, and a lack of declining production over the cropping cycle (Ferreira et al., 2001). Following many years of field testing, required reviews by the US Animal Plant Health Inspection Service (APHIS), the Environmental Protection Agency (EPA), and the Food and Drug Administration (FDA), and the completion of regulatory and licensing procedures, transgenic Rainbow and SunUp papaya seeds became commercially available in May, 1998.

IV. GM Papaya Adoption and Impacts in Hawaii

The transgenic Rainbow variety, preferred by growers, was adopted immediately and widely after it became available. A recent study of GM papaya adoption by Puna-area growers demonstrates some striking results (C. Gonsalves, 2001). Within the first full year of release (1999), 256 of 262 Hawaiian papaya growers – *fully 98 per cent of growers* – had registered with the Papaya Administrative Committee[2] to receive the transgenic seed. Traditional time paths of agricultural-technology adoption last over many years; that nearly the entire population of Hawaiian papaya growers demonstrated enough initial interest to register to receive seeds is unprecedented. Of 93 randomly surveyed Puna-area farmers (more than one-half of the 171 registered growers), 73 per cent were growing GM papaya within one year. For those growing GM papaya, the *average* time between release of the seeds and planting was just two months. Forty-four per cent of Puna-area papaya acreage was in transgenic papaya within one year. By the second year (2000), this had increased to 56 per cent of bearing acreage (Hawaii Department of Agriculture, 2002).

The experience of GM papaya in Hawaii is one of remarkably rapid adoption. The reasons for this unprecedented rate of adoption are several. First, the severity of the disease and its devastating impacts on yields – PRSV commonly results in total loss of the crop – made farmers open to, even desperate for, new technologies that would effectively address the disease. The likelihood of experiencing devastating crop failure was (and remains) very high if a grower does not adopt the technology. The development of the GM papaya thus addressed a problem that farmers themselves found to be critically important to their livelihoods. Table 1 shows the reasons given by surveyed farmers for adopting the GM papaya. Resistance to PRSV was overwhelmingly the major factor behind adoption; other major factors are closely related – expected gains in yields and profits and a decrease in production risk. In addition, producers were highly familiar with ongoing research and development efforts leading to the development of the GM papaya. Publicity surrounding its development and commercialisation throughout the 1990s complemented growers' longstanding participation in farmer organisations: the Papaya Administrative Committee (established in 1971), the Hawaii Papaya Industry Association (1965), and the Big Island Papaya Growers Association (1964). An initial distribution of about 2500 pounds of free GM papaya seeds and the promotional and educational campaign that surrounded seed distribution contributed to rapid adoption.

Survey results demonstrate differential effects according to the size distribution holding. Small growers – those farming from one–six acres of papaya – had experienced the greatest decrease in non-GM papaya acreage through 1999, a 71 per cent decline in aggregate acreage (C. Gonsalves, 2001). Total acreage declines among other size groups ranged from 9 per cent (medium-sized growers), to 23 per cent (small-to-medium producers) and 37 per cent (large growers). But small growers nonetheless experienced the *highest* adoption rate of GM papaya, as measured by proportion of total papaya acreage devoted to the GM crop (Table 2). This is not wholly surprising: those growers who had experienced the greatest losses would naturally benefit disproportionately from a new technology that ensured a return to sustainable production. Nonetheless, the relative size-neutrality of initial GM papaya adoption is an important result, with important implications for developing countries.

Although initial adoption rates were high among all grower groups, growers remained sensitive to the traditional production risks stemming from a new

Table 1. Farmers' reasons for planting GM papaya

Reason for planting/wanting to plant GM papaya	No.	(%)
Resistance to papaya ringspot virus	87	(96)
To decrease production risk	17	(19)
To experiment with GM papaya	15	(16)
Expect higher profits	12	(13)
Expect higher yields	9	(10)
Puna area closer to home	6	(7)
Because neighbors are doing it	5	(5)
Fruit has longer expected shelf life	5	(5)
GM plants flower and bear fruit earlier	3	(3)
GM plants use less fertiliser	1	(1)
Other	7	(8)

Source: C. Gonsalves (2001).
Note: Sums and percentages are greater than 100 per cent due to multiple responses.

technology. All groups of growers continued to farm large quantities of non-GM papaya as well as GM papaya following initial release of seeds (Table 2). Remarkably though, within one year of initial seed release, there was a 26 per cent increase in total papaya acreage (GM and non-GM) of surveyed farmers *above the papaya acreage existing prior to the onset of PRSV* (Table 3). Thus, the adoption of the GM seeds had more than made up for the devastating impacts of PRSV on papaya acreage during the 1990s.

Following the initial period of rapid adoption (1998–2000), the proportion of Hawaiian papaya acreage dedicated to the transgenic Rainbow variety declined. After peaking at 56 per cent of total papaya acreage in 2000, Rainbow variety bearing acreage decreased to 47 per cent in 2001, and 40 per cent in 2002 on the island of Hawaii (Hawaii Department of Agriculture 2002). The reasons are not hard to identify. Japan, the largest traditional export market for Hawaiian papaya, has continued to maintain import restrictions on GM foods, including GM papaya. This is a critically important factor in the Hawaiian papaya market, as the industry is heavily dependent on exports, most of which are destined for Japan. In response, at least one major Hawaiian packer that exports solely to the Japanese market has developed an extensive 'farm to market' protocol designed specifically to meet Japanese standards. This involves a rigorous system of product tracking, testing (to avoid importation of genetically modified fruit), and product monitoring and tracing

Table 2. Initial adoption patterns of transgenic papaya, Puna area, Hawaii*

Farm Size	Ave. no. of acres (T + NT)	No. of growers surveyed	Ave. no. acres of T planted	% total acres of T planted	% of growers W/some T	Ave. time before planting (mos.)
Very large	162	4	97	45	75	1.3
Large	28	21	13	47	100	2.2
Medium	11	45	6	38	73	2.2
Small	3	20	1.8	61	70	1.9

Source: C. Gonsalves (2001).
Notes: T = transgenic papaya; NT = non-transgenic; *based on farmer surveys in Puna area, following first phase of adoption.

Table 3. Historical and current papaya acres of surveyed Puna area farmers

Farm type	No. farmers	Historical acres[a] (before PRSV)	Nontransgenic (NT)	Transgenic (T)	Total (NT + T)
Growing just NT	22	396	458	0	458
Growing just T	15	123	0	114	114
Growing both NT and T	53	871	545	679	1224
Other	3	33	0	0	0
Total	93	1423	1003	793	1796

Source: C. Gonsalves (2001).
Notes: [a]Total papaya acres of surveyed farmers prior to PRSV infections (which very between 1992 and 1997).

all the way from the individual papaya tree through the entire packing process, to the retail market in Japan. Production, product handling and distribution protocols are carefully followed in order to meet Japanese import market requirements, though these add substantially to production and marketing costs.

For papaya farmers, Japanese import restrictions and the uncertainties they create have led to three strategies: (1) produce exclusively non-transgenic papaya for both domestic and international (primarily Japanese) markets and try to obtain the highest price premium possible, but assume significant production risks and added costs of meeting Japanese import standards; (2) produce GM papaya for the domestic US market and selected importers such as Canada, obtaining a discounted price, but greatly reducing production risks and added costs associated with exports to Japan; or (3) 'hedge their bets' and produce and market both GM and non-GM papaya.

As a result of Japanese import restrictions, a segmented market for Hawaiian papaya has developed, in which large price premia may be gained for non-GM papaya in the Japanese market. Because marketing of GM papaya is mostly limited to the US market – after January, 2003, the Canadian export market was opened to GM papaya – GM papaya experiences a significant price discount relative to non-GM papaya. The non-GM price differential in excess of GM prices has been estimated at approximately $.15 per pound in 2000 and $.10 per pound in 2001 (Sebastiano, 2003). As a proportion of the price received by growers these differentials are significant: 47 per cent of price in 2000 and 36 per cent of price in 2001. These price differentials – which are analogous to, but higher than, the price differentials arising in other GM crops where identity preservation and segmented markets also matter – is a primary factor influencing production decisions of growers and marketing decisions of packers. For example, in 2002, the average Japanese papaya import price was $1.29 per pound, while the average US import price (excluding domestic Hawaiian production) was only $0.34 per pound. As a result of Japanese import barriers and their effects on relative prices, many producers have at least partially shifted back to non-GM papaya. A recent study of international trade in papaya (Sebastiano, 2003) confirms that the principal limiting constraint to the success of GM papaya is neither technical nor economic, but institutional; Japan's import restrictions make a significant impact on product and market requirements, which are in turn translated throughout all levels of the production, packing, distribution and trading system.

V. Implications for Developing Countries

One implication concerns the initial adoption of this technology – so rapid and so widespread compared to more typical pattern in which many years are required for most varieties to become extensively adopted. To begin, there are the demand-side characteristics of the product that are highly favourable. Papaya in the markets served by Hawaiian growers has benefited from surging consumer demand for fresh horticultural products and, in the US, growing immigration from countries where papaya is a traditional food. Given its high vitamin A content and the prevalence of vitamin A deficiencies, the papaya can be expected to continue to be an attractive food for many households in low-income countries. Indeed, the Center for Science in the Public Interest has ranked papaya first in nutrition among fruits (Liebman, 2001).

Despite these promising demand-side factors, rapid adoption in Hawaii is largely due to supply-side factors. Although it is primarily a commercial crop in Hawaii, in developing countries, papaya is typically produced in home gardens, is easily grown by poor households, and except in commercial operations, only hand labour is typically required. GM papaya seeds are wholly substitutable in production for non-GM seeds, requiring no additional labour or mechanical input.[3] There is still need for pesticide use in papaya production to avoid other insect and disease problems, so there is not a major impact on pesticide use as has been documented with *Bt* cotton, for example. In addition, early adopters liked the Rainbow variety because of their perceptions about the quality of the product, and the fact that it bore fruit earlier than the traditional Kapoho variety (C. Gonsalves, 2001; D. Gonsalves, 2004). From both demand and supply side perspectives, the GM papaya and its seed appear to be 'appropriate technologies' for developing country households.

More important, however, the devastating loss caused by the papaya ringspot virus makes the opportunity cost to farmers of *not* using the technology extremely high. Plant viruses decrease yields and crop values significantly worldwide; aggregate loss in crop value has been estimated at roughly $60 billion dollars annually (Flasinski et al., 2001). The risk of total crop loss, plus the fact that alternative management practices simply do not solve the problem, means that farmers are desperate for solutions. This is not a situation in which the more modest (but still significant) yield gains attributable to use of *Bt* cotton, for example, might be in part offset by the increased cost of seeds. Once established, transgenic papaya production costs per hectare are virtually the same as for traditional varieties (Gonsalves, 2001; Sebastiano, 2003). Indeed, our survey research shows that the availability of the GM technology actually brought papaya farmers who had quit the industry back into production. Papaya does require replanting after as few as three–four years, but the turnaround time between planting and the first commercial harvest is very short, 10–12 months; thus, there is not the lengthy time between planting and first harvest, typical of perennial crops, during which the producer has no cash revenues to cover the initial investment. Thus the underlying economics of the GM crop appear to be highly favourable,[4] prospectively, in developing countries, based on projections from its initial adoption in Hawaii.

In terms of research, intellectual property, and biosafety factors that may constrain farmer adoption of other GM crops, the GM papaya in Hawaii also holds some important lessons. Although not finally commercialised until 1998, development of the GM papaya was the culmination of a research effort that extends back 20 years and initially addressed the potential of cross protection for PRSV (D. Gonsalves, 2004). Subsequent to early success in developing GM papaya resistant to the Hawaii strain of PRSV, efforts began in the 1990s to develop papaya resistant to the strains of PRSV found in other countries. Suitable varieties have already been developed for Brazil, Jamaica, Venezuela and Thailand, and in all but Venezuela, field trials have been undertaken (in Brazil, on a limited basis). The potential availability of GM papaya for smallholders in developing countries is one of the factors behind development of the Papaya Biotechnology Research Network for south-east Asia (Thailand, Malaysia, Vietnam, Philippines, Indonesia), established by the International Service for Acquisition of Agri-biotech Applications (ISAAA) in 1998. Thus it is clear that focused, targeted research on a critical problem of direct relevance to growers, although it may take considerable time, can be successful.

In terms of intellectual property, the development and commercialisation of the GM papaya was the product of a unique collaboration of the public, land-grant university sector (Cornell University and the University of Hawaii) with the private sector. The Papaya Administrative Committee successfully negotiated licensing agreements with the Monsanto Company, Asgrow Seed Company, Cambia Biosystems LLC, and the Massachusetts Institute of Technology, completing this task in 1998. These included limitations-of-use and compliance provisions. The fact that the papaya is not a major globally traded crop (versus maize, soybeans and cotton, for example), that none of the patent holders held a strong commercial interest, and that licensing agreements only extended to the Hawaii-grown crop all contributed to the resolution of IPR issues and commercialisation of the crop (Swain and Powell, 2001).

Regarding biosafety constraints, it took a three-year effort (1995–98) to deal with the regulatory constraints of the USDA's Animal and Plant Health Inspection Service, US Environmental Protection Agency and the Food and Drug Administration, to successfully commercialise the crop. Delays would presumably be longer in developing countries, where regulatory procedures and protocols are often missing and must be developed.

A common argument holds that educational and promotion efforts to familiarise producers with a new technology and its characteristics can have a rapid payoff. Early and widespread promotion of papaya technology made its potential benefits known widely to Hawaiian farmers. The existence and involvement of the Papaya Administrative Committee and its close involvement with the development, deregulation, and commercialisation of GM papaya – and thus the high degree of familiarity of growers with the GM technology – clearly played a major role in its immediate and widespread adoption. Although the initial distribution of GM papaya seed was costless to growers, survey results showed that 86 per cent of growers indicated a willingness to purchase the seed.

Finally, our data show that there appear to be no significant scale economies to the adoption of transgenic papaya – other than those that exist in papaya production more generally – so that smallholders are able to benefit significantly. Smallholders were indeed represented disproportionately among the early adopters in Hawaii. Our results appear to confirm the argument that transgenic virus resistance, in general, may actually convey its greatest value for subsistence farmers, who are often unable to afford other means of virus protection (Flasinski et al., 2001). In the debate over incidence of benefits versus costs of GM crops in developing countries, the question of 'who benefits' has loomed large. In this case, demonstration of scale neutrality is an important result, again confirming that the development of transgenic technologies appropriate to the context and problems faced by small and medium-sized farmers can be achieved.

In sum, many of the factors that constrain adoption of GM crops in developing countries appear to have been surmounted in the case of GM papaya adoption in Hawaii; moreover, many of the conditions and characteristics present in Hawaii apply, in varying extents, to low-income tropical countries: (a) consumer demand

and nutritional attributes, important in Hawaii, are likely to be even more relevant to low-income households in tropical developing countries, where papaya is frequently part of the diet and nutritional constraints are more serious; (b) the effects of papaya ringspot virus have been equally devastating in many tropical countries as in Hawaii, creating a need for the technology; (c) the underlying technology is ‘appropriate’ in that it is consistent with farmers’ management practices and capacities and requires no large capital investments (unlike many mechanical and irrigation technologies); (d) variable production costs are not excessive (indeed appear no more than non-GM papaya) and thus do not present obstacles to adoption as they often do for mechanical technologies, fertilisers, pesticides, and other purchased inputs; (e) intellectual property ownership is already or potentially in the public domain (these are being negotiated currently in several developing countries); and (f) the lack of scale economies associated with the technology should assure benefits to smallholders. Research and field trials on GM papaya in developing countries are underway, and commercial introduction in at least some of these countries is anticipated in the near future.

Notwithstanding these positive characteristics, there are cautionary notes from the Hawaii experience relevant to developing-country producers. Institutional factors – in this case, Japanese import prohibitions – can ‘trump’ technological and economic factors. Major expansion of Hawaiian papaya acreage will have to await changes in these import constraints. Second, product and process standards – such as the HACCP standards adopted by the UN Codex Commission and being increasingly adopted internationally – are science-based, often technologically advanced, and information-intensive. Sophisticated and complex systems of monitoring, testing, quality control, and documentation are required. The harmonisation of international standards will increasingly require that developing-country producers and marketers remain current with international standards in testing, quality assurance, and relevant scientific underpinnings. For many developing countries, this will be a challenge. In addition, market impacts, although beyond the scope of this paper, can represent a significant limiting factor to extensive technology adoption, especially in those cases where product price declines result from aggregate yield and production impacts in the early stages of adoption.

Finally, there is an additional, and ironic, twist in the story of the Hawaiian papaya worth noting. Import restrictions in the Japanese market – Hawaii’s major export destination – are currently the major factor limiting the growth of transgenic papaya production in Hawaii. Yet Japanese imports of fresh papaya have continued to surge. The country that is largely filling this gap in import demand – given that only non-transgenic papaya exports from Hawaii are permitted in Japan – is the Philippines. Between 1998 and 2002, Japanese papaya imports from the Philippines increased well over 40 per cent, such that by 2002, for the first time, they exceeded imports from Hawaii (JETRO, 2004). Who is the major Philippine papaya exporter? It is the Dole Company, the world’s largest producer and marketer of fresh fruit and vegetables, which is reported to be eager to capture the Japanese market for imported papayas (*Davao Sun Star*, 2003). In recent years, Dole has sourced papaya from over 600 hectares and 1500 growers in the Davao area of the Philippines. In the interest of addressing alleged food safety problems, Japanese import restrictions effectively penalise small and medium-sized Hawaiian transgenic papaya producers and, in turn, directly foster increased market share and potential market dominance by a major transnational company with global reach. This result is not only highly ironic, given the commonly heard concerns of biotechnology opponents, but is testimony to the truly globalised nature of food markets and the unintended spillover effects that changes in policies and regulatory frameworks can effect.

Notes

1. Revision of paper presented at the Cornell Conference on 'Transgenics and the Poor: Science, Regulation and Development Policy', November 2003, Ithaca, NY.
2. The Papaya Administrative Committee was the organization authorised by the US Congress to administer the Federal Marketing Order for Hawaiian papaya under the oversight of the USDA. The PAC and the Federal Marketing Order have recently been terminated.
3. Unlike the evidence of large-scale labour savings in the use of *Bt* cotton, for example, early evidence does not appear to show significant labour savings.
4. This does not take into account market price effects due to greatly increased aggregate yields, which are outside the scope of the individual farmer.

References

Beadle, A. (2000) Intellectual property rights issues and developing countries: a private sector perspective, in M. Qaim, A. F. Krattiger and J. von Braun (eds), *Agricultural Biotechnology in Developing Countries: Towards Optimizing Benefits for the Poor*, pp. 281–8 (Boston: Kluwer Academic Publishers).

Bennett, R. M., Ismael, Y., Kambhampati, U. and Morse, S. (2004) Economic impact of genetically modified cotton in India, *AgBioForum*, 7, pp. 96–100.

Byerlee, D. and Fischer, K. (2001) Accessing modern science: policy and institutional options for agricultural biotechnology in developing countries, *World Development*, 30, pp. 931–48.

Dashwood, R. and Bertram, J. (1994) Role of papayas in human health, paper presented at the 30th Annual Hawaii Papaya Industry Association Conference, Kihei, Hawaii, 23–24 September 1994.

Davao Sun Star (2003) Davao papayas may soon re-enter Japan market, 28 June, www.sunstar.com.ph/static/dav/2003/06/28/index.html

De Kathen, A. (2000) Managing biosafety capacity development: technical and political aspects, in M. Qaim, A. F. Krattiger and J. von Braun (eds), *Agricultural Biotechnology in Developing Countries: Towards Optimizing Benefits for the Poor*, pp. 39–68 (Boston: Kluwer Academic Publishers).

Ferreira, S. A., Pitz, K. Y., Manshardt, R. M., Zee, F., Fitch, M. and Gonsalves, D. (2001). Virus cost protein transgenic papaya provides practical control of papaya ringspot virus in Hawaii, *Plant Disease*, 86, pp. 101–5.

Flasinski, S., Aquino, V., Hautea, R., Kaniewski, W., Lam, N., Ong, C., Pillai, V. and Romyanon, K. (2001) Value of engineered virus resistance in crop plants and technology cooperation with developing countries, in *Proceedings from the 5th International Conference on Biotechnology, Science and Modern Agriculture: A New Industry at the Dawn of the Century*, Ravello, Italy, 15–18 June 2001, pp. 123–4.

Food and Agriculture Organization of the United Nations (FAO) (2000–03) *FAO Production Yearbook* (Rome: FAO).

Food and Agriculture Organization of the United Nations (FAO) (2004) *The State of Food and Agriculture 2003–2004: Agricultural Biotechnology: Meeting the Needs of the Poor?* (Rome: FAO).

Gonsalves, C. (2001) Transgenic virus-resistant papaya: analysis of farmer adoption in the Puna area of Hawaii, M. A. thesis, State University of New York Empire State College, New York.

Gonsalves, D. (1998) Control of papaya ringspot virus in papaya: a case study, in R. K. Webster, G. Shaner and N. K. Van Alfen (eds), *Annual Review of Phytopathology*, pp. 415–37 (Palo Alto, CA).

Gonsalves, D. (2004) Transgenic papaya in Hawaii and beyond, *AgBioForum*, 7, pp. 36–40.

Gonsalves, D., Ferreira, S. Manshardt, R., Fitch, M. and Slightom, J. (2000) Transgenic virus resistant papaya: new hope for controlling papaya ringspot virus in Hawaii, APSnet (The American Phytopathological Society), accessed at: http://www.apsnet.org/education/feature/papaya/Top.htm

Hawaii Department of Agriculture, Hawaii Agricultural Statistics Service (2001–03) Papaya acreage survey results, accessed at: http://www.nass.usda.gov/hi/fru.PDF

Huesing, J. and English, L. (2004) The impact of *Bt* crops on the developing world, *AgBioForum*, 7, pp. 84–95.

Huang, J., Pray, C. and Rozelle, S. (2002) Enhancing the crops to feed the poor, *Nature*, 418, pp. 678–84.

International Council for Science (2003) *New Genetics, Food and Agriculture: Scientific Discoveries – Societal Dilemmas* (Paris: ICSU).

JETRO (2004) *Marketing Guidebook for Major Imported Products*, accessed at: http://www.jetro.go.jp/ec/e/market/mgb/data_e/1/10/pdf

Kent, L. (2004) What's the holdup? Addressing constraints to the use of plant biotechnology in developing countries, *AgBioForum*, 7, pp. 63–9.

Lele, U. (2003) Biotechnology: opportunities and challenges for developing countries, *American Journal of Agricultural Economics*, 85, pp. 1119–25.

Liebman, B. (2001) Fresh fruit – papaya a day?, *Nutrition Action Healthletter*, Center for Science in the Public Interest, Washington, DC.

Lohr, L. (2001) Factors affecting international demand and trade in organic food products, Working Paper, faculty series FS 00-02, Department of Agricultural and Applied Economics, University of Georgia, Athens, GA.

Nuffield Council on Bioethics (2003) *The Use of GM Crops in Developing Countries* (London: Nuffield Council on Bioethics).

Paarlberg, R. (2005) From the Green Revolution to the Gene Revolution, *Environment*, 47, pp. 38–40.

Pingali, P. L. and Traxler, G. (2002) Changing locus of agricultural research: will the poor benefit from biotechnology and privatization trends?, *Food Policy*, 27, pp. 223–38.

Roberts, D., Josling, T. E. and Orden, D. (1999) A framework for analyzing technical trade barriers in agricultural markets, USDA Economic Research Service, Technical Bulletin No. 1876, March.

Sebastiano, J. N. (2003) An economic analysis of the adoption of a genetically modified variety: the case of the Hawaiian papaya, unpublished M. S. thesis, Cornell University, New York.

Swain, S. and Powell, D. A. (2001) Papaya ringspot virus resistant papaya: a case study, University of Guelph Technical Report, accessed at: http://www.plant.uoguelph.ca/safefood/gmo/papayarep.htm

US Department of Agriculture, Economic Research Service (1998) Biotechnology in Agriculture Confronts Agreements in the WTO, Agriculture in the WTO, WRS-98-4/December.

US Department of Agriculture, Foreign Agricultural Service (2002) *Foreign Agricultural Trade Statistics* (Washington, DC: US Department of Agriculture).

World Health Organization (2001) Nutrition/micronutrient deficiencies: combating vitamin A deficiency, accessed at: http://www.who.int/nut/vad.htm

Supplying Crop Biotechnology to the Poor: Opportunities and Constraints

CARL E. PRAY & ANWAR NASEEM

I. Introduction

In the late 1980s and early 1990s, rapid advances in molecular biology and related fields led many scientists to predict that their application in agriculture would result in significant productivity gains and would usher in the 'Gene Revolution'. It was felt that tools of biotechnology would allow scientists to develop novel crop and animal technologies that would not have been possible through conventional methods. Genetically modified (GM) crops would not only overcome the yield plateau that had been reached for many major crops in the period after the Green Revolution, but also could be engineered to be higher quality and more nutritious.

Early scientific excitement led to major investments by the private sector in biotech – especially in the US and Europe. These investments were driven, in part, by changes in patent law in industrialised countries which began to allow patenting of biotechnology tools and products, including living organisms. Since the mid to late 1980s, firms have been able to protect biological innovations through patents and hence appropriate more of the returns to their research investments. The prospects of greater appropriability – and greater profits – led them to invest even greater amounts in biotechnology research.

The promise of biotechnology also led some governments in both industrialised and developing countries to visualize potential for dramatically increasing agricultural production. This view resulted in major government investments in research on basic biology, as well as more applied medical and agricultural biotechnology research in the US, Europe, China, India, Brazil and elsewhere.

With almost 10 years of experience with commercialisation of genetically modified crops, it is now widely acknowledged that the impacts of biotechnology have been unbalanced and mixed. While GM crops have increased yields and productivity in places where they have been adopted, the majority of crops that have been genetically modified, such as soybean, are planted largely in industrialised countries, particularly the US. Little or no biotech research has been conducted on crops – such as cassava, white maize, rice, and millet – that are planted by resource-poor farmers in developing countries. Further, the traits that have been introduced in GM crops to date tend to largely favour the existing farming practices of industrial agriculture, rather than meet the needs of the poor.

The concentration of GM crop research and adoption around a few countries, crops and traits seem surprising at first glance, particularly in light of growing evidence that GM crops have increased farm-level productivity and profits for farmers who have adopted them. If GM crops are so productive, why has their adoption not been more widespread? Part of the answer lies in the limited demand for such products by final users, namely consumers. In particular, European consumers have been reluctant to accept the technology, which many regard as unsafe and unnecessary, leading to a virtual moratorium by the European Union on importation or planting of GM crops. In a ripple effect, some developing-country governments have also imposed limits on GM crops because they fear that they will be unable to export products to European markets. Intense lobbying and anti-biotechnology campaigns by environmental and consumer groups have also caused many governments to take a slow and cautious approach towards adoption of GM crops.

While demand-side concerns regarding GM crops may influence a farmer's decision not to adopt, it is still the case that a significant number of the crops grown in developing countries – especially those grown by poor farmers – seldom enter the global market and are unlikely to become a trade issue. It is also doubtful that informed consumers in poor countries would resist consuming GM crops, especially considering that the majority of consumers in developing countries spend a large share of their income on food, and productivity-increasing technology may lower food prices significantly. Limited adoption as a result of limited demand therefore is not the complete story.

We argue that a number of related factors affecting the supply of technology have also been significant constraints to widespread adoption of GM crops. For example, the strengthening of intellectual property rights in developed countries and large investments in biotech made by private firms have resulted in a number of key technologies being concentrated amongst a few multinational firms. Multinational firms are unwilling to make the necessary investments in biotechnology research relevant to developing country agriculture due to limited market potential, fear of piracy of their intellectual property and the high cost of meeting regulatory requirements. Taken together, this has meant that research on crops important to poor farmers yields low private returns and hence provides limited incentives for private firms to invest. This supposed market failure necessitates government intervention that would either use the taxpayer's money to invest directly in public-sponsored research or create the necessary incentives to encourage private investment. However, neither has occurred; public investment in agricultural research remains low, and most developing countries do not have the appropriate institutional capacity to encourage higher private investment.

This paper considers supply-side constraints limiting adoption of GM crops and their implication for public- and private-sector research. In the next sections, using data on research expenditures and field trials of GM crops we review the nature of current research, emphasising the focus of research, where it is being conducted and by whom. Data on field trials support the argument that crops and traits most important to the poor have been largely neglected. Section three reports farm-level impacts of GM crops in those developing countries where they have been adopted. This evidence indicates that significant gains have accrued to adopting farmers. In light of this evidence, the next two sections examine the reasons for limited adoption in developing countries. Section four begins by reviewing the determinants of private and public research: how might these might explain variation in investments across countries? Section five discusses the impact that intellectual property rights (IPRs), costs of doing research and meeting regulatory mandates, and monopoly power of biotech multinational companies have on research and the supply of crops to farmers. Section six discusses research priorities for the poor and how they can be attained through correct incentives for private firms and public research institutes. Section seven concludes the paper.

II. Agricultural Biotech Research Expenditures and Characteristics

Information on how much biotech research is being conducted globally is quite limited. We first review available estimates of research expenditure, then look at data on field trials of GM crops, which is perhaps the only way to measure and characterise applied biotech research by private firms.

Biotech Research: Expenditures

Current state. Precise estimates of worldwide plant biotech research are not available, though there are preliminary estimates of expenditure (Byerlee and Fischer, 2001). We have used these estimates to construct Table 1. About 90 per cent of the biotech research expenditure occurs in industrialised countries. The public sector tends to place more emphasis on basic research, but private firms are doing a remarkable amount on such upstream research as mapping genomes and identifying the function of specific genes. The private sector conducts a large amount of applied research to develop new GM crop varieties – considerably more than the public sector. In total there is more private than public research but the public sector still plays a large role, which sometimes is overlooked in the debates about biotech research.

In developing countries much less money is spent on biotech research. The data on biotech research expenditure is incomplete, especially for the private sector. National Agricultural Research Systems (NARS) spend the most money, and foreign aid

Table 1. Estimated expenditure on crop biotechnology research in 2000

	Biotech R&D expenditure (million $/year)	Biotech as % of agricultural R&D
Industrialised countries		
Private sector (seed/chemical, biotech MNCs)	1000–1500	40
Public sector	900–1000	16
Developing countries		
Public (from own resources)	100–150	5–10
Public (from foreign aid donors)	40–50	NA
CGIAR Centers	25–50	8
Private firms	NA	NA
World total	2065–2730	

Source: Byerlee and Fischer (2001).

donors contribute about a third of the money spent by NARS. Cohen (2005) documents extensive research in LDCs on transgenic plants. More recent evidence indicates that estimates of public research in developing countries are low. Huang and his colleagues estimated that China spent over US$100 million on agricultural biotech research in 1999 (Huang et al., 2002). The International Agricultural Research Centers (IARCs) spend about 8 per cent of their budget on biotechnology research (Morris and Hoisington, 2000), but the total is considerably less that of NARS. Some private biotech research is conducted in developing countries, mostly by multinationals although local research institutes also do some. For example, in Brazil and South Africa local private sugarcane research institutes have fairly large biotech research programs while in India Maharashtra Hybrids and several other seed companies have research programs. The total investment of these private efforts is unknown but is undoubtedly less than the public sector is investing in biotech research in developing countries.

Trends. Both public and private investments in biotech research grew rapidly in the US and Europe until around 2000 when Table 1 was constructed. Public research in biotechnology as a share of total research and in real expenditure has continued to grow in industrialised countries. Even in countries such as France and Germany, where there is considerable opposition to genetically modified crops, governments continue to invest in biotech research.

Private sector biotech research growth in industrialised countries was fuelled at first by venture capital and IPOs by small companies and then by established seed and agricultural chemical companies. Since the late 1990s these companies consolidated into a few major companies that dominate agricultural biotechnology.

Since 2000, research funded by the major private firms has declined according to their annual reports and other published information. Monsanto reduced its research expenditure – 85 per cent of which is on biotech and plant breeding – from US$588 million to $550 million between 2000 and 2001 (Monsanto, 2002). Syngenta's total agricultural research, of which about 21 per cent is biotech and 15 per cent plant breeding, declined from $745 million to $723 million (Syngenta,

2002). Recently, Syngenta announced that they were closing their Torrey Mesa Research Institute which led their effort to sequence the rice genome (Vogel, 2002). In 2001, Savia, the large Mexican conglomerate, cut the research of its biotech subsidiary, DNAP, in half. Dow Agrosciences also reported declining research. The only firms that have announced plans to increase their biotech research are Bayer, which purchased Aventis to get into the biotech business, and BASF, which just started to participate in biotech research.

Public sector agricultural biotech research in developing countries is growing. The largest investments are in Brazil, China, India, South Africa, Egypt and a few other large countries. India may be spending as much as $30 million and Brazil about $15 million annually on agricultural biotech (Pray, 2001). China appears to be leading the growth in public sector biotech research, spending US$112 million in 1999 and planning to increase that amount by 400 per cent by 2004 (Pray et al., 2002).

Governments that have made substantial investments in public-sector agricultural biotechnology research in the developing world – Brazil, China and India – have several things in common. First, they have a strong scientific base – Brazil, China, India, and South Korea account for three quarters of all science that is conducted in developing countries (Byerlee and Fischer, 2001). Second, they have identified science and biotechnology in particular as an engine of economic growth in both agriculture and medicine. Third, the agricultural sector is a large and important component of their economies, and public agricultural science has had substantial success in promoting rapid agricultural growth. Thus, potential payoffs from biotech research are high and the cost of biotech research is relatively low because of the agricultural and medical research capacity already in place. Politicians in these countries have seen the explosive growth of information technology and its contributions to their economies and hope for similar growth through medical and agricultural biotechnology.

However, enthusiasm for agricultural biotech research will be lost if payoffs to research are long in coming – so far the only obvious payoffs are public *Bt* cotton varieties in China. In Brazil public and private varieties of GM soybeans have been a huge success, but approval of GM food crops is stalled by concerns about exports and biosafety. Although transgenic soybeans have been approved for commercial use by the government, they are still tied up in its courts. In the meantime illegal GM soybeans are being produced by local breeders in Brazil or being imported from neighbouring Argentina and have been planted on up to four million ha of land. In India GM cotton was approved for planting but other GM crops have been held up by regulatory concerns. An obvious conclusion is that efforts to establish an effective, transparent, and efficient biosafety regulatory system are essential.

Biotech Research: Field Trials

While biotech research expenditure in industrialised countries is fairly evenly divided between public and private research, production of technology is almost entirely in the hands of the private sector. All plant biotechnology that has been commercialised in the world, with the exception of China, was developed by the private sector.

Field trials of GM crop varieties are an important indicator of the spread of agricultural biotechnology research worldwide. These trials are an essential early step in bringing agricultural biotechnology to commercial markets. After much research and many glasshouse trials, a private company or public institution puts its new crop variety through a battery of tests to ensure that it does not have an adverse effect on agricultural production, the environment, or animal and human health. To do this, the firm must submit an application to the government for permission to conduct a GM crop release into the environment.[1] After crops have been approved for environmental and food-safety trials and tests have been successful, the company or institution submits an application to the government for deregulation (or general release) of the crop. Once the crop has been deregulated, the company can commercially release the crop for sale to farmers in some countries. In other countries the GM crop variety also must undergo mandatory trials required of all new varieties.

Field trials of GM crops are measured by the total number of applications approved in each country by its appropriate governmental regulatory department. In our study we have used the US counting system in which one application is for one event (a gene or series of genes transplanted in one variety) no matter how many locations this event is tested in.[2]

Trends in Field Trials

Countries. Field trial data collected from 1987–2000 indicates rapid growth in biotech research after 1993 (see Figure 1). The US has always been the leader in number of field trials. No other industrialised countries' field trial research grew at a similar pace, especially since 1993 when the number of trials in the US accelerated dramatically. In 1997, field trials in other industrialised countries peaked and have declined slightly since then. There was a significant reduction or plateau in GM crop research in Western European countries, Japan, Argentina and New Zealand.

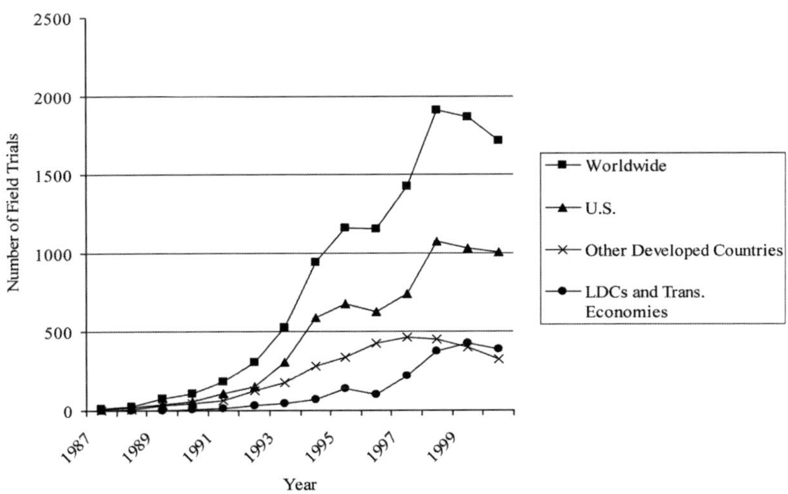

Figure 1. Number of GM crop field trials worldwide

Field trial data shows that developing countries have been slow to take up applied research on GM crops, although in recent years the percentage has increased to 23 per cent. For the entire period approximately 15 per cent of all field trials were in developing or transitional countries (Table 2). The low numbers reflect lack of private firm interest in small, poor countries and difficulties their governments are having in establishing bio-safety regulatory systems. In recent years the number of field trials in those countries has increased. Field trial activity grew rapidly in part due to increase in the number of countries conducting them. In 1999 field trials of GM crops were approved for the first time in Lithuania, Czech Republic, Moldova, Romania, Slovakia, Slovenia, Turkey and the former Yugoslav Republic. In 2000 Kenya and the Philippines were added to the list; by 2000 at least 58 countries had reported biosafety field trials (Pray, Courtmanche and Govindasamy 2002). In developing countries that have been conducting field trials for some time, the annual number of approvals proved to be more erratic than in developed nations. There was a 100 per cent increase in the number of field trials approved in South Africa between 1999 and 2000 (from 55 to 110 approved applications), but others such as Argentina and New Zealand have stopped field trials in certain years while re-evaluating their biosafety system.

Research institutions. The private sector continues to dominate research and development of new GM crops. Figures 2(a) and 2(b) show the division of GM crop research by public and private sectors. Less than 20 per cent of the applications for field trials approved worldwide were submitted by universities or NARS/IARCs. Additionally, many of those that have been submitted by the public sector have been submitted jointly with a private-sector or multinational life-science partner. The transitional economies of Europe report the lowest public sector participation: only 7 per cent of field trials are submitted by the public sector.

In industrialised countries, multinational firms[3] are responsible for most of the field trials (70 per cent), with single-country firms[4] accounting for 10 per cent of all field trials. Universities are conducting 12 per cent of the GM crop research followed finally by the national and international agricultural research centres conducting 8 per cent of all field trials in developed countries. In developing countries, national and international agricultural research centres are conducting 16 per cent of field trials with universities accounting for a mere 4 per cent; research institutes rather than universities conduct most agricultural research in developing countries (Pray and Umali-Deininger, 1998). Within the transitional economies of Europe, 79 per cent of all field trials are being conducted by multinational corporations, followed by a mere 1 per cent by local firms and 7 per cent of the research by national agricultural research centres. None of the trials are being conducted by universities.[5]

Crops. Since 1987, when the first field trials were approved, there have been more than 11,000 field trials of 81 GM crops approved worldwide. The crop most widely tested worldwide is corn (with 35 per cent of the field trials) followed by canola (11 per cent of all field trials) and potato (10 per cent of field trials). Soybean trials rank fourth tied with cotton at 7 per cent each in approved field trials worldwide (see Figure 3).

Most subsistence crops have not been the subject of much applied biotech research. The number of field trials of some important subsistence crops, such as

Table 2. Field trials by crop and region (1987–2000)

	Corn	Canola	Potato	Soybean	Cotton	Tomato	Sugar beet	Tobacco	Wheat	Rice	Other	Total
Total number	3881	1242	1088	782	723	654	394	308	232	189	1610	11105
US and Canada	2749	826	770	552	407	494	118	194	190	102	1087	7489
Europe, Australia, New Zealand and Japan	452	366	227	20	72	89	237	61	23	36	316	1901
Transitional	61	17	27	7	2	2	33	6	1	0	9	165
LDC	619	33	64	203	242	69	6	47	18	51	198	1550
Percent of all crops	35	11	10	7	7	6	4	3	2	2	14	100
US and Canada	37	11	10	7	5	7	2	3	3	1	15	100
Europe/NZ/Aus/Japan	24	19	12	1	4	5	13	3	1	2	17	100
Transitional	37	10	16	4	1	1	20	4	1	0	6	100
LDC	40	2	4	13	16	5	0	3	1	3	13	100

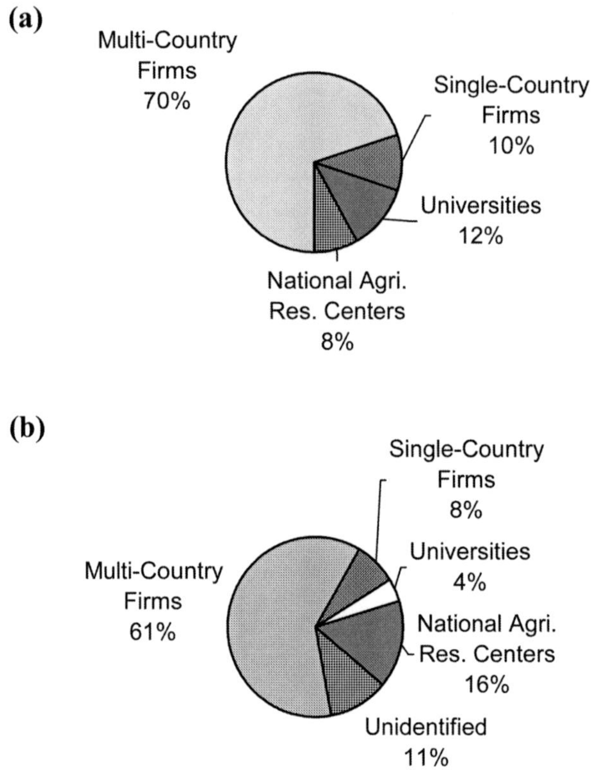

Figure 2. (a) Public/private research field trials of GM crops in industrialized countries (1987–2000), (b) Public/private research field trials of GM crops in developing countries (1987–2000)

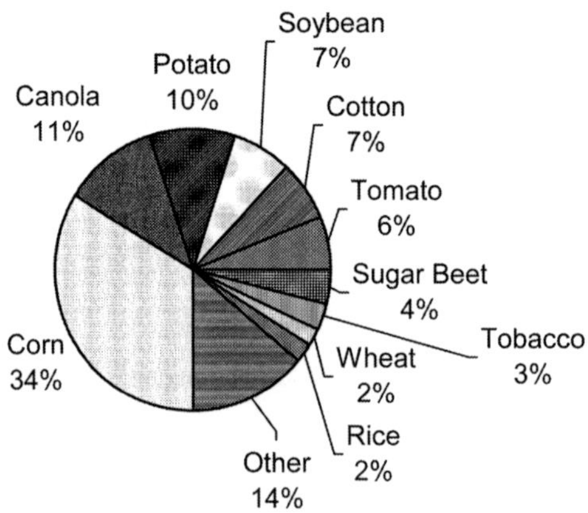

Figure 3. Ten major GM crops under field trials worldwide (1987–2000)

wheat and rice, has accelerated, however, in recent years. Wheat and rice were the ninth and tenth most frequently tested crop worldwide, with 232 and 189 field trials approved, respectively. This may mean that these crops are starting to receive some much-needed attention. Other GM crops that have been the focus of recent research are cotton, sugar beet, tobacco, alfalfa, melon, grasses, trees, squash and sunflower. Notably, in 2000, cassava was tested for the first time worldwide in Colombia and was to be approved for testing in the US in 2001. Yam, another important subsistence crop, has not yet reached the field trial test stage, but other 'orphan commodities' such as bananas, sweet potato, and lentils have all been approved for field testing in one or more countries.

Table 2 details the amount of research on different crops by region. In both less-developed countries (LDCs) and developed countries (DCs), corn is the most field-trial-tested crop. After corn, the most frequently field-tested crops are canola and potato in DCs and soybean and cotton for LDCs. In LDCs, however, rice and wheat, which are the most important food crops in developing countries, were the subject of only 3 per cent (rice) and 1 per cent (wheat) of all field trials.

Traits. The characteristics or traits of GM crops being researched globally primarily focus on herbicide tolerance and insect resistance. As shown in Figures 4(a) and 4(b), 29 per cent of all field trials are for herbicide tolerance. In industrialised countries the second most important trait tested is insect resistance (21 per cent of all field trials). In developing countries insect resistance is the most widely tested (37 per cent) trait. Another major difference between industrialised and developing countries is that industrialised countries put much more emphasis on product quality (more than 16 per cent) than developing countries (6 per cent). In both parts of the world considerably fewer field trials have been done on agronomic traits, such as increased yield.

III. Impact of GM Crops on Agricultural Productivity

Controversies surrounding biotechnology and the role of multinationals have generated studies of farm level impacts of these crops. All genetically modified crops that have been commercialised so far, except in China, have been developed and marketed by private firms. The results of these studies are summarised in Table 3.

Roundup Ready (RR) soybeans did not increase yields in Argentina, but did reduce pesticide costs, which increased farmers' profits. Results of *Bt* cotton studies were pretty consistent – *Bt* cotton leads to increases in yields, decreases in pesticide use and increases in profits in all four countries. The only difference was that in low-pesticide-use countries – Argentina, Mexico and South Africa – use of *Bt* cotton led to increases in total cost of production while in China it reduced the cost of production.

In addition to financial benefits for farmers, several of these studies report positive impacts in replacing dangerous pesticides and improving farmers' health. Surveys of 400 farmers in northern China showed the adoption of *Bt* cotton has led to dramatic reductions in the use of dangerous organophosphate pesticides and an FAO class II pyrethroid, reducing the probability of pesticide poisonings significantly (Hossain et al., 2004). A study of hospital records in one region of South Africa where small holders adopted *Bt* cotton showed a marked decline in cases of pesticide poison after *Bt* cotton was adopted (Ismael et al., 2002). The adoption of RR soybeans in Argentina led to a change in the type of herbicide used from very toxic (Toxicity Class II) to much safer (Toxicity Class IV) herbicides.

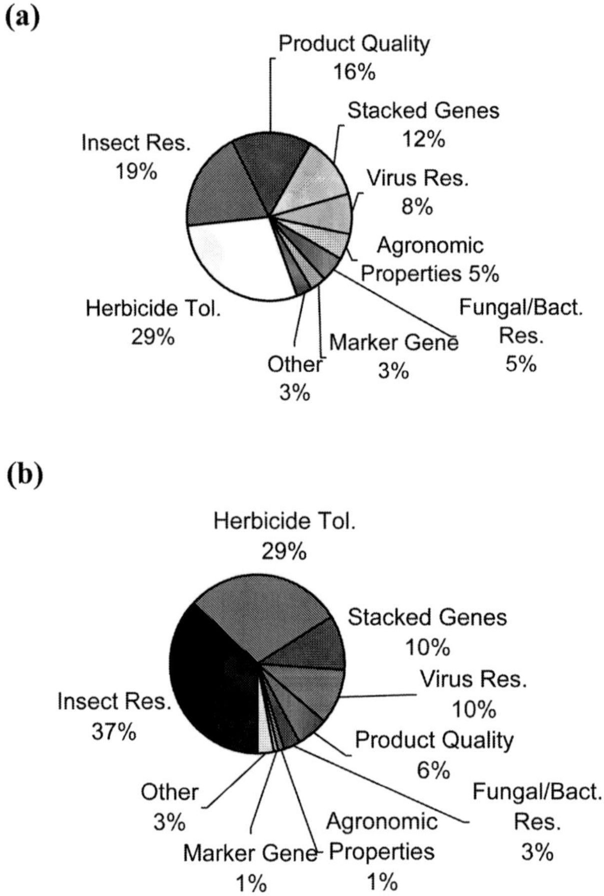

Figure 4. (a) GM crops traits tested in industrialised countries (1987–2000), (b) GM crops traits tested in developing countries (1987–2000)

Reduction in poisonings not only improves the well-being of farmers by reducing sickness, but also has economic consequences: labour productivity increases and health costs are reduced. There is also evidence of environmental benefits from adoption of GM varieties. Adoption of RR soybeans is associated with large increases in low-tillage cultivation, which has spread to millions of ha in Argentina (Qaim and Traxler, 2002). This type of technology lowers pollution by tractors which pass over the land less and greatly reduces erosion by wind and water. In China there is evidence from agricultural extension agencies (Pray et al., 2001) and scientists (Wu, 2002) that *Bt* cotton has greatly increased the number of insect species found in *Bt* cotton fields.

IV. Determinants of Agricultural Research

Research as an engine of growth for the agricultural sector and the economy as a whole is now accepted by economists and a growing number of policy makers.

Table 3. Impact of private biotech research on farmers, seed/biotech and pesticide companies profits

Country	Crop	Farmers	Consumer	Technology suppliers	Total surplus
				Millions of US $s	
Argentina	RR soy	303.20	4.30	27.60	335.10
	Bt cotton	0.77		1.70	2.47
Mexico	*Bt* cotton	2.35		0.43	2.78
S. Africa	*Bt* cotton	0.49		0.28	0.77
China	*Bt* cotton	33.29		4.36	37.65
India	*Bt* cotton	4.15		2.08	6.23
				% of total surplus	
Argentina	RR soy	90	1	8	
	Bt cotton	31		69	
Mexico	*Bt* cotton	84		16	
S. Africa	*Bt* cotton	63		37	
China	*Bt* cotton	88		12	
India	*Bt* cotton	67		33	

Sources: Qaim and Traxler (2002), Qaim and de Janvry (2002), Traxler et al. (2005), Ismael et al. (2002), Pray et al. (2001), Qaim (2003).

Given that adoption GM crops has, by and large, increased yields and productivity, why is there such limited R&D on developing-country crops? Analyzing determinants of both public and private research, we suggest that thin markets, lack of appropriability of research returns and high cost of undertaking research have discouraged private-sector research. Public sector support for research, on the other hand, is constrained by limited funds, research capacity, and, increasingly, lobbying pressure from environmental groups discouraging such research.

Determinants of Private Research

Demand. For private firms agricultural research is an investment meant to increase profits; otherwise, they will not invest in research for very long. Profitability of agricultural R&D investments is determined by costs and returns to the research activity, which are in turn influenced by several factors. Returns to private research improve in the presence of sizable expected demand for the products of research, the availability of exclusion mechanisms to appropriate part of the benefits from the new product or process, and a favourable business environment that permits efficient operations. Potential demand for inputs and consumer products developed through research, and thus market size, varies among regions depending on the size of the population, the purchasing power of the prospective buyers, local agro-climatic conditions, and sectoral and macroeconomic policies that influence input and output prices. For example, in 2000, the size of the global crop protection market was estimated to be $28 billion (Syngenta, 2003); consequently, the first generation of traits was designed to capture a portion of this market by either substituting for, or enhancing the productivity of, existing chemicals. Further, firms introduced these traits into crops with large markets, thereby enhancing their ability to extract rents.

Sectoral policies. Changes in the incentive environment affect demand for research services and the speed at which countries can adopt new agricultural innovations. Macroeconomic and sectoral policies alter the relative profitability of agricultural activities that in turn affects the expected profitability of adopting agricultural innovations, as well as the capacity of different segments of the farm community to acquire new technologies. The effectiveness of agricultural support services delivery (public and private), in particular agricultural extension, and rural infrastructure (roads, markets, irrigation) will also influence significantly the types and range of technologies introduced and the speed of adoption by farmers. Finally, bilateral and multilateral trade agreements reshape trading rules and influence market access and thus potential market size.

Government policies that affect the local business environment directly influence returns to private research. Examples of such policies are government marketing of inputs that reduce the market share of private firms and licensing and investment regulations that favour smaller firms over larger firms (Pray and Ramaswami, 2001).

Appropriability. Appropriability is an important precondition for private for-profit firms' participation in agricultural research. If firms cannot capture (appropriate) some of the social benefits of their research, they will stop investing. To capture some benefits from innovation, the innovating firm must be able to prevent imitators from using the innovation. The ability to do this is a function of the characteristics of the technology, the laws on intellectual property and their enforcement, the structure of the industry which is producing the technology and the industry that is using it. The legal means of protection against unauthorised use includes patents, plant breeder's rights, and trademarks. Firms also control use by keeping inventions or key parts of their inventions secret, which in some countries is protected by trade secrecy law. These legal means, however, give only limited protection in developing countries.

Inventors can also protect inventions biologically: putting new characteristics into hybrid cultivars or utilizing technical means to prevent copying (genetic use restriction techniques (GURTs) also called 'terminator' technologies). Hybrids and GURTs increase the size of the market for new cultivars because farmers cannot reproduce seed on their own farm and for subsequent planting. In the case of hybrids, seeds will grow but the crop will yield 15 to 20 per cent less. This is usually sufficient incentive for farmers to purchase new seeds each year. In genetic use restriction techniques, some of the proposed techniques (none are in commercial use yet) would use genetically engineered crops, which would produce sterile seed unless the seed had been treated with a specific chemical. Farmers then could not reproduce the seed for replanting. This type of GURT caused considerable controversy because poor farmers tend to save much of their seed; companies are accused of planning to exploit poor farmers by forcing them to purchase seed every year. Some GURT variations would produce fertile seeds, but subsequent generations would lack the patented trait – e.g., insect resistance – unless seeds were treated with a chemical.

In summary, the degree of appropriability is a function of the strength of intellectual property laws, the degree to which government agencies can enforce the laws, the structure of industry that reduces the cost of enforcing IPRs, and the technical capacity of firms to protect their varieties through the use of hybrids or GURTs.

Costs. Private research investments are also determined by potential costs of the agricultural research program and associated risks (Pray and Echeverría, 1991). The cost of research is the combination of quantity and price of research inputs, the number of years that these inputs will be employed to develop a new technology, and the stock of knowledge in this area of science. Such costs decrease with the supply of research inputs, the presence of a favourable business environment, the stock of existing knowledge and technology, and available human capital for conducting research activity. Research costs increase in the presence of anti-competitive markets or regulatory requirements.

The supply of research inputs and thus their price depends on the availability and accessibility of research tools and knowledge, many of which are produced by the public sector. For example, private breeders, to add desirable traits to new private varieties, may use improved populations of crop germplasm developed by public-research programs as parent material. Advances in biotechnology knowledge have led to a significant increase in private investment in agricultural research in the US and Europe over the past two decades. Greater private sector R&D implies that the marginal cost of applied agricultural research will decline as firms take advantage of economies of scale and scope. However, concentration of key research inputs amongst a few firms raises the possibility that cost of conducting research for those who do not have access to such technologies will increase.

The domestic supply and quality of human capital, a key input to the research activity, influences the level of research investments. In the Philippines, the availability and low cost of local research personnel encouraged some multinational firms to transfer research programs to teams of Filipino scientists (Pray, 1987). The domestic supply of skilled personnel is heavily dependent on the level and composition of public and private expenditures on education.

Several aspects of the business environment affect research costs and the productivity of research costs. Industrial policy can influence the degree of market concentration, the intensity of competition, and the prices of research inputs and outputs. Various government incentive programs, such as government contracts for new products and processes, grants and concessional loans, technical information services, and tax incentives, reduce research costs. Indirectly, the development of capital markets makes it easier for firms to raise funds for research (for example, venture capital). Bilateral and multilateral agreements also improve trade opportunities by facilitating access to intermediate technologies.

Regulation such as excessive product testing regulations and seed certification procedures can greatly increase costs of commercialising research output. Regulations to ensure that products developed using biotechnology are environmentally benign and safe for human consumption are necessary to gain consumer acceptance, but have greatly increased the cost of developing and releasing transgenic plant varieties. For example, seed companies have spent US$1.6 to 1.8 million to obtain regulatory approval for *Bt* cotton in India (Monsanto, personal communication, 2003). In the US the average cost for complying with the regulatory requirement for a single GM product is $5–6 million (Suresh, 2003). This is more than the annual research budgets of most Indian seed companies. As a result, only the largest companies can afford to attempt to commercialise genetically modified crops.

Determinants of Public Research

The basic justification for government expenditure on agricultural research is that private firms will under-invest in research when the output of that research has the characteristics of a public good – that is, the outputs of research are often non-rival and non-excludable. Because of the public-goods nature of research, social benefits are much higher than private benefits, hence the justification for public intervention. While many public investments have high social benefits, public investment will be justified only if the return is higher than other forms of public investment. A number of recent reviews of the returns to research have shown that public investments, indeed, has high payoffs – between 40 to 50 per cent depending on the methodology used (Gollin and Evenson, 2002). Considering that private companies and governments usually can obtain credit at interest rates below 10 per cent and public rates of return on other types of government investment are considerably lower than 40 per cent, these rates of return to research are very high; they suggest that there is considerable underinvestment in agricultural research in both the industrialised and developing world.

While studies that show high social rates of return to research investments may convince economists, most policy makers who actually decide on investments do not appear to have been convinced to make investments based solely on high social rates of return. Rather, public investments in agricultural research respond to many of the same forces that influence the amount and direction of private research (Hayami and Ruttan, 1985). Public research increases when there are changes in basic research that increase the possibility of an innovation or reduce its cost. In private-sector research models this is referred to as an increase in technological opportunity. The discovery of dwarfing genes in rice and wheat created the opportunity for plant breeders around the world to create many new green-revolution varieties that would respond to higher doses of fertiliser and good water conditions. These opportunities led to major increases in public-sector plant-breeding research around the world. Likewise, tools of biotechnology have created a major shift outward in the innovation possibility curve for plant breeding research; some governments have again responded by increasing their investments in research.

It has also been shown that changes in the demand for technology by farmers and consumers induce research to develop a particular type of technology. For example, in Asia, population increases pushing against limited land increased land prices, led government scientists to develop technology in the form of high-yielding varieties that allowed farmers to substitute biological technology and inexpensive fertiliser for land. Private firms did not attempt research to fulfil this demand because there was no way to make sufficient money selling new rice varieties to pay for the research. Public scientists in Japan did attempt this research because consumers in Japan put pressure on the Japanese government to provide cheaper food and Japanese farmers had run out of land for expansion. The government put money into research that eventually delivered fertiliser-responsive rice varieties.

In the case of biotechnology, some of the key factors have been demand for ways to protect plants against pests and disease for which no genetic resistance seemed to exist; for more efficient and labour saving ways of controlling pests; for ways to reduce the cost of pesticide use; and for methods to improve the quality of fruits, vegetables and edible oils.

V. Implications of Greater Private-Sector Role in Biotechnology Research

The large role of the private sector in biotechnology research has raised a number of concerns. These concerns are primarily based on perceptions that (a) IPRs on key technologies in the hands of a few private firms will limit access of such technology to developing-country researchers; and (b) monopoly power of biotech firms enables them to extract excessive rents from farmers. With declining support of public research, it is felt that poor people's crops and poor regions will be neglected. This section examines the empirical evidence that is available on these issues and considers their implications.

IPRs and the Question of Technology Access

Large biotech firms in the US and Europe through their own research and through mergers and acquisitions have amassed major portfolios of proprietary biotechnology tools, genes, and information. Much of this knowledge and these techniques can be obtained only from the company that owns it through a license for research and/or for commercialisation. The main concern is that companies may try to charge licensing royalties that local firms and public research institutes in developing countries could not afford. In some cases, owners may not be interested in licensing because of the possibility of liability problems or bad public relations if something goes wrong.

Another concern is that many of the major breakthroughs in biotechnology developed and patented by public-sector biotechnology institutes are likewise not available. The universities that developed them often licensed them exclusively to private firms, which then have rights to sublicense them. For small biotech firms these patents are the main assets of the firm. They may not have time and resources to market these technologies to developing countries. They will be very reluctant to license the technology in countries with weak intellectual property rights where they could lose control of their technology. Finally, if they do market the technologies, they may be too expensive for the public sector.

Although most of the discussions on access to biotechnology have focused on the negative impact of intellectual property rights, IPRs can increase access to biotechnology to developing country scientists. IPRs give owners of these technologies – multinationals, universities, and individuals – increased incentive to market their technologies. Prior to IPRs, the main incentive for sharing the technology was humanitarian; now firms also have a financial incentive. The other way in which IPRs make technology more accessible is through publication of patents, making information about how technology works publicly available. Patents of most industrialised countries can be searched on the web anywhere there is an internet connection and computer. Some non-profit organisations, like CAMBIA in Australia, provide guidance on how to search for patents as well as access to some sets of patents.

Many biotech labs in developing countries are currently using proprietary biotechnology. Descriptions of some of these technologies have been published, some genes are in commercial use and can be obtained through reverse engineering, and some techniques have become available through unauthorised routes. In most developing countries these techniques either cannot be or have not been patented; there is nothing illegal about use of these technologies for research or commercialisation of plant varieties based on these techniques. The complexity of developing a transgenic plant means that many patented technologies may be required to produce a GM plant with the desired trait. A now famous study of Golden Rice found that it was based on 70 patented technologies held in 30 different institutions (Kryder et al., 2000). These patents applied in both industrialised and developing countries. However, in some major rice consuming countries none of these patents exist and only a few are applicable (Pardey et al., 2003).

No comprehensive survey of patented genes and research tools in use in developing countries has been published, but a recent survey of scientists in Latin America by ISNAR indicates that their use is extensive. They found that government scientists were using patented techniques extensively without knowing they were patented in some cases or not caring about it in others (Salazar et al., 2000).

If genes and research tools are protected by patents, if the LDC scientists or farmers hope to export agricultural products developed using these techniques, innovators who wish to commercialise the technology need to license them from their owners. A recent study found that in most developing countries this was not a major problem because they export little back to markets where they might be blocked (Binenbaum et al., 2003). Consumer acceptance of GM crops is likely to be a higher export hurdle than objection based on violation of intellectual property.

The main problem in accessing these technologies in developing countries is not companies that are unwilling or unable to share their technologies due to IPRs, but the lack of research capacity to use these techniques (Pardey et al., 2003). Countries often do not have the scientific capacity to know which technologies would be most useful, or how to use them even if they were to get access. Far too many developing countries fall into this category.

Access to patented technology, however, remains a legitimate and serious concern for many countries with large NARS. Many technologies are difficult to obtain without going to the owners of the technology. The technologies may require biological material, complementary proprietary technology, and information not disclosed in the patents to make them work. Even if the technology could be used without permission from the owners, public or private institutes may decide that they will not use it without obtaining proper permissions. This decision may be motivated by desire of the local institute to have longer-term collaboration with the owner of the technology or by desire to export products produced by this technology. In some developing countries, IPRs on plants and biotechnology are growing stronger and the local research institute may have a legal obligation to obtain permission.

Discussions with researchers from national institutes and IARCs indicate that IPRs have slowed access to biotechnology in developing countries. Foundations and aid donors are so concerned about these problems that they are funding a number of institutions such as the International Service for the Acquisition of Agri-Biotech Applications (ISAAA), which is a decade old, and the newly established African Agricultural Technology Foundation (AATF) to address them.

Monopoly Power of Biotech Multinationals and Economic Exploitations

The concentration of plant biotech research and patents in the hands of a few multinationals has caused fears of market power and monopoly pricing. There are concerns that multinationals will take over the seed industry from local firms and use their market power to increase the price of biotechnology to capture all benefits from research.

Several researchers have recently estimated the distribution of benefits from adopting GM crops (see Table 3 for a summary). In nearly all studies surveyed, farmers captured most of the benefits (see Table 3). With the exception of *Bt* cotton in Argentina, the share of farmers in total benefits to society (measured as farmers' profits plus the profits of the monopoly supplier of technology) was far greater than the benefits to the suppliers of the technology. The seed/biotech company obtains more than the farmers only in the Argentina *Bt* cotton case. However, if the benefits to farmers who are using black-market seed in Argentina were to be included, most of the benefits would go to farmers (Qaim and de Janvry, 2002). Biotech companies capture about a third of the benefits as illustrated in the case of South Africa and India (Gouse et al., 2005; Qaim 2003). In other cases – China and RR soybeans in Argentina – farmers capture between 80 and 90 per cent of the benefits (Pray et al., 2001; Qaim and Traxler, 2002).

The difference in share of benefits clearly reflects differences in firms' ability to protect their technology from unauthorised copying. This ability derives from a combination of intellectual property law, enforcement of regulation, institutional structure of the industry, and characteristics of seed technology. Table 4 describes combinations of IPR, institutions, and technical characteristics that can lead to strong appropriability for firms. It then presents examples of these characteristics and the share captured by firms from empirical studies. The weakest appropriability is at the top of the table. The top row describes the situation during the Green Revolution. High-yielding wheat and rice varieties were not protected by intellectual property and could be easily duplicated by farmers. During the Green Revolution, a few small firms made limited profits from producing and distributing seed, even though research and seed distribution were financed by the government. Another case in which innovators have not been able to capture much benefit is Brazil where transgenic herbicide-resistant soybeans are illegal but are very easy for farmers and small seed companies to produce and sell. Moving down Table 4 seed companies start to make money from plant-breeding research when they develop hybrids. In India, private firms produced pearl millet and sorghum hybrids based on public inbred lines. This allowed the firms to capture some benefits from the farmers.

In the middle of the Table 4, China has patent laws, plant breeders' rights, and trademarks, which protect Monsanto and the Chinese government *Bt* varieties from being copied. The biosafety regulations are stringent, and it is time-consuming to obtain permission for commercialising a new product, which should restrict competition. However, due to problems of enforcement, most of the *Bt* cottonseed that farmers planted was produced by farmers or small seed companies and government research institutes that did not apply for biosafety committee approvals or pay royalties to the owners of varieties or genes (Pray et al., 2005). Thus, GM seed is sold for very low prices; Monsanto and the Chinese firm are able to capture only a small share of benefits whereas farmers, in the short run at least, obtain most of the benefits.

Near the bottom of Table 4 the highest technology fees, seed prices and share of benefits that accrue to firms are for *Bt* cottonseed in Argentina, South Africa, and

Table 4. Weakest to strongest intellectual property protection on plants or traits

IPR legislation	Institutional – enforcement and competition	Biological/technical	Examples (biotech & non-biotech)	Biotech/seed company share of economic benefits
None on plants	Public supplies seed	Pure-line varieties, open pollinated varieties	Green revolution rice and wheat in Asia	Small seed co. make limited profits
GM technology illegal	Small seed cost and farmers supply seed	GM crops, open pollinated varieties	RR soybeans in Brazil and Paraguay	Small seed co. make limited profits
None on plants	Public supplies seed in competition with private	Hybrids	Pearl millet, Sorghum in India	6% to 19%
Plant variety protection	Weak enforcement, competition between public & private research	Pure-line varieties	*Bt* cotton in China	5% to 10%
Just passed PVP – no protection yet	Biosafety regulations	Hybrids	*Bt* hybrid cotton in India	30%
Patents & PVP	Strong enforcement, monopoly on trait, control seed	Pure-lines, farmers can't replant seed	*Bt* cotton S. Africa	35% to 40%
Patents & PVP	Strong enforcement, monopoly on trait, control seed	Hybrids farmers can't replant seed	RR corn US, RR canola Canada,	50%
Patents on plant varieties		GURTs	Not available yet	Unknown

Source: Authors.

India. In Argentina, *Bt* cotton farmers sign contracts agreeing not to reproduce seed, and Monsanto has contracts with the limited number of ginning companies to buy all cotton seed. Nevertheless, experts estimate that more black-market *Bt* seed is planted than legal seed. In South Africa, farmers also agree not to replant their own seed, and Monsanto has contracts with the two gins in the country to ensure that none of the seeds go back to farmers. In addition they have patents on the genes and plant variety certificates for the varieties. In India, the *Bt* gene is in a hybrid, which farmers tend to purchase each year to get higher yields. In all these countries, it is now very expensive, time-consuming and risky for companies or government agencies proposing new *Bt* strains or new *Bt* varieties to move through the biosafety regulatory process. This indirectly acts as an entry barrier to new firms and creates a de jure monopoly for Monsanto and its partners for a number of years in all three countries.

There are several important results from Tables 3 and 4. First, when there are strong IPRs, complemented by contracts and regulations, as is the case with *Bt* cotton in Argentina and GM crops in South Africa, companies that supply the technology capture a larger share of the benefits. The second important result is that even in developing countries where companies have strong intellectual property rights such as South Africa, farmers capture most of the benefits from the GM crops. A final result is that strong intellectual property rights can induce private firms to do some research and technology transfer even in cases like South African cotton where the size of the crop and market is very small.

VI. Research Policy for the Poor

Evidence indicates that biotechnology research has not been focused to meet the needs of the poor farmers in developing countries; a number of policy recommendations have been suggested. (see Lipton, 2001; Byerlee and Fisher, 2002; Naylor et al., 2004). This section discusses policy measures to provide incentives for both private and public research to focus on crops harvested by the poor. We also discuss ways that biotechnology could be accessed to benefit the poor.

Incentives for the Private Sector Research in Generating and Targeting the Poor

While not all private sector research is expected to directly benefit the poor, it is important that incentives exist for firms to undertake some research for the poor. Private sector research, even when not focused on the poor, has been shown to create scientific capacity and spill-overs in developing countries. Biotech research in the private sector is producing knowledge, research tools, genes, and GM varieties that can be useful to the poor. Such research includes the rice-genome research financed by Monsanto and Syngenta and functional genomics research on rice by smaller biotech firms such as Metahelix in India. Research of this type will probably require public-sector plant breeding to produce actual varieties for the poor; however, with some changes in appropriability the private sector could play a role. Other examples include the *Bt* gene developed by Monsanto that has been incorporated into cotton for small farmers in Africa and Asia and recently into white maize in South Africa. In these cases private firms have been willing to commercialise technology that can benefit the poor.

A private firm's investments in research or commercialisation of a new technology will depend on the expected size of the market, their ability to appropriate some of the gains from the crop, and the costs of research and commercialisation. Government actions that increase expected markets, increase appropriability, or reduce costs of research could induce greater private sector investment.

There are a number of policies that can increase the market size of an innovation. By removing unnecessary regulations on the export of new conventional and transgenic varieties between developing countries, private firms would have a bigger market for the products of their research. For example, the South African seed company, Pannar, could supply more technology throughout Africa, and Chinese seed companies could supply biotechnology to India, Vietnam and Bangladesh. Where national research institutions are weakest, basic government interventions may be needed to put in place the physical infrastructure such as transportation and communication, and the institutional infrastructure such as law and order and enforceable contract law that is required for markets to work.

In countries with more advanced NARS, appropriability can be strengthened not only through stronger IP laws and their enforcement, but also through government research programs. The evidence on the role of IP laws and complementary institutions in inducing research was discussed in Section IV. Government research has increased appropriability in the past by developing methods of producing hybrid seed in crops that had been pure-line varieties. For example, the Indian government and International Rice Research Institute (IRRI) research in developing hybrid rice has induced more private plant breeding research in India (Pray and Ramaswami, 2001).

Governments can reduce the cost of research by increasing the production of skilled scientists by strengthening public research universities. Government research institutes can identify and provide useful new germplasm for private plant-breeding programs. Reducing restrictions on importing research inputs such as chemicals can reduce research costs. Reducing restrictions on foreign direct investment can allow more resources for research and technology transfer. Small local firms may also need subsidies from the government or assistance from government universities and research institutes to access proprietary research tools.

The major government policy that restricts the private role in spreading biotechnology is the regulatory system. If there is no regulatory system for GM varieties, companies will not do biotech research on the problems of a country. If the regulatory system is unnecessarily strict, companies may have to spend millions of dollars on biosafety research, which will reduce their incentive to invest. Finally, if the regulatory process is very time-consuming and expensive, the only institutions that will able to afford commercialisation of GM crops will be large multinationals. American and European firms typically expect to spend $10 million for each new transgenic product – develop the portfolio of health, environment, and agricultural biosafety information – that they will then use in most countries.

In addition to providing commercial incentives for private research to assist the poor, governments may be able to induce some private research and technology transfer by providing good publicity and money in the form of prizes to firms that develop and spread technology to the poor. Tax incentives or better investment opportunities could be provided for private firms that invest in the needs of the poor.

In the US and elsewhere, private foundations and charitable organisations such as the Rockefeller Foundation have been established and grown in part because of tax incentives.

One final possibility for providing incentives to the private sector to do more research is a major prize program for agricultural technology that reduces poverty or food insecurity (Masters, 2003). This competition would focus on crops of major importance to the poor; both public and private institutes could compete; and the amounts would be large enough to make competition worthwhile. The money could come from a combination of government and private foundations. The recent announcements of a $250 million program by the Bill and Melinda Gates Foundation to fund research on diseases that cause millions of deaths in developing countries offers a possible model for funding such a program (Bill and Melinda Gates Foundation, 2005).

Role of Public Research

We have demonstrated why private firms will not develop many of the traits or new varieties needed by the poor. Most of the major biotech companies have been reducing the number of crops that they work on as research budgets shrink. Monsanto, for example, now focuses their research on four crops: maize, soybean, canola, and cotton. There is consensus among economists that public research must be part of the answer. The public agricultural research systems of developing countries are an essential ingredient for harnessing benefits for the poor. The IARCs are also starting to play an important role in biotech research, just at the time when their budgets are being slashed.

The key issue is the difficulty of obtaining funding for biotechnology and conventional research on problems of poor people. Public agricultural research in general and especially the IARCs are facing declining financial support. In the competition for declining funds, the poor face considerable difficulty. Almost by definition the poor do not have well organised groups that can lobby for their interests when government research resources are being allocated. Yet there are NGOs, charities, foundations and some donors who are specifically focused on the poor. These groups need to be mobilised to provide political support for government funding of research on poor peoples' problems although some of these groups, particularly some of the NGOs, are very sceptical of or opposed to biotech research.

Another issue is that an expensive, unpredictable, and opaque biosafety regulatory regime can be even more of a constraint to commercialisation of technology developed by public research than technology from the private sector. Public institutes often have less money to finance ecological, health, and agricultural trials to meet regulatory requirements than the private sector. This may be one of the reasons that there are so few field trials by the public sector. Governments must find ways to fund these trials if they want public research to provide any GM crops to help the poor or compete with the private sector. Finally, the public-research sector can be most efficient if they can find ways to access some of the proprietary technology available in small and large biotech firms and not have to re-invent the wheel.

Options For Accessing Proprietary Biotechnology to Benefit the Poor

A recent review of options for accessing biotechnology highlighted possibilities for partnerships between public NARS, local seed companies, global companies and the CGIAR (Byerlee and Fischer, 2001). This section summarises ways in which public research institutes or local firms can obtain patented biotech genes and tools.

First, the public sector in developing countries may be able to purchase a technology or negotiate free access and make it available to local scientists free of royalty. For example, a consortium of public research institutes in Asia, led by IRRI, purchased the rights to a *Bt* gene from a small Japanese biotech company. However, few key technologies will be available for purchase. In some cases private firms will agree not to enforce their patents on their technology in a poor country.

Material transfer and licensing agreements are a second possibility. Material transfer agreements (MTAs) define the conditions for research use and leave the conditions for commercialisation to a future date. Initially this method is cheaper, but there is risk that the company that does the research will not be allowed to commercialise the technology. Licensing agreements, on the other hand, specify the conditions for commercialisation of a technology, the payments, and the sharing of benefits. These are probably the most common form of authorized technology and know-how transfer.

Alliances and joint ventures (JV) are the third possibility. In JVs both parties agree to provide specific assets and to share the benefits; JV contracts typically include MTAs and licensing agreements on technology. There is a growing consensus that partnerships between the public and private sector will be needed to effectively use biotechnology for the poor in developing countries (Byerlee and Fischer, 2001; Pingali and Traxler, 2002).

Last, it may be efficient and legal for scientists to use specific technologies that are easily copied or fully revealed in patent disclosures, as long as the patent on the invention has not been filed in the country of use. The drawback of such a strategy is that commercial products that are developed using proprietary technology without permission could be blocked from being exported to a country where the technology is patented. The loss of export markets could be an important reason not to use such proprietary technologies.

These options are not mutually exclusive. Developing country policy-makers and scientists will have to use a combination of these options to access proprietary technology that meets their research needs.

VII. Conclusions

Dominance of the private sector in agricultural biotechnology research and commercialisation has raised concerns that benefits may not reach the poor in developing countries. In this paper, a review of trends in public and private agricultural research shows that private research on agbiotech is more than twice that done by the public sector and that the majority of it is in industrialised countries. Where the products of biotechnology have been adopted in developing countries, the evidence shows that farmers and consumers are the primary beneficiaries of GM crops. This is the case even in countries that have very strong intellectual property rights that are reinforced by technical and other institutions to prevent unauthorised copying.

Given that adoption of GM crops has potential for increasing social welfare in developing countries, this paper considers constraints that have limited both adoption and research. The private sector, which is doing much of the research in industrialised countries and owns many of the research tools, has little incentive to undertake research in developing countries due to low private returns. Although such market failure emphasises the need for public sector involvement, many NARS are constrained by declines in funding and private ownership of key technologies. To overcome these constraints, it is suggested that policy makers create greater incentives for private-sector research. These include commercial incentives such as more efficient biosafety regulations and stronger IPRs; tax incentives and government subsidies for research; and financial prizes for research and technology for the poor. A transparent and efficient regulatory process will ensure that any risks from adopting GM crops are minimised and are not at the expense of higher research costs.

Private-sector research by itself will not meet all needs of the poor. The role of the public sector is to induce private research and to conduct research that will benefit those neglected by the private sector. Sustainable public biotech research for the poor requires the development of groups who will lobby for the poor, which can be difficult. Nevertheless, with leadership from local anti-poverty groups and donors who are committed to reducing poverty, effective local support for public research to reduce poverty might be built. International support for the IARC biotech research programs is also essential.

In closing, it must be emphasised that biotechnology research by itself cannot eliminate poverty and food insecurity. It can be most effective in delivering benefits if other ingredients for sustainable economic growth are present – namely good institutions and governance, sound macroeconomic policies and human capital to take advantage of the technology.

Notes

1. These are widely known as crop field tests, *field trials* or environmental releases.
2. Field-trial data cannot be interpreted as representative of all biotechnology research being conducted in a country. It represents only the most applied R&D and only a piece of a country's agricultural research.
3. Multinational firms are defined as companies that have applied for permission to conduct field trials in more than one country. More than 54 multinationals have conducted GM crop field trials worldwide.
4. Single-country firms are defined as companies that have conducted field trials in only one country thus far; there are 205 single-country firms.
5. Due to data limitations, 13 per cent of the companies or institutions conducting the field trials in transitional economies could not be identified, but that percentage most likely consists of multinational corporations.

References

Binenbaum, E., Nottenburg, C., Pardey, P. G., Wright, B. D. and Zambrano, P. (2003) South-north trade, intellectual property jurisdictions, and freedom to operate in agricultural research on staple crops, *Economic Development and Cultural Change,* 51(2), pp. 309–35.

Byerlee, D. and Fischer, K. (2001) Accessing modern science: policy and institutional options in developing countries, *IP Strategy Today*, No. 1 International Institute Inc, Ithaca, NY.

Cohen, J. (2005) Poorer nationals turn toward publicly developed GM crops, *Nature Biotechnology*, 23(1), pp. 27–33.

Bill and Melinda Gates Foundation (2005) 'Bill Gates Urges World Leaders to Seize Historic Opportunity to Improve Health', Press Release May 16, 2005. Available at http://www.gatesfoundation.org/GlobalHealth/Announcements/Announce-050516.htm

Gollin, D. and Evenson, R. (2002) Technology in agricultural development: contributions of public sector research programs, unpublished paper, Economic Growth Center, Yale University. New Haven.

Gouse, M., Pray, C. E. and Schimmelpfennig, D. (2005) The distribution of benefits from *Bt* cotton adoption in South Africa, *AgBioForum*, 7(4), pp. 187–94.

Hayami, Y. and Ruttan, V. W. (1985) *Agricultural Development: An International Perspective*. Baltimore: Johns Hopkins University Press).

Hossain, F., Pray, C. E., Lu, Y., Huang, J., Fan, C. and Hu, R. (2004) GM cotton and farmer's health in China: an econometric analysis of the relationship between pesticide poisoning and GM cotton use in China, *International Journal of Occupational and Environmental Health*, 10(3), pp. 307–14.

Huang, J., Roselle, S., Pray, C. E. and Wang, Q. (2002) Plant biotechnology in China, *Science*, 295, pp. 674–7.

Ismael, Y., Bennett, R. and Morse, S. (2002) Benefits from *Bt* cotton use by smallholder farmers in South Africa, *AgBioForum*, 5(1), pp. 1–5.

Kryder, R. D., Kowalaski, S. P. and Krattiger, A. F. (2000) The intellectual and technical property components of pro vitamin-A rice (Golden Rice™): a preliminary freedom-to-operate review, ISAAA Brief No. 20, ISAAA, Ithaca, New York.

Lipton, M. (2001) Reviving global poverty reduction: what role of genetically modified plants?, *Journal of International Development*, 13, pp. 823–46.

Masters, W. A. (2003) Research prizes: a mechanism to reward agricultural innovation in low income regions, *AgBioForum*, 6(1/2), pp. 1–74.

Monsanto (2002) *Annual Report*, accessed at:http://monsanto.com/monsanto/content/media/pubs/2002/Monsanto_2002_Annual_Report.pdf

Morris, M. and Hoisington, D. (2000) Bringing the benefits of biotechnology to the poor: the role of the CGIAR Centers, in M. Qaim, A. F. Krattiger and J. von Braun (eds), *Agricultural Biotechnology in Developing Countries: Towards Optimizing the Benefits for the Poor*, pp. 327–55 (Boston: Kluwer Academic Publishers).

Naylor, R. L., Falcon, W. P., Goodman, B. M., Jahn, M. J., Sengooba, T., Tefera, H. and Nelson, R. J. (2004) Biotechnology in the developing world: a case for increased investments in orphan crops, *Food Policy*, 29(1), pp. 15–44.

Pardey, P., Wright, B., Nothenberg, C., Binenbaum, E. and Zambrano, P. (2003) Intellectual property and developing countries: freedom to operate in agricultural biotechnology, Biotechnology and Genetic Resource Policies, Brief No. 3, IFPRI: Washington, DC.

Pingali, P. L. and Traxler, G. (2002) The changing locus of agricultural research: will the poor benefit from biotechnology and privatization trends?, *Food Policy*, 27(3), pp. 239–46.

Pray, C. E. (1987) Private sector research and technology transfer in Asia, in V. Ruttan and C. E. Pray (eds), *Policy for Agricultural Research*, pp. 411–32 (Boulder, Westview Press).

Pray, C. E. (2001) Public/private sector linkages in research and development: biotechnology and the seed industry in Brazil, China and India, *American Journal of Agricultural Economics*, 83(3), pp. 742–7.

Pray, C. E., Bengali, P. and Ramaswami, B. (2005) Costs and benefits of biosafety regulation in India: a preliminary assessment, *Quarterly Journal of International Agriculture*, 44(3), pp. 267–89.

Pray, C. E., Courtmanche, A. and Govindasamy, R. (2002) The importance of intellectual property rights in the international spread of private sector agricultural biotechnology, presented at 6th International Conference convened by ICABR on Agricultural Biotechnologies: New Avenues for Production, Consumption and Technology Transfer, Ravello, Italy.

Pray, C. E. and Echeverría, R. G. (1991) Determinants and scope of private sector agricultural research in developing countries, in P. G. Pardey, J. Roseboom and J. Anderson (eds), *Agricultural Research Policy: International Quantitative Perspectives*, pp. 243–364 (Cambridge: Cambridge University Press).

Pray, C. E., Huang, J., Hu, R. and Rozelle, S. (2002) Five years of *Bt* cotton in China – the benefits continue, *The Plant Journal*, 31(4), pp. 423–30.

Pray, C. E., Huang, J., Ma, D. and Qiao, F. (2001) Impact of *Bt* cotton in China, *World Development*, 29(5), pp. 813–25.

Pray, C. E. and Ramaswami, B. (2001) Liberalization's impact on the Indian seed industry: competition, research, and impact on farmers, *The International Food and Agricultural Marketing Review*, 2(3/4), pp. 407–20.

Pray, C. E. and Umali-Deininger, D. (1998) The private sector in agricultural research systems: will it fill the gap?, *World Development*, 26(6), pp. 1127–48.

Qaim, M. (2003) *Bt* cotton in India: field trial results and economic projections, *World Development*, 31(12), pp. 2115–27.

Qaim, M. and de Janvry, A. (2005) *Bt* cotton and pesticide use in Argentina: economic and environmental effects, *Environment and Development Economics*, 10, pp. 179–200.

Qaim, M. and Traxler, G. (2002) Roundup ready soybeans in Argentina: farm level, environmental, and welfare effects, presented at 6th International Conference convened by ICABR on Agricultural Biotechnologies: New Avenues for Production, Consumption and Technology Transfer, Ravello, Italy.

Salazar, S., Falconi, C., Komen, J. and Cohen, J. I. (2000) The use of proprietary biotechnology research inputs at selected Latin American NAROs, Briefing Paper 44, ISNAR, The Hague, The Netherlands.

Suresh, N. (2003) Transgenic supporter, *Biospectrum*, 21 July, accessed at: http://www.biospectrumindia.com/content/shakers/103072101.asp

Syngenta (2002) *Annual Report*, accessed at: http://www.syngenta.com/en/downloads/ar2002_en.pdf

Syngenta (2003) *Annual Report*, accessed at:http://www.syngenta.com/en/downloads/ar2002_en.pdf

Traxler, G. S., Godoy-Avila, J., Falck-Zepeda, J. and Espinoza-Arellano, J. (2003) Transgenic cotton in Mexico: economic and environmental impacts, in N. Kalaitzandonakes (ed.), *Economic and Environmental Impacts of First Generation Biotechnologies*, pp. 183–202 (New York: Kluwer).

Vogel, G. (2002) Retreat from Torrey Mesa: a chill wind, *Ag Research Science*, 298, p. 2106.

Wu, F. (2002) *Bt* or not Bt? Tools for regulatory decisions concerning genetically modified corn, doctoral dissertation, Engineering and Public Policy, Carnegie Mellon University, Pittsburgh, Pennsylvania.

Agroecological Alternatives: Capitalising on Existing Genetic Potentials

NORMAN UPHOFF

I. Introduction

The most direct way to reduce poverty is to raise the productivity of those factors of production controlled by the poor: first of all, their labor, but also their knowledge and skills, and for many though not all, small areas of land. Increased factor productivity of land, labor, capital and water can have second-order benefits for the poor, urban as well as rural, by lowering the price of food and other things on which the poor spend most of their meager incomes. Current strategies for agricultural development have relied on the genetic improvement of crops (and livestock), on mechanisation to reduce labor input and expand the scale of production units, and on greater use of inputs, chemical fertilisers, agrochemical biocides, and water. These practices have been faulted for being inaccessible to many smaller farmers and for offering little to those who own no land. However, they have had some large and favorable impacts for many of the poor over and above their generation of some additional employment on- and off-farm. Since 1960, there has been about a 30 per cent increase in per capita food production worldwide; the real price of food has declined by almost half (Halweil, 2002). Such a decline may have affected adversely some small farmers whose gains in productivity were less than others', but overall, this technological change has improved the nutritional and income positions of the poor in comparison to counter-factual alternatives.

II. Looking Ahead

This said, we must acknowledge that yield increases through conventional strategies – taking cereals as indicative of sector performance – have been slowing since the 1990s and are no longer keeping pace with population growth (Table 1). A number of factors contribute to prospects for a less-favorable agricultural future.

- *Price effects:* Economists point out, correctly, that some of the recent decline in growth rates is attributable to the lower prices that have resulted from the successes of the Green Revolution. Lower prices discourage expansion of cultivated area and irrigation facilities. Simultaneously, shifts in factor productivity have been taking place. Land and water resources offering the highest productivity for lowest cost have now been developed; marginal returns from further investments should be expected to be less, unless there are significant improvements in technology. Other biophysical constraints, less tractable to policy intervention, intersect with economic considerations.
- *Genetic potentials:* Nature's genetic cornucopia may no longer supply comparable gains through conventional breeding. This slowdown in gains, complemented by new opportunities created by advances in biotechnology, has made more attractive the possibilities of transgenic innovation, reinforced by trends that are slowing the pace of the Green Revolution.

Table 1. World total and per capita grain production, fertiliser use and exports of pesticides, with growth rates by decade, 1961–2000

	World grain production (million tonnes)	World fertiliser use (million tonnes)	World pesticide exports (billion 2002 USD)	World grain production (kg per capita)
1961	805	31	1.32	261
1970	1087	69	2.42	293
1980	1418	117	7.88	319
1990	1779	138	9.51	337
2000	1864	135	11.82	307
More recently	1874 (2003)	138 (2001)	10.91 (2002)	298 (2003)
Changes by decade:				
1961 to1970	+386	+38	+1.10	+32
1970 to 1980	+331	+48	+5.46	+26
1980 to 1990	+361	+21	+1.63	+26
1990 to 2000	+85	−3	+2.31	−30

Source: Worldwatch Institute's Vital Signs database, derived from UNFAO, USDA, International Fertilizer Industry Association and other sources.

- *Mineral fertilisers:* These have contributed as much as genetic improvements to the last century's production increases, but are encountering diminishing returns, reducing their value to farmers (Uphoff, 2006). Even if the market price of fertiliser does not rise in coming decades, government subsidies are being reduced because of fiscal constraints and/or new policy directions. Moreover, given probable changes in global supply-demand balances in petroleum, if only because of rising incomes, the cost of fertilisers to farmers is more likely to increase than remain the same even as their marginal productivity diminishes. Table 1 illustrates a slowdown in the rate of increase in world grain production during the 1990s; fertiliser use declined, indicating a decline in marginal returns on a global scale.[1] Reduction or substitution of reliance of plants on nutrient amendments to the soil has not been a focus of significant transgenic research and development to date.
- *Agrochemical use for pest and disease control:* This grew even more rapidly than the use of fertilisers during the 1970s, thanks in large part to public-sector subsidies. Tracking trends in global agrochemical use is difficult because price effects can obscure changes in volume; moreover, many different kinds of chemicals need to be aggregated (insecticides, herbicides, fungicides, nematocides, molluscicides, and so forth). One current estimate (Pretty and Hine, 2005) is that 2.5 billion kg of active ingredients were applied in 1998–99. In the absence of comprehensive data on agrochemical use, the volume of world trade in pesticides is one indicator of global trends.

  The level of exports reported (by value) in Table 1 for 2002 was 13 per cent below their peak level of $12.6 billion in 1996. Actually, according to Pretty and Hine (2005), agricultural use by volume peaked in 1981, but the value of exports continued to rise because of price increases, peaking about 15 years later. The value of imports continues to rise in developing countries. Agrochemical use can create a kind of treadmill where more and more need to be applied to maintain an edge in what can become a losing battle.[2] Moreover, there are concerns about hazards to human and environmental health from agrochemicals (Kishi, 2005). These considerations have spurred interest in biotechnology alternatives that can reduce dependence on chemicals or that favor less toxic and harmful versions.
- *Water constraints:* Water scarcity is growing globally; the agricultural sector's claims on fresh water are challenged by growing urban and industrial needs (Postel, 1999). Finding ways to get 'more crop per drop' is becoming imperative for rich and poor farmers alike as both quantity and quality problems in the water sector loom larger for agriculturalists.
- *Influences of climate:* Farmers are now confronting climate change as a large but ominous unknown factor. While they can adjust to gradual warming by changing cropping systems, there is no protection against climate variability. Extreme weather events and greater variance are devastating to agricultural production, especially with monoculture and genetic homogeneity. Engineering plants to be more tolerant of drought or flooding may offer some relief, but improving the water absorption and retention qualities of soil systems – by increasing soil organic matter and through different, better methods of soil and crop management – could have large benefits.

III. An Agroecological Approach to Meeting Food Demand and the Needs of the Poor

The longer-run benefits and sustainability of current 'modern' agricultural strategies are being questioned for many reasons; here I want to focus on an agroecological strategy for meeting world food needs in ways that are more accessible to the poor, cost-effective and environmentally beneficial. This presentation proceeds from the premise (fact) that by modifying management practices, it is possible to get more productive phenotypes from existing genotypes. Phenotypes are the actual organisms that result from the interaction between genetic potential (genotype) and the biophysical environment in which this potential gets expressed. We don't consume genotypes; we nourish ourselves with phenotypes.

Research and experience are demonstrating significant untapped opportunities in agroecological innovation. Specifically, the production of rice, the most widely consumed staple food, can be greatly increased – and productivity of the factors of production being devoted to rice can be raised – by modifying plants' growing environment without altering their present genetic endowments. Such gains can be achieved relatively quickly and at low cost, and are likely to be more accessible to the poor than are higher-tech innovations. Moreover, they are associated with agricultural practices that improve rather than degrade their biophysical environment. Further, they are not encumbered by the costs and complexities of monitoring, testing, regulation, and enforcement of biosafety regimes (Herring, this issue).

Agroecological approaches to agricultural development do not make genetic improvements unimportant or unnecessary. Transgenic interventions can produce a variety of benefits, and may indeed be needed to deal with certain problems of crop production and protection. However, there are understandable concerns about the ways in which much of the current transgenetic research is being undertaken, driven largely by private investments and incentives that need not take public and environmental interests into account systematically and transparently. And there is justifiable concern about the true costs of such innovations, including regulation and monitoring costs currently required of transgenic modifications (Herring, 2007).

Even so, some transgenic innovation carries important potential – e.g., plants resistant to the papaya ring spot virus (see Gonsalves et al., 2007). Research is currently being done at Cornell University on the transfer of a gene that increases crop plants' production of trehalose, a common sugar that is produced naturally in a wide variety of organisms, from bacteria and fungi to certain insects. Trehalose confers greater resistance to drought, cold and salinity while improving yield. Millions of poor households in rural areas support themselves on drought-prone or otherwise marginal lands; widespread benefits for the poor are certainly possible. Trehalose is already used in many food and health products, and thus not a good candidate for disparagement as a 'Franken-compound'. Cornell is placing this innovation in the public domain so that improved crop varieties can be freely available in resource-poor parts of the world where the need is greatest (Garg et al., 2002). This example makes clear that genetically-modified organisms (GMOs) represent a diverse set of phenomena, about which broad generalisations, either pro or con, are untenable.

When evaluating contributions of transgenic research and resulting crops to food security and other needs of the poor, we should consider whether these objectives can be achieved in any other ways. Are there more cost-effective means to reduce poverty? Can this goal be achieved with fewer adverse impacts on the biophysical environment?

Given the experience to date with the system of rice intensification (SRI), developed in Madagascar about 20 years ago and now beginning to gain acceptance around the world (Stoop et al., 2002; Uphoff, 2003), I would answer these questions with a definite 'yes', at least regarding rice, which is the staple crop that has the largest impact on smallholder and poor households around the world (Dixon and Gulliver, 2001). The lessons learned from SRI could contribute also to improving the production of other crops in cost-effective and sustainable ways.

IV. The System of Rice Intensification and Its Spread

SRI can be characterised as a 'civil society' innovation (Lines and Uphoff, 2005). The person most responsible for its creation was Father Henri de Laulanié, a Jesuit priest who spent 34 years working with farmers in Madagascar to improve rice production. Having been educated at a leading French agricultural school, he brought basic scientific knowledge to his quest. But his sources of knowledge for devising SRI were farmer practices and his own observations of rice plants (Laulanié, 1993).

The practices that Fr. Laulanié combined produce a different, more productive phenotype from rice genotypes; these methods for plant, soil, crop and nutrient management utilise available genetic potentials more fully. These practices include:

- Transplanting *very young* seedlings very carefully, while still only 8 to 12 days-old and certainly less than 15-days-old, before their fourth phyllochron of growth. This preserves their full genetic potential for tillering and root growth.[3]
- Spacing the plants *widely apart* in a square pattern (25 × 25 cm or wider), planted *singly* rather than in clumps of four–six plants so that there is no competition among roots and plenty of room for canopy growth. These practices help farmers achieve 'the edge effect' throughout the whole field.
- *No continuous flooding* during the plants' growth period, applying small amounts of water regularly or alternating wet and dry field conditions to maintain a mix of aerobic and anaerobic soil conditions. After flowering, the recommendation is that a thin layer of water be kept on the field, although some farmers find alternate wetting and drying of fields throughout the crop cycle to be feasible and even beneficial.
- Controlling weeds by *weeding with a rotating hoe,* which aerates the soil at the same time that it destroys weeds by incorporating them back into the soil, where their nutrients can be recycled.
- *Applying compost* to the field to increase the soil's organic matter content. This is optional, as improved yields can come with fertiliser or even with no nutrient amendments, at least for some time as nutrient reserves in the soil are mobilised through microbial activity.

By changing the ways rice plants, soil, water and nutrients are managed, SRI farmers are usually able to attain rice yields to seven–eight tonnes/hectare (t/ha),

about double the present world average of 3.8 t/ha. With proper and sustained use of these methods, yields can go beyond 15 t/ha, which has been thought by researchers to represent a kind of 'yield ceiling' for rice (Khush and Peng, 1996). A simple summary of SRI production levels in a number of countries is presented in Table 2, comparing SRI yields, average and maximum, with those obtained by standard production methods under similar conditions.[4]

Since the data in this table come from diverse sources and different ecosystems, between-country differences are not easily interpreted; within-country comparisons represent real differences in that yields were measured by the same methods, and usually the comparisons were of different cultivation methods on the same farms. The differences are large and consistent enough that it is unlikely any measurement

Table 2. Average and maximum yield with SRI vs. comparison yields, in tonnes per hectare from 13 countries (N and ranges shown in parentheses)

Country	Data sources (no. of farmers)	Ave. comparison yield [t/ha]	Average SRI yield [t/ha]
Bangladesh			
(2002–04)	4 evaluations funded by IRRI/ Bangladesh (1,278 on-farm trials)	5.61 (4.7–6.71)	7.08 (4.7–8.69)
Cambodia			
(2000–02)	26 reports at national workshop (1,665 farmer experiences)	2.7 (1.3–6.0)	4.1 (3.1–12.0)
(2001–03)	Evaluation of 120 farmers using SRI for 3 years	1.34 (0.4–4.8)	2.75 (0.3–8.9)
(2001–04)	Survey of 400 SRI farmers and 100 SRI farmers, all randomly selected	1.63	2.29
China			
(2000–02)	7 on-station trials with hybrid varieties	10.9 (10.0–11.8)	12.4 (9.7–15.8)
Cuba			
(2003)	27 on-farm trials in 8 municipalities	4.3 (1.6–7.2)	7.3 (3.0–13.0)
Gambia			
(2001)	On-farm trials (10)	2.3 (2.0–2.5)	7.1 (6.8–7.4)
India:			
AP (2003–04)	On-farm trials (51)	7.13 (3.9–10.3)	9.67 (5.4–15.0)
AP (2004)	On-farm trials (476)	5.48	7.92
TN (2003–04)	On-farm trials (100)	5.65	7.23
Indonesia			
(2000–02)	2 on-farm trials and 5 on-station trials	5.0 (4.1–6.7)	7.4 (6.2–8.4)
Madagascar			
(2000–02)	11 on-farm trials (3,025) 3 on-station trials	2.6 (1.5–3.6)	7.2 (4.2–10.35)
Myanmar			
(2001–03)	228 Farmer Field School trials	<2.0	4.95 (2.0–15.3)
Nepal			
(2002–03)	12 sites with Farmer Field School trials	FP: 4.3 (2.5–6.5) IP: 6.0 (3.8–8.5)	8.3 (5.5–11.1)
(2004)	On-farm comparison trials (22)	3.37 (3.0–5.0)	7.85 (5.0–13.5)

(continued)

Table 2. (*Continued*)

Country	Data sources (no. of farmers)	Ave. comparison yield [t/ha]	Average SRI yield [t/ha]
Philippines (2002–04)	Results of 75 on-farm and on-station trials	3.0 (2.0–3.6)	6.6 (4.95–7.6)
Sierra Leone (2001–02)	8 on-farm trials (160)	2.5 (1.9–3.2)	5.3 (4.9–7.4)
Sri Lanka (2000–02)	6 on-farm trials (275) and 2 on-station trials	3.6 (2.7–4.2)	7.8 (7.–13.0)
(2003)	Survey of 60 SRI farmers and 60 non-SRI farmers	3.84	5.52

Notes: Comparisons of SRI in China and India are with the best modern practices using highest-yielding varieties and agrochemical inputs; in China, average rice yields are 6.3 t/ha; and in India, 3.01 t/ha in Andhra Pradesh, and 3.42 t/ha in Tamil Nadu. Sources unless given below are from Uphoff et al. (2002).
Bangladesh: Final evaluation report on verification and refinement of the system of rice intensification (SRI) in selected areas of Bangladesh (SP 36 02), report by A. M. Muazzam Husain et al. to IRRI/Bangladesh, Dhaka, June 2004.
Cambodia: Cambodia SRI yield data, from report to national workshop, Prey Veng, and ecological system of rice intensification (SRI) impact assessment (2001–03), reports of Center for Study and Development of Cambodian Agriculture (CEDAC), Phnom Penh, January 2003 and May 2004; Jürgen Anthofer et al., Evaluation of the system of rice intensification (SRI) in Cambodia, February–April 2004, report for GTZ, Phnom Penh. Farmers were randomly sampled in 5 randomly-selected villages in 5 provinces; farms < 0.3 ha were excluded from yield calculations.
China, Gambia, Indonesia, Madagascar, Sierra Leone, Sri Lanka: Data from country reports in Uphoff et al. (2002).
Cuba: Review of results and progress of system of rice intensification in Cuba, report from Dr Rena Perez, based on data reported by provincial rice coordinators at meeting, January 2004.
India: Andhra Pradesh: System of rice intensification: an innovative technology to produce 'more crop per drop', PP presentation by Dr A. Satyanarayana, director of extension, Andhra Pradesh, to IWMI-Tata Fourth Annual Water Meeting, Anand, Gujarat, India, February 2005.
Tamil Nadu: On-farm evaluation of system of rice intensification in Tamiraparani command area, Tamil Nadu, India, PP presentation by Dr T. M. Thiyagarajan, dean, College of Agriculture, Tamil Nadu Agricultural University, Killikulam, to World Rice Research Conference workshop on SRI, Tsukuba, Japan, November 2004.
Myanmar: The results of SRI in Myanmar (Summary), report from Humayun Kabir, agricultural advisor, Metta Development Foundation, Myitkyina, November 2003.
Nepal: Report on farmer field school demonstration plots, monsoon season 2002, and report on farmer field school demonstraiton plots, monsoon season 2003, reports from NEDECO-SMIP Team, Sunsari-Morang Irrigation Project, Biratnagar, March 2004 (FP = farmer practice, and IP = improved practice, using fertiliser and other recommended external inputs); and performance of the system of rice intensification in Morang District, 2004, report by Rajendra Uprety, District Agricultural Development Office, Morang, November 2004.
Philippines: Data reported at national SRI workshop, University of the Philippines, Los Baños, October 2004, compiled by Robert Verzola, coordinator, SRI-Pilipinas Network.
Sri Lanka: 2003 data from Namara et al. (2004). Farmers were randomly selected in 2 districts. Most 'SRI farmers' were not yet using full set of SRI practices, so the full potential of SRI was not evaluated in this study.

errors would invalidate the conclusion that SRI methods raise productivity. The data in Table 2 should be considered as indicative of the range, magnitude and pattern of observed SRI effects. Actual results can vary widely.

These yield increases are achieved with more effort and skill on the part of farmers, but also with lower costs of production, usually by about 20 per cent (Uphoff, 2005). This makes them more appropriate for small farmers who have easier access to labor than to cash.

- There is no need to use *new seed varieties*. While high-yield varieties and hybrids have produced the highest yields with SRI methods, all varieties have responded positively. In Sri Lanka, traditional varieties have given yields of 8–13 t/ha (Dissanayake, 2002).
- Nor need to apply *chemical fertilisers*. Fertiliser can augment yield with SRI, but the highest yields are achieved with compost, made from any decomposed biomass.
- Or *agrochemicals* for crop protection. Farmers have found SRI plants to be more resistant to pest and disease attacks, so spraying is not worth the cost.
- Water applications can be *reduced by a quarter to a half* compared to rice crops that are continuously irrigated. Less water is used throughout the growth cycle.
- Only *10–20 per cent as much seed* is needed because SRI seeding rates are 5–10 kg/ha, compared with the 50–100 kg of seed more usually sown per hectare.

SRI thus allows farmers to:

(a) Increase *factor productivity*, raising output per unit of land, labor, capital and water simultaneously. One normally expects tradeoffs among factor productivity gains, but SRI changes these parameters.

(b) Increase *incomes* as the costs of production are lowered at the same time that yields are raised. This boosts farmers' profits by more than their yield increases and enables them to escape the cost-price squeeze that adversely affects so many producers around the world.

V. Relevance for the Poor

As SRI gains acceptance from farmers, it is coming under attack from some established scientists, as reported in Surridge (2004; see articles by Dobermann, 2004; Sheehy et al., 2004; Sinclair, 2004).[5] This is not the place to replay the critiques and rebuttals.

The initial argument against SRI, which this author encountered when trying to get scientists to take it seriously enough to evaluate it, was that the yield increases reported were spurious or idiosyncratic. The growing number of reports, some quite detailed, as summarised in Table 2, have shown consistent patterns, though exact levels of production vary, not surprisingly with any biologically-based production process. Farms are not like factories, which produce the same output from specified inputs every time. Barrett et al. (2004) did detailed econometric analysis of data from a sample of 108 Madagascar farmers who were using both SRI and conventional methods, thereby controlling for inter-farmer and inter-farm differences. They found a 37.4 per cent increase in factor productivity associated solely with SRI, with an additional 49.5 per cent attributable to farmers' skillful use of these methods.

Some skeptics have pointed to SRI's slow acceptance and spread in Madagascar, and to instances of disadoption by some farmers after trying it, as evidence that its methods are too complicated or too difficult for farmers to accept despite evident productivity benefits. Of greater concern here is the suggestion that although or because SRI is more labor-intensive (at least initially), the poor may find it difficult to adopt. Even though poor households have relatively more labor than other factors of production, they may not be able to afford to invest their labor in SRI because they need immediate, daily returns from labor for their subsistence.

These claims were examined by Moser and Barrett (2003) analysing SRI use in five villages, three in the rain forest area around Ranomafana National Park where SRI was introduced by the NGO *Association Tefy Saina* (ATS) in cooperation with the Cornell International Institute for Food, Agriculture and Development (CIIFAD) between 1994 and 1999, and two on the high plateau. A significant rate of disadoption was found, as high as 40 per cent, once technical support was withdrawn. Moreover, in this sample, poorer households were less likely than richer households to adopt SRI because of its increased labor requirements, particularly for careful transplanting and more frequent weeding. These activities need to be undertaken during the 'hunger season' when poor households are experiencing 'liquidity crises'. They cannot afford to invest more labor in their own rice production, even if this would give them higher returns some months later, if this would reduce immediate off-farm income opportunities.

These conclusions were probably correct for the villages and farmers studied, but they are not necessarily representative of Madagascar, let alone of farmer responses to SRI elsewhere.[6] In Cambodia, for example, a more positive attitude has been taken toward SRI by local institutions and farmers. SRI users have grown from 28 farmers in 2000 to almost 17,000 in 2004, to >55,000 farmers as of March 2006. Disadoption has not been reported by the NGOs and donor projects disseminating SRI in Cambodia. When the NGO leading SRI efforts interviewed 120 farmers who had been using SRI for three years, a majority (57%) said that the system requires less labor (Tech, 2004). An evaluation commissioned by the German development agency GTZ, studying 400 randomly-selected SRI farmers in five provinces, compared with 100 non-SRI farmers randomly selected in the same villages, found no difference in the labor expended per hectare by the two groups of farmers. Indeed, more experienced SRI users spent less time using its techniques than when they used conventional practices. Moreover, the SRI pattern of labor demand was judged to be more favorable for poorer farmers and for women (Anthofer, 2004).

In 2003, the International Water Management Institute (IWMI) evaluated SRI use in Sri Lanka, interviewing 60 randomly-chosen SRI users in two districts, and 60 non-users.[7] This study found that SRI was adopted proportionally more by both poorer and richer farmers compared to middle-income farmers, with no significant difference between poorer and richer farmers in terms of adoption (Namara et al., 2004). The same conclusion was reached in an evaluation of SRI in a village in Sichuan province, where the number of SRI users expanded from 7 in 2003 to 398 in 2004. Li et al. (2005) similarly found no significant difference by economic level in adoption rates. The IWMI study observed further that 'the probability of disadoption among poor farmers is significantly lower than that among rich farmers. This implies that once the poor farmers adopt SRI, they have a higher probability of continuing with it than rich farmers' (Namara et al., 2004). Another IWMI evaluation of SRI practice, conducted in very poor tribal communities in West Bengal, India found that the number of SRI users had grown from 4 to 150 over three seasons and labor input per hectare actually went *down* by 9 per cent with SRI (Singh and Talati, 2005). So while the barriers to adoption that Moser and Barrett discussed need to be kept in mind, experience elsewhere is suggesting that there are no intrinsic barriers for SRI to be taken up by the poor.

Barrett et al. (2004) found from their data set with 108 farmers that SRI yields and labor productivity are more variable than with conventional practice, implying that SRI entails more risk, which could be a deterrent especially for the poor. However, data from the IWMI study in Sri Lanka showed just the opposite: that the risk of net losses from cultivating rice was greatly reduced by SRI methods (last two columns in Table 3). In Sri Lanka, the major risk comes from drought, which is compensated for by the deeper root systems that SRI develops. The evaluation done for GTZ in Cambodia comparing SRI and non-SRI users concludes, from econometric analysis of data from 500 farmers, that economic risks were significantly lower for those farmers who practiced SRI (Anthofer, 2004).

SRI farmers in Sri Lanka reported one of the most attractive outcomes was reduced cost of production (rupees/kg); they can reduce or even eliminate purchases of mineral fertiliser and agrochemicals, while reducing also cost of seed. Cambodian farmers reported the same (Anthofer, 2004), saying they appreciated not having to make cash outlays at the start of the growing season, when their cash reserves were depleted. These considerations are especially important for poor farmers, who lack access to credit or wish to avoid debt traps, having no asset buffer to protect them if their crops fail.

As noted already, the main requirement for successful SRI cultivation, apart from knowledge and skill, is – at least initially – more labor input. Farmers are expected to monitor their crop and irrigation, making adjustment as needed; more weeding is necessary since fields are not kept flooded to suppress weeds. Several analyses have indicated about a 30 per cent increase in labor input initially. However, Barrett et al. (2004) found that by the fourth and fifth years of SRI cultivation, when techniques had been mastered, four per cent and 10 per cent *less* labor was required than with conventional rice growing methods. In the IWMI Sri Lanka study, total hours of labor/hectare recorded with SRI were 10 per cent less in the dry (*yala*) season and 6 per cent less in the wet (*maha*) season compared with conventional cultivation (Namara et al., 2004: Table 10).

More important for poor households, which depend mostly on returns to labor as their primary factor of production, the IWMI data showed a labor productivity gain with SRI (kgs of rice/hr of labor) of 50 per cent in the dry season and 62 per cent in the wet season (Namara et al., 2004). The IWMI India study found a 9 per cent reduction in labor requirements per hectare, and Chinese farmers told researchers from China Agricultural University that labor saving was the SRI benefit that they appreciated most – more than higher yield, greater profitability or water saving (Li et al., 2005).

Table 3. Gross and net revenue with SRI vs. conventional cultivation methods, with cost of production and incidence of net losses from production, calculating labour costs according to three different levels, Sri Lanka, 2003

Incidence of Net Loss (%)			Gross revenue (rupees/ha)			Net revenue (Gross Revenue – Costs)			Cost of production (rupees/kg)		
Season	Conv.	SRI	Increase	Conv.	SRI	Increase	Conv.	SRI	Decrease	Conv.	SRI
At zero wage rate, assuming all labour is unpaid family labour:											
Yala	27017	34899	29%	16032	29371	83%	7.6	3.6	111%	15%	2%
Maha	23871	36720	35%	14373	34575	141%	6.8	2.2	209%	0	0
At farm wage rate, assuming all labor is paid the typical agricultural laborers' wage:											
Yala	27017	34899	29%	8417	15992	90%	13.3	11.1	17%	45%	7%
Maha	23871	36720	35%	7490	16285	117%	11.9	8.7	27%	25%	2%
At non-farm wage rate, assuming that all labor is paid at the off-farm employment wage, representing the opportunity cost for agricultural labor:											
Yala	27017	34899	29%	4215	10223	143%	15.5	15.0	3%	48%	10%
Maha	23871	36720	35%	4369	13374	206%	14.0	11.4	19%	37%	3%.

Source: Namara et al. (2004: Table 15). The incidence of net loss is the percentage of 60 farmers who suffered a net economic loss from their rice production given the level of production resulting from the methods used (SRI or traditional) for each of three levels of labor remuneration.

SRI is starting to spread in places as different and difficult as Cuba, Myanmar and Sierra Leone. Networks to promote evaluation and dissemination have been established in Bangladesh, China, Indonesia, Nepal, Philippines and Thailand, linking NGOs, universities, government research agencies, farmer groups, and concerned individuals. Informal networks are active and effective in Cuba, Madagascar, and Sri Lanka. Interested individuals and organisations are working voluntarily to better understand and promote SRI in still other countries such as Guinea, India, Mozambique, Pakistan, Peru and Senegal.

VI. Accounting for the Productivity of SRI

Scientific understanding of SRI is still incomplete because agricultural scientists have only recently begun to take an interest. However, researchers in China, Indonesia, India, Madagascar and at Cornell University have started making SRI scientifically comprehensible. Until 1999, Madagascar was the only country in the world where SRI was known and used. Today there are confirming results from at least 22 countries across Asia, Africa and Latin America; SRI methods are starting to spread because of farmer interest and enthusiasm.

One possible explanation for how SRI achieves these improvements is by accessing and utilising the vast and free supplies of oxygen and nitrogen in the atmosphere. With more aerobic soil conditions, there is greater growth of plant roots and more biological nitrogen fixation, as well as more mineralisation of nutrients like phosphorus by soil organisms (Turner and Haygarth, 2001). Green Revolution technologies have done nothing to enhance the abundance, diversity and activity of these organisms, whereas agroecological approaches attend to the bacteria, fungi and other microbes as well as to soil fauna such as earthworms that support plant growth and health in symbiotic ways (Pinton et al., 2001). Symbiosis between plants and microorganisms extends back 400 million years (Margulis and Sagan, 1997).

Adding compost augments the plants' own infusion of carbon and other compounds into the soil around their roots through processes of root exudation and rhizodeposition. Although these processes are well-known in plant science, they have received relatively little attention in the literature, and almost none in rice research, perhaps because rice plants when kept continuously submerged have only shallow root systems, which degenerate due to hypoxia. By the flowering stage, as many as three-fourths of the rice plants roots are lost (Kar et al., 1974).

Plants share some of the sugars, amino acids and other growth-supporting substances that they create through photosynthesis and by other biochemical processes with the billions of soil microorganisms that live in the rhizosphere, the few millimeters of soil that surround plant roots like a glove. These organisms support plants' growth in a symbiotic manner. In return for the carbon and other substances that they receive from root exudation, microbes enhance soil nutrient availability and uptake and provide other benefits such as induced systemic resistance to disease (Neumann and Römheld, 2001; Doebbelaere et al., 2003; Bonkowski, 2004).

The roots of SRI plants are several times larger than those of rice plants grown with conventional practices, which create anaerobic soil conditions through continuous flooding. Figure 1 shows two rice plants of the same age and variety, planted at the

Figure 1. Two rice plants of same variety (VN 2084) and age (52 days) being grown in Cuba with conventional practices (left) and SRI practices (right). Picture courtesy of Dr Rena Perez

same time in adjacent plots but managed differently. The one on the left was raised in a flooded nursery with dense plant population following conventional practice; in Cuba, transplanting is normally done when seedlings are 50 to 55 days-old. The plant on the right benefited from being carefully transplanted at nine days into well-drained but moist soil, with wide spacing and enhanced soil organic matter. Figure 1 shows the augmentation of root growth that is possible, from identical genetic potentials, when growing conditions are made more favorable. These conditions concurrently increase the abundance and diversity of soil organisms.

What makes the inputs of land, labor, capital and water more productive? In recent years, there is a growing understanding of how to take advantage of biological processes linking plants, soil organisms, air, nutrients and water that have been 'eclipsed' by the emphasis put on chemical inputs over the past century and on genes in this one (Uphoff et al., 2006). Much remains to be studied and evaluated, however. The following points are suggestive of where agroecological approaches are likely to open new possibilities in the future:

- SRI practices can tap enhanced supplies of O_2 and N_2 that increase soil microbial populations and their provision of nutrients to the growing crop through biological nitrogen fixation (BNF), phosphorus solubilisation, and other biological processes (Magdoff and Bouldin, 1970; Turner and Haygarth, 2001).
- Under aerobic soil conditions, plant roots get beneficially 'infected' by mycorrhizal fungi whose hyphae (filaments) greatly expand the volume of soil from which nutrients such as phosphorus and zinc can be extracted (Martin et al., 2001).

- When root exudation is increased, larger populations of bacteria can be supported on the roots' surface, which are in turn 'grazed' by protozoa, which require a lower carbon-to-nitrogen ratio than that of bacteria, so 'excess nitrogen' is excreted onto the roots (Brimecombe et al., 2001; Bonkowski, 2004).

In these and other ways, plants by interacting with soil biota that they help to nourish can acquire more nutrients for their own use, in particular by exuding various organic compounds into the soil that surrounds their roots (Pinton et al., 2001). Many gains in agriculture could come from knowing more about how microorganisms enhance plant growth, for example, how rhizobia bacteria that live as endophytes inside rice plant roots enhance both crop yield and grain protein content per hectare (Yanni et al., 2001). Rhizobia fix nitrogen in the nodules of leguminous plants, but they can also benefit non-leguminous plant species such as rice by producing growth-promoting hormones, for example, auxins and cytokinins, that aid roots (Khalid et al., 2006).

A weakness of modern agriculture, and its current incarnation in biotechnology, is that opportunities for synergy are missed by focusing on just one species at a time, ignoring the interdependencies that have evolved productively over several hundred million years. A large scientific literature supports agroecological strategies that focus on and utilize interactions among species across plant, animal and microbial kingdoms. Much about these processes is still unknown because research investments to date in soil chemistry and soil physics have been many multiples those in soil biology. Yet enough is known now to propose that this domain warrants more extensive scientific investigation (Uphoff et al., 2006). Arguments about the importance and urgency of investing heavily in genetic modification look different when other opportunities for achieving productivity gains possibly more simply, cheaply and quickly are taken into account.

VII. Looking Beyond Rice

The Green Revolution was able to increase food production substantially by making changes in crops' genetic potential, mostly through conventional plant breeding. Its strategy was to raise outputs by increasing external inputs, particularly mineral fertiliser, but also chemical biocides, together with more irrigation water, because the new plant varieties were bred to be more responsive to such inputs. SRI works instead with whatever genetic potential already exists and decreases or eliminates synthetic inputs; it relies more on human knowledge, labor and skill than external inputs. The essential components of the SRI approach are: expanded *root growth*, through a set of new management practices, and mobilisation of the benefits of *soil biota*, including (a) greater abundance and diversity of soil microorganisms; and (b) greater numbers and activity of soil fauna, particularly earthworms.

The SRI paradigm reflects principles often grouped under the broad heading of agroecology. This paradigm parallels that developed over the past 20 years by French researchers supported by the Center for International Research on Agricultural Development (CIRAD) who have worked closely with colleagues in Brazil, Vietnam, Madagascar and other countries. The cropping systems that they have evolved from experimentation in a variety of agroecosystems, from the humid tropics to temperate climates, are based on direct seeding and use of mulch (Séguy et al., 2006). This is an advance upon what has become known as 'no-till' or 'zero-till' cultivation. The soil is not disturbed by plowing; additionally, it is kept protected by some vegetative cover, living or dead. In warm climates, this lowers soil temperatures and nurtures more subterranean biodiversity and biological activity.

A range of such systems have demonstrated that productivity can be greatly enhanced, as with SRI, by promoting root growth and soil biotic abundance and diversity by changing the way that plants, soil, water and nutrients are managed together, achieving greater synergy from symbiotic processes. CIRAD programs in Brazil have shown that soybean production in direct-seeded, mulch-based cropping systems can be increased by about 50 per cent with a 50 per cent reduction in chemical fertilisation. Over time, as soil conditions improve as a result of these practices, the profitability of such operations increases. Rain-fed rice, which usually produces one–two t/ha, has with such practices produced yields of eight t/ha, with improvement in grain quality (Séguy et al., 2006). Elsewhere in Latin America, agroecologically-grounded practices have raised the yields of maize, beans and potatoes by 100 to 200 per cent, and even more, in the impoverished hillside and montane environments in Central America and the Andes (Bunch, 2002; Ruddell, 2002).

VIII. New Technologies and the Poor

This essay has explored opportunities to achieve some of the goals of transgenic research through understanding and better utilising capabilities already existing in genomic endowments and their surrounding ecosystems. An agroecological approach does not study and promote single species like rice or soybeans or chickens by themselves but rather considers the functioning species as parts of biological communities that include microflora and fauna in soil systems. This perspective suggests that there may not be as much urgency as some have claimed for pursuing biotechnology 'to help the world's poor'. In this twenty-first century, both biotechnological and agroecological advances have contributions to make to improving their situation.

Agroecological methods require skill and knowledge, which can be acquired more easily by the poor than more land or capital. Water is becoming more and more scarce in a world where infertile soils forfeit large amounts of annual rainfall that could be absorbed and stored, but instead is lost through runoff that also imposes the costs of erosion and lost nutrients. Soil that is crusted, low in organic matter and diverse life forms, loses its capacity for water retention. Although water is considered a scarce commodity, in fixed supply, agroecological practices that improve soil fertility can loosen the growing constraint of water supply. Water saved is in effect water produced. The poor in particular can benefit from water harvesting and conservation methods as part of agroecological production strategies that enhance the life of the soil (Toulmin et al., 1996).

Transgenic research is an important current issue that should be considered, even contested, from a variety of perspectives – scientific, economic, political, social, ethical, and of course environmental (Pretty, 2001). In fact, not all 'natural' innovations may pass such tests; no blanket judgments should be made about either biotechnology or agroecology for research and application. There are no powerful or well-financed interests supporting agroecology at present, as there are for biotechnology, from the White House to corporate boardrooms to prestigious scientific advocates. Considering prospective cost-effectiveness and environmental impacts, it could be argued that equal support is warranted for the two domains.

Transgenic work could be directed more toward enhancing natural processes and potentials. Some of the initial transgenic products such as Roundup-Ready maize and soybeans have been essentially an extension of the Green Revolution paradigm, continuing and even reinforcing dependence on chemical inputs that can have adverse environmental impacts rather than enhancement of biological processes. *Bt* crops that build resistance to certain pests into the genome move away from chemical dependence. Enhancement of nutritional value enabling plants to produce more of certain vitamins or mineral elements in their edible portions, could have particular value for the poor, provided the costs of production and purchase increase little or not at all. Such transgenic improvements may not be attractive to investors, motivated more by profit than altruism. Explicitly pro-poor transgenic improvements probably need special funding provisions or channels if they are to have more than a marginal place on the research agenda. In the mean time and in any case, agroecological innovations could have substantial and quick payoffs for the world's present poor and for future generations who depend on a robust and healthy environment for long-term survival.

Notes

1. In countries where fertiliser has been under-utilised, positive returns are attainable; in high-application countries, marginal returns are diminishing. In China, the incremental amount of rice produced from 1 kg of nitrogen fertiliser has dropped from 15–20 kg before the Green Revolution to about 5 kg now (Peng, 2004).
2. Pimentel (1997) reports that since 1945, the use of insecticides has gone up 10-fold while the percent of crops lost to insect damage has increased from 7 to 13 per cent. Interest in integrated pest management (IPM) has been growing because of increasing evidence that pesticides make crops more vulnerable to insect and disease by weakening natural biological controls that keep pests and disease vectors in check (Chaboussou, 2004).
3. On phyllochrons, see Nemoto et al. (1995). Transplanting young seedlings is recommended to reduce or avoid trauma to young plant roots; however, it is not a requirement. Direct seeding is one of the many adaptations being made as knowledge about SRI spreads; experiments are being done by farmers in Cambodia, Cuba, India and Sri Lanka. Reducing labor requirements will make the method more accessible, especially for larger-scale use, since these can be a constraint on adoption of SRI (Moser and Barrett, 2003).
4. When SRI is introduced to farmers, they are told there is no need to change their varieties to get higher yields, so there should be little if any genetic difference that could account for the differences in yield seen between the two right-hand columns; the main change with SRI was in management practices, using fewer purchased inputs.
5. A letter responding to criticisms reported in Surridge (2004), written by the director of extension for the Indian state of Andhra Pradesh, clearly refuted critics' objections on the basis of demonstrated SRI performance in his state, now spreading to other states (Satyanarayana, 2004).
6. Three of the study villages were located around Ranomafana National Park. These are very 'traditional' populations, culturally disposed to respect and perpetuate 'the ways of the ancestors' and slow to accept change. Disadoption when outside support was withdrawn was unfortunate but not very surprising. Two of the study villages were located on the high plateau, which is more 'modernised'. Data from an irrigation management project there funded by the French development

agency showed that, with very little extension activity, the area under SRI increased in five years' time from 34 to 548 ha, with little evidence of disadoption. Average SRI yield was 8.55 t/ha compared to 3.77 t/ha using 'improved' (modern input) practices and 2.36 t/ha with typical practice (Hirsch, 2000).

7. Few of the 'SRI users' in the sample were using the full set of recommended SRI practices; full use could further improve yields. Farmers' profitability with SRI at least doubled. Among non-SRI users, 87.5 per cent had heard about SRI, mostly from other farmers; and 25 per cent said that they intended to practice SRI in the future. When asked about the general attitude toward SRI in their communities, 69 per cent reported that it was favorable, only 9 per cent said it was unfavorable, while 22 per cent offered no opinion (Namara et al., 2004).

References

Anthofer, J. (2004) The potential of the system of rice intensification (SRI) for poverty reduction in Cambodia, paper presented at the 2004 Deutscher Tropentag, Berlin, accessed at: http://www.tropentag.de/2004/abstracts/full/399.pdf

Barrett, C. B., Moser, C., McHugh, O. and Barison, J. (2004) Better technologies, better plots, or better farmers? Identifying changes in productivity and risk among Malagasy rice farmers, *American Journal of Agricultural Economics*, 86(4), pp. 869–88.

Bonkowski, M. (2004) Protozoa and plant growth: the microbial loop in soil revisited, *New Phytologist*, 162(3), pp. 617–31.

Brimecombe, M., DeLeij, F. A. and Lynch, J. M. (2001) The effect of root exudates on rhizosphere microbial populations, in R. Pinton, Z. Varanini and P. Nannipieri (eds), *The Rhizosphere: Biochemistry and Organic Substances at the Soil-Plant Interface*, pp. 95–140 (New York: Marcel Dekker).

Bunch, R. (2002) Increasing productivity through agroecological approaches in Central America: Experiences from hillside agriculture, in N. Uphoff (ed.), *Agroecological Innovations: Increasing Food Production with Farmer Participation*, pp. 162–72 (London: Earthscan).

CEDAC (2003) Report of First SRI Working Group Meeting, Kampot, 21 March 2003, Centre pour les Etudes et Developpement d'Agriculture Cambodgien (CEDAC), Phnom Penh.

Chaboussou, F. (2004) *Healthy Crops: A New Agricultural Revolution* (Charnley, UK: Jon Carpenter).

Dissanayake, S. (2002) A policy-maker's perspective on SRI from Sri Lanka, in N. Uphoff et al. (eds), *Assessments of the System of Rice Intensification: Proceedings of an International Conference, Sanya, China, April 1–4, 2002*, pp. 29–31 (Ithaca, NY: Cornell International Institute for Food, Agriculture and Development).

Dixon, J. and Gulliver, A. (2001) *Farming Systems and Poverty: Improving Farmers' Livelihoods in a Changing World* (Rome: FAO).

Dobermann, A. (2004) A critical assessment of the system of rice intensification (SRI), *Agricultural Systems*, 79(3), pp. 261–81.

Doebbelaere, S. J., Vanderleyden, J. and Okon, Y. (2003) Plant growth-promoting effects of diazotrophs in the rhizosphere, *Critical Reviews in Plant Science*, 22(2), pp. 107–49.

Garg, A., Kim, J.-K., Owens, T. G., Ranwala, A. P., Yang, D. C., Kochian, L. V. and Wu, R. J. (2002) Trehalose accumulation in rice plants confers high tolerance levels to different abiotic stresses, *Proceedings of the National Academy of Sciences* (USA), 99(25), pp. 15898–903.

Gonsalves, C., Lee, D. R. and Gonsalves, D. (2007) The adoption of genetically modified papaya in Hawaii and its implications for developing countries, *Journal of Development Studies*, 43(1), pp. 177–91.

Halweil, B. (2002) Farming in the public interest, in C. Flavin, H. French, G. Gardner, S. Dunn, R. Engelman, B. Halweil, L. Mastny, A. Platt McGinn, D. Nierenberg, M. Renner and L. Starke (eds), *State of the World 2002*, pp. 51–74 (New York: Norton).

Herring, R. J. (2007) Stealth seeds, bioproperty, biosafety, biopolitics, *Journal of Development Studies*, 43(1), pp. 130–57.

Hirsch, R. (2000) La Riziculture Malgache Revisitée: Diagnostic et Perspectives (1993–99), report to Agence Française de Développement, Départment des Politiques et des Etudes, Antananarivo.

Kar, S., Varade, S. B., Subramanyam, T. K. and Ghildyal, B. P. (1974) Nature and growth pattern of rice root system under submerged and unsaturated conditions, *Il Riso* (Italy), 23(2), pp. 173–9.

Khalid, A., Arshad, M. and Zahir, Z. A. (2006) Phytohormones: microbial production and applications, in N. Uphoff et al. (eds), *Biological Approaches to Sustainable Soil Systems*, pp. 207–20 (Boca Raton, FL: CRC Press).

Khush, G. S. and Peng, S. (1996) Breaking the yield frontier of rice, in: *Increasing Yield Potential in Wheat: Breaking the Barriers*. Proceedings of a workshop held in Ciudad Obregon, Sonora, Mexico, CIMMYT, Mexico, DF, pp. 36–51.

Kishi, M. (2005) The health impact of pesticides: what do we know?, in J. Pretty (ed.), *The Pesticide Detox*, pp. 23–38 (London: Earthscan).

Laulanié, H. de (1993) Le système de riziculture intensive malgache, *Tropicultura* (Brussels), 11(1), pp. 110–4.

Li, X., Xu, X. and Li, H. (2005) A socio-economic assessment of the system of rice intensification (SRI): a case study of Xinsheng Village, Jianyang County, Sichuan Province, report for the College of Humanities and Development, China Agricultural University, Beijing, February.

Lines, G. and Uphoff, N. (2005) A remarkable civil society contribution to food and nutrition security in Madagascar and beyond, in M. Schulz and U. Kracht (eds), *Food and Nutrition Security in the Process of Globalization* (Münster, Germany: Lit-Verlag).

Magdoff, F. R. and Bouldin, D. R. (1970) Nitrogen fixation in submerged soil-sand-energy material media and the aerobic-anaerobic interface, *Plant and Soil*, 33(1), pp. 49–61.

Margulis, L. and Sagan, D. (1997) *Microcosmos: Four Billion Years of Evolution from Our Microbial Ancestors* (Berkeley: University of California Press).

Martin, F. M., Perotto, S. and Bonfante, P. (2001) Mycorrhizal fungi: a fungal community at the interface between soil and roots, in R. Pinton, Z. Varanini and P. Nannipieri (eds), *The Rhizosphere: Biochemistry and Organic Substances at the Soil-Plant Interface*, pp. 263–96 (New York: Marcel Dekker).

Moser, C. M. and Barrett, C. B. (2003) The disappointing adoption dynamics of a yield-increasing, low external-input technology: the case of SRI in Madagascar, *Agricultural Systems*, 76(3), pp. 1085–100.

Namara, R. E., Weligamage, P. and Barker, R. (2004) Prospects for adopting system of rice intensification in Sri Lanka: a socioeconomic perspective, Research Paper 75, International Water Management Institute, Colombo.

Nemoto, K., Morita, S. and Baba, T. (1995) Shoot and root development in rice related to the phyllochron, *Crop Science*, 35(1), pp. 24–9.

Neumann, G. and Römheld, V. (2001) The release of root exudates as affected by the plant's physiological status, in R. Pinton, Z. Varanini and P. Nannipieri (eds), *The Rhizosphere: Biochemistry and Organic Substances at the Soil-Plant Interface*, pp. 41–93 (New York: Marcel Dekker).

Peng, S. (2004) Improving fertilizer-nitrogen use efficiency of irrigated rice: progress of IRRI's RTOP project in China, paper presented at the International Conference on Sustainable Rice Production, China National Rice Research Institute, Hangzhou, October.

Pimentel, D. (ed.) (1997) *Technologies for Reducing Pesticide Use* (New York: John Wiley).

Pinton, R., Varanini, Z. and Nannipieri, P. (eds) (2001) *The Rhizosphere: Biochemistry and Organic Substances at the Soil-Plant Interface* (New York: Marcel Dekker).

Postel, S. (1999) *Pillar of Sand: Can the Irrigation Miracle Last?* (New York: Norton).

Pretty, J. N. (2001) The rapid emergence of genetic modification in world agriculture: contested risks and benefits, *Environmental Conservation*, 28(3), pp. 248–62.

Pretty, J. N. and Hine, R. (2005) Pesticide use and the environment, in J. Pretty (ed.), *The Pesticide Detox*, pp. 1–22 (London: Earthscan).

Ruddell, E. (2002) Raising smallholder crop and livestock production in Andean mountain regions, in N. Uphoff (ed.), *Agroecological Innovations: Increasing Food Production with Farmer Participation*, pp. 173–86 (London: Earthscan).

Satyanarayana, A. (2004) Rice, research, and real life in the field (letter), *Nature*, 429, p. 803.

Séguy, L., Bouzinac, S. and Husson, O. (2006) Direct-seeded tropical soil systems with permanent soil cover: learning from Brazilian experience, in N. Uphoff et al. (eds), *Biological Approaches to Sustainable Soil Systems*, pp. 323–62 (Boca Raton, FL: CRC Press).

Sheehy, J. E., Peng, S., Dobermann, A., Mitchell, P. L., Ferrer, A., Yang, J., Zou, Y., Zhong, X. and Huang, J. (2004) Fantastic yields in the system of rice intensification: fact or fallacy?, *Field Crops Research*, 88(1), pp. 1–8.

Sinclair, T. (2004) Agronomic UFOs waste valuable scientific resources, *Rice Today* (Los Baños, Philippines: International Rice Research Institute), 3, p. 43.

Singh, S. K. and Talati, J. (2005) Impact of the system of rice intensification (SRI) on rice yields: results of a new sample study in Purulia District, India, paper presented to the Fourth Annual IWMI-Tata Meeting on Water Management, Anand, India, 24–26 February, Ahmedabad, International Water Management Institute, India Program.

Stoop, W., Uphoff, N. and Kassam, A. (2002) A review of agricultural research issues raised by the system of rice intensification (SRI) from Madagascar: opportunities for improving farming systems for resource-poor farmers, *Agricultural Systems*, 71(1), pp. 249–74.

Surridge, C. (2004) Feast or famine?, *Nature*, 428, pp. 360–1.

Tech, C. (2004) Ecological system of rice intensification (SRI) impact assessment, CEDAC Field Document, February, Phnom Penh, Center for Study and Development of Cambodia Agriculture.

Toulmin, C., Reij, C. and Scoones, I. (eds) (1996) *Sustaining the Soil: Indigenous Soil and Water Conservation in Africa* (London: Earthscan).

Turner, B. and Haygarth, P. (2001) Phosphorus solubilization in rewetted soils, *Nature*, 411, p. 258.

Uphoff, N., Fernandes, E. C. M., Yuan, L. P., Peng, J., Rafaralahy, S. and Rabenandrasana, J. (eds) (2002) *The System of Rice Intensification: Proceedings of an International Conference, Sanya, China, April 1–4, 2002* (Ithaca, NY: Cornell International Institute for Food, Agriculture and Development), accessed at: http://ciifad.cornell.edu/sri/

Uphoff, N. (2003) Higher yields with fewer external inputs? The system of rice intensification and potential contributions to agricultural sustainability, *International Journal of Agricultural Sustainability*, 1(1), pp. 38–50.

Uphoff, N. (2005) Features of the system of rice intensification (SRI) apart from yield, report to SRI Internet Network, accessed at: http://ciifad.cornell.edu/sri/

Uphoff, N. (2006) Opportunities for overcoming productivity constraints with biologically-based approaches, in N. Uphoff et al. (eds), *Biological Approaches to Sustainable Soil Systems*, pp. 693–713 (Boca Raton, FL: CRC Press).

Uphoff, N., Ball, A., Fernandes, E. C. M., Herren, H., Husson, O., Laing, M., Palm, C., Pretty, J. N., Sanchez, P.A., Sanginga, N. and Thies, J. (eds) (2006) *Biological Approaches to Sustainable Soil Systems* (Boca Raton, FL: CRC Press).

Yanni, Y. G., Rizk, R. Y., Abdel Fattah, F. K., Squartini, A., Corich, V., Giacomini, A., de Bruijn, F., Radamaker, J., Maya-Flores, J., Ostrom, P., Vega-Hernandez, M., Hollingsworth, R. I., Martinez-Molina, E., Mateos, P., Velasquez, E., Wopereis, J., Triplett, E., Umali-Garcia, M., Anama, J. A., Rolfe, B. G., Ladha, J. K., Hill, J., Mujoo, R., Ng, P. K. and Dazzo, F. B. (2001) The beneficial plant growth-promoting association of *Rhizobium leguminosarum* bv. trifolii with rice roots, *Australian Journal of Plant Physiology*, 28, pp. 845–70.

INDEX

Page numbers in *italics* represent Tables. Page numbers in **bold** represent Figures.